Quantum Computing Unveiled

Aimed at advanced undergraduate and graduate-level students, this textbook covers the core topics of quantum computing in a format designed for a single-semester course. It will be accessible to learners from a range of disciplines, with an understanding of linear algebra being the primary prerequisite. The textbook introduces central concepts such as quantum mechanics, the quantum circuit model, and quantum algorithms, and covers advanced subjects such as the surface code and topological quantum computation. These topics are essential for understanding the role of symmetries in error correction and the stability of quantum architectures, which situate quantum computation within the wider realm of theoretical physics. Graphical representations, examples, and exercises are included throughout the book and optional expanded materials are summarized within boxed "Remarks." Lecture notes have been made freely available for download from the textbook's webpage, with instructors having additional online access to selected exercise solutions.

Yidun Wan is a tenured Full Professor of Physics at Fudan University. He has a diverse educational background, with degrees in computer science, economics, and physics. He gained his PhD in theoretical physics from the University of Waterloo and the Perimeter Institute for Theoretical Physics. His research interests include topological orders, quantum information and computation, cosmology, and quantum gravity. He has taught a broad range of courses, including General Relativity, Quantum Computation, Differential Geometry, Mathematical Methods in Physics, and Scientific English Writing.

Quantum Computing Unveiled

A Concise Course with Topological Extensions

Yidun Wan

Fudan University, Shanghai

Shaftesbury Road, Cambridge CB2 8EA, United Kingdom

One Liberty Plaza, 20th Floor, New York, NY 10006, USA

477 Williamstown Road, Port Melbourne, VIC 3207, Australia

314–321, 3rd Floor, Plot 3, Splendor Forum, Jasola District Centre,
New Delhi – 110025, India

103 Penang Road, #05–06/07, Visioncrest Commercial, Singapore 238467

Cambridge University Press is part of Cambridge University Press & Assessment,
a department of the University of Cambridge.

We share the University's mission to contribute to society through the pursuit of
education, learning and research at the highest international levels of excellence.

www.cambridge.org
Information on this title: www.cambridge.org/highereducation/isbn/9781009656078

DOI: 10.1017/9781009656061

First published 2026

A catalogue record for this publication is available from the British Library

A Cataloging-in-Publication data record for this book is available from the Library of Congress

ISBN 978-1-009-65607-8 Hardback

Additional resources for this publication at www.cambridge.org/wan-quantumcomputing

To Zheng, Yitong, Yisong, and my parents

Contents

Preface

Quantum computation is one of those topics that never ceases to intrigue. It's an intellectual puzzle, a practical tool, and a glimpse into what might just be the future of computing. Teaching this subject at Fudan University for a one-semester course, I've seen firsthand how students' eyes light up when they begin to grasp the strange, beautiful logic of quantum mechanics.

Yet diving into quantum computation can feel like jumping into an ocean without knowing how to swim. The field is vast and complex, and many textbooks on the subject seem determined to cover everything under the sun. That's why I decided to write "Quantum Computing Unveiled: A Concise Course with Topological Extensions." I wanted to create a book that provides a solid grounding without overwhelming you with information.

The idea was simple: Cover what's essential and make it accessible within the span of a single semester. A book that doesn't just sit on the shelf, collecting dust, but one that can be actively used in a classroom. And not just for the experts but for graduate and advanced undergraduate students eager to explore this fascinating field.

Now, I must admit, I couldn't resist adding a bit of my personal touch. There's a chapter on topological quantum computing, a subject I find fascinating. Before we dive into that, though, we explore topological orders and surface codes. It's essential to understand these underlying concepts, as they pave the way for the innovative thinking that topological quantum computing represents. These chapters are more than a sequential progression; they're a logical unfolding of ideas, one leading naturally into the other, offering a cohesive understanding of a complex subject.

Teaching quantum computation is a bit like leading an expedition into uncharted territory. There are familiar landmarks, unexpected obstacles, and plenty of opportunities for discovery. In this book, I've tried to lay out a clear path, starting with the foundational concepts and gradually building up to more complex topics like quantum algorithms, error correction, and yes, topological quantum computing.

Along the way, I've done my best to keep things engaging. Quantum computation shouldn't be a dry, abstract subject. It's filled with surprising insights, puzzling paradoxes, and practical applications that could change the world. I've tried to capture some of that excitement in these pages.

I know that quantum computation can be daunting. I've seen students struggle with the material, wrestle with the concepts, and then suddenly have that "aha" moment when it all clicks. That's what I hope to achieve with this book – a guide that doesn't just tell you about quantum computation but helps you understand it, question it, and may even fall in love with it, just a little bit. So here it is, "Quantum Computation Unveiled: A Concise Course with Topological Extensions."

While quantum information, particularly quantum communication and cryptography, often appears in books on quantum computation, this text will not address these topics. First, quantum computation is both a subset and an application of quantum information theory, emphasizing the effective manipulation of quantum information. Second, key aspects of quantum information theory, such as quantum error correction and quantum decoherence, will be covered in the context of constructing quantum computers. Additionally, quantum information is an extensively developed field with numerous branches, each deserving comprehensive treatment in its own right. Covering all these areas would require an entire book for each branch, making it reasonable to focus this work specifically on quantum computation. Maintaining a clear focus on quantum computation ensures a more coherent narrative and avoids diluting the core subject with material that, though related, is vast and adequately covered elsewhere.

This being a concise book, instructions for its use are straightforward: For a 14- to 16-week course, three 45-minute lectures per week, the entire book can be covered. For shorter courses, chapters can be selected accordingly. If students are already well versed in quantum mechanics, Chapter 3 can be skipped. For those with a computer science background, Chapter 2 may be unnecessary. If the course aims to provide a primer on quantum computation, Chapters 8 and 9, which cover surface codes and topological quantum computation, may be beyond its scope. In the case of a single-quarter course, Chapters 2, 3, 8, and 9 can all be omitted.

Each chapter includes exercise problems strategically placed within the content, encouraging readers to solve them as they go to reinforce, clarify, and deepen their understanding. Solutions to selected exercise problems will be available online for instructors.

This book does not contain any appendices. Essential mathematics and physics are integrated into the main text. Useful but not immediately necessary extensions appear in boxed remarks that can be skipped on a first reading.

Whether you're an instructor looking for manageable course material or a student taking your first steps into the quantum realm, I hope you find this book both enlightening and enjoyable. Let's embark on this journey together. The quantum world is waiting, and there's so much to explore.

Acknowledgements

I wish to partially dedicate this work to the memory of my late friend and collaborator, Professor Ray Laflamme, who passed away during this book's final preparation. His profound intuition in physics, unwavering courage in the face of adversity, and invaluable guidance during my time as an Affiliate Member at the IQC from 2018 to 2023–all of which profoundly shaped my academic journey–are deeply remembered. I am eternally grateful for his mentorship, which began during my PhD studies at the Perimeter Institute and University of Waterloo, where he served on my dissertation committee, and for his generosity as a host and a brilliant mind. His legacy continues to inspire this work and the entire field of quantum information.

I am deeply grateful to my wife, Zheng, whose unwavering support made this book possible and allowed my career as a theoretical physicist to grow beyond a distant aspiration. The all–round support from my parents, who applauded every small success of mine, provided a steady source of motivation.

I owe a debt of gratitude to Professor Yong-Shi Wu at the University of Utah, an illustrious physicist, for his unwavering encouragement and support since my PhD years, as well as for our inspiring discussions – not only during the writing of this book but throughout my career so far. I also thank Professor Mikio Nakahara, a renowned physicist and author of widely used textbooks who retired from Kindai University in Japan, for offering me my first postdoc position – an opportunity that shifted my research from quantum gravity to quantum computation. I'm indebted to Professor Xiao-Gang Wen, a celebrated physicist at MIT and a pioneer in topological phases of matter, for his seminars and the postdoc offer that opened the door for me to the fascinating realm of topological order. I deeply appreciate the opportunity to work with Professor Seigo Tarucha, a distinguished experimentalist now retired from the University of Tokyo. During my time as his postdoc, I had the privilege of studying surface code in its early stages. Working with these physicists broadened and deepened my understanding of quantum computation, topological order, and physics in general, ultimately making this book a reality.

I would also like to thank my PhD advisor, Professor Lee Smolin, an acclaimed physicist and a founder of loop quantum gravity. His acceptance of me as a PhD student preserved my career and ideal of becoming a theoretical physicist. My gratitude extends to Professors Xiaoyi Bao and Liang Chen, renowned experts in optics at the University of Ottawa, for accepting me – despite my having no background in physics – as their Master's student. This opportunity allowed me to switch from computer science to physics and realize my childhood dream of being a physicist, while also allowing my wife and me to avoid a long-distance relationship.

My sincere thanks go to Professor Anthony Zee, an eminent theoretical physicist and inspiring author, for guiding my decision to write a textbook and influencing my approach. I must also thank Professor Lucien Hardy, renowned physicist in quantum information and quantum foundations at the Perimeter Institute, for reading an early draft of the book and offering encouraging feedback, and Professors Huangjun Zhu at Fudan University, Yuting Hu at Hangzhou Normal University, and Chenjie Wang at the University of Hong Kong for their careful proofreading of several chapters.

This book took shape alongside my teaching responsibilities for the quantum computation course at Fudan University. Balancing these commitments was made easier by a dedicated team of students (hired through Fudan University Grant No. IAH1512004: Seven Major Series of One Hundred Premium Textbooks) who transformed my initial sketches into most of the figures and maintained a comprehensive Zotero library of references. I am profoundly grateful to Yu Zhao, Yingcheng Li, Yanyan Chen, Siyuan Wang, Ruohan Yang, and Jiaqi Guo for their invaluable assistance. I also benefited from the keen eyes of students enrolled in the course, including Hengzhun Chen, Chengmeng Wang, Yiwen Wu, and Chenye Li, who identified numerous typos and errors in the draft.

I would like to thank Professor Zixiang Hu at Chongqing University for inviting me to teach a lecture series covering most of the book. These lectures helped make the book more approachable for students and researchers of diverse backgrounds.

As a first-time textbook author working outside the traditional Western academic sphere, approaching Cambridge University Press could have been daunting. Cambridge's clear, concise author guidance and the patient, supportive expertise of Commissioning Editor David Liu eased any apprehensions and made the publishing journey a positive experience. I am deeply grateful to Managing Editor Jane Adams, Editorial Assistant Roxanne Daruwalla, Copy Editor Tom Moss Gamblin for their meticulous editing of the manuscript, ensuring linguistic precision and stylistic coherence, and Senior Project Executive Vigneswaran Gurumurthy for his kind assistance in proofreading. Their invaluable support extended to coordinating revisions, managing timelines, and addressing countless logistical details throughout the publication process.

I acknowledge the hospitality of the Perimeter Institute, where I completed segments of this book during my stints as a visiting fellow. During one visit, I also taught a minicourse (on Chapters 8 and 9), which helped me further refine certain details in those chapters.

1   Overview

Although this book focuses on quantum computation, the architecture of quantum computing – in particular, the quantum circuit model, which is by far the most popular quantum computing scheme – has not diverged significantly from that of classical computing. To better appreciate the necessity and facilitate the understanding of quantum computing from the ground up, we shall first briefly review the core concepts of classical computing. In this chapter, we will explore the fundamental differences between classical and quantum computers and discuss the compelling reasons for the development of quantum computing technology. We will introduce you to various types of quantum computers, different models of quantum computation, and the current state of the art in this rapidly evolving field. Additionally, we will examine the concept of quantum information, elucidating its intrinsic relationship with quantum computation.

1.1   Classical Computers and Quantum Computers

We shall first take a tour through the evolution of computing, starting from classical computers to quantum computers. We'll discuss their capabilities and limitations, and why the transition from classical to quantum is so revolutionary.

1.1.1   Classical Computers

We begin by delving into the well-established world of classical computing. We'll explore how classical computers work, their historical evolution, and the challenges they face as we move towards a future dominated by quantum technologies.

What is a Computer?

What is a computer? Imagining a sleek laptop or a massive server, you're not wrong – but there's more to it. As shown in Figure 1.1, a computer is a "black box" that gobbles up input, munches on it using some algorithms, and then spits out an output [1]. Inside that box, you've usually got a memory stashing all your data and a control unit acting like a maestro, orchestrating the entire algorithmic symphony [2].

Figure 1.1 A computer is an information processor.

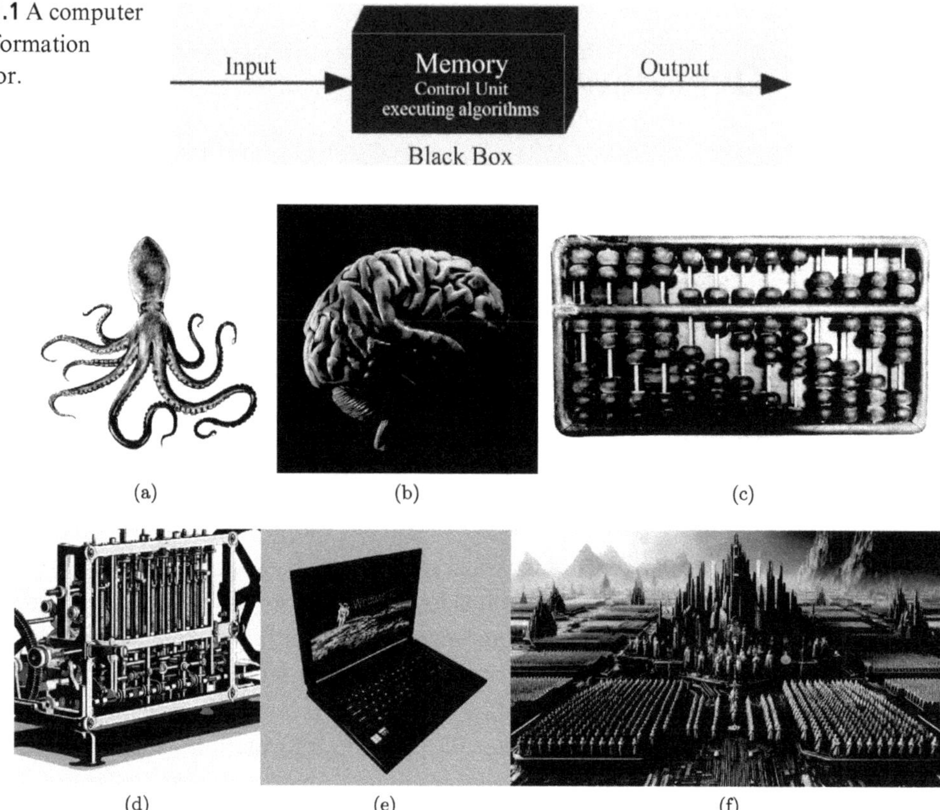

Figure 1.2 Examples of computers: (a) an octopus, (b) a human brain, (c) a Chinese suanpan, (d) Babbage's analytic machine, (e) a laptop, (f) a 3D human–matrix computer in "The Three-Body Problem."

Computers

One of the world's first "computers" was a bunch of beads on sticks, aka the Chinese suanpan [3]. And here's a curveball: Even your brain is a sort of computer, an amazing organic one at that [4]. Fast-forward to the steampunk era, and we find Charles Babbage messing around with his mechanical Difference Engine [5]. Babbage is often regarded as the father of the computer. It's also worth mentioning Babbage's student, Ada Lovelace, born in 1815, who first recognized that Babbage's machine had applications beyond pure calculation, and published the first algorithm intended to be carried out by such a machine [6]. As a result, she is often regarded as the first computer programmer. Ever read *The Three-Body Problem* by Cixin Liu? That sci-fi novel imagines a whole new concept of a computer made up of ... humans [7]! See Figure 1.2 for illustrations of these examples.

One Computer Doesn't Fit All

Some computers are specialists, like calculators that are whizzes at math but can't help you browse the web. These are what we call special-purpose computers [8].

Figure 1.3 The first modern computer in the world – ENIAC. Credit: The original uploader was TexasDex at English Wikipedia; transferred from en.wikipedia to Commons by Andrei Stroe using CommonsHelper. https://creativecommons.org/licenses/by-sa/3.0/.

Then there are computers, like your laptop, which can do a bit of everything – these are the general-purpose ones [9].

A Pocket-Sized History Lesson

The first general-purpose electronic digital computer was called ENIAC, and it was so big that it could fill an entire room [10]. See Figure 1.3.

As a graduate student in computer science at the University of Pennsylvania, the author had the unique privilege of walking by ENIAC – displayed behind a glass wall – almost every day from 2000 to 2002. It's a vivid illustration of how far we've come, a humbling reminder of our technological roots. Seeing those vacuum tubes, switches, and wires wasn't just a step back in time; it was a daily inspiration. So, what propelled us from room-sized giants like ENIAC to the sleek laptops we use today? Enter Moore's Law.

Moore's Law is the idea that the number of tiny transistors you can pack onto a chip will double about every two years [11]. While it's not a law like gravity, it's been a pretty good bet for the past several decades, turning room-sized giants into pocket-sized miracles [12]. Figure 1.4 provides a road map of the key milestones in the development of integrated circuits. Remarkably, Moore's Law has held true empirically for several decades, with some of the most advanced integrated circuits containing up to 50 billion transistors as of 2024.

1.1.2 Quantum Computers

Now, get ready to meet the new prodigies of the computational world: quantum computers. Brace yourselves as we explore how these marvels leverage the quirks of quantum mechanics to achieve feats unimaginable for their classical counterparts.

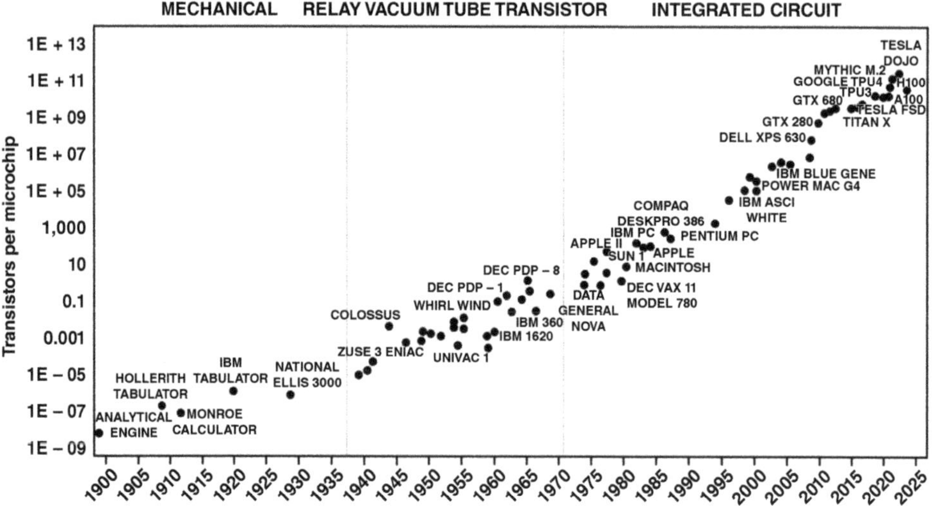

Figure 1.4 Moore's Law up to 2025.

What Makes a Computer "Quantum"?

If you're thinking, "Hey, don't modern computers already use quantum mechanics in transistors?", you're onto something. But here's the twist: That doesn't make them quantum computers [13]. The game changer? Quantum logic. Unlike classical computers, which use Boolean logic, quantum computers are all about quantum superposition, entanglement, and the wild wonders of quantum mechanics.

One Quantum Computer Doesn't Fit All

There's more than one kind of quantum computer. Some are specialists, often called quantum simulators, tailored to simulate other quantum systems [14]. Then there are the general-purpose ones, the Swiss Army knives of quantum computing [13].

Models of Quantum Computation

In the realm of quantum computing, a "model" serves as a conceptual framework that defines how quantum information is processed and manipulated. Just as classical computers can be understood through models like the Turing machine, quantum computers also have their own set of models – blueprints, if you will – that help us understand their inner workings. Here is a list of four important models:

- **Quantum circuit model**: This is the quantum cousin to the classical circuit model of computing, which underpins modern computers. Bits are transformed into "qubits," and Boolean logic gates morph into quantum gates [13]. This is the most mainstream model nowadays because it's the most straightforward and works similarly to its classical cousin.

- **Adiabatic quantum computing (AQC):** This model is all about evolving from a simple Hamiltonian[1] H_I to a complex Hamiltonian H_F. The magic formula?

$$H(t) = (1 - t/T)H_I + (t/T)H_F. \tag{1.1}$$

  Stick to the adiabatic route, and voilà, you end up at H_F's ground state, where your problem's solution lies [15].

- **Holonomic quantum computing:** Here, quantum gates are realized through non-Abelian geometric phases. These are phases that the quantum states pick up during their adiabatic evolution. It's like a car turning on a winding mountain road – the curve of the path is what counts [16].

- **Topological quantum computation:** Forget circuit boards; in this model, you're braiding anyons around one another to perform computations. Now, what are anyons? These particles don't quite fit into our neat categories of bosons and fermions. They live in two dimensions and have more exotic statistics. When you swap anyons, the quantum state of the system can pick up a phase factor that is not simply 1 or -1, but some fraction thereof. In topological quantum computing, these anyons are manipulated to perform quantum operations. Their states are topologically protected, which makes them less susceptible to certain kinds of errors [17, 18].

Don't fret if some of these terms seem intimidating. We'll break down all this jargon in the chapters to come. So hang tight, and let's enjoy the quantum ride together!

The Path Forward: A Focus on Circuit and Topological Models

Now that we've glimpsed the colorful landscape of quantum computing models, you might be wondering where we go from here. Buckle up because we're going to deep-dive into the quantum circuit model and topological quantum computation in this textbook. While holonomic and adiabatic quantum computing have their merits, they're not too far removed from the circuit model in terms of their underlying principles. But topological quantum computation? That's an entirely different beast. In topological quantum computation, there aren't even physical qubits; instead, we work directly with logical qubits embedded in the Hilbert space of anyons. Don't worry if these terms sound overwhelming now; we'll demystify all the quantum jargon in the chapters to come.

Quantum Computing Platforms

After understanding the theoretical models that serve as the backbone of quantum computing, it's time to meet the real-world implementations or "platforms." These are the physical systems designed to bring those conceptual models to life.

Before we dive into the different platforms, let's get one thing straight: all of them are trying to manipulate something called a "qubit." You can think of a qubit as the

---

[1] Chapter 3 will explain such concepts.

quantum version of the bits in your classical computer. Unlike regular bits, which can be either a 0 or a 1, qubits can exist in a superposition – essentially, they can be both 0 and 1 at the same time. This superposition allows quantum computers to perform complex calculations much faster than classical computers for certain tasks. But enough about that for now; you'll get the full rundown in the chapters to come.

With this in mind, let's meet a few mainstream platforms for universal quantum computing.

- **Superconducting qubits**: Picture a circuit where electricity flows without any loss – yes, that's "superconducting." In this almost magical circuit, electrical currents can be in a superposition of flowing clockwise and counterclockwise at the same time [19]. The foundational work demonstrating this macroscopic quantum behavior was pioneered by John Clarke, Michel H. Devoret, and John M. Martinis, whose contributions to macroscopic quantum tunneling and energy quantization in electrical circuits were recognized with the 2025 Nobel Prize in Physics. And how do we control it? Through precise microwave pulses, which IBM and Google are getting pretty good at.
- **Ion-trap systems**: Imagine ions floating in a vacuum chamber, perfectly still thanks to electric fields. How to control them? Lasers come into play, interacting with these ions to place them into different quantum states. These systems offer long coherence times,[2] so they're stable and reliable [20, 21].
- **Neutral atoms in optical lattices**: Think of a 3D grid made of lasers, each cell hosting a neutral atom. The atoms can be in superpositions of different energy levels. To control them, you adjust the laser parameters, effectively playing quantum chess [22].
- **NMR-based platforms**: Think of specific atoms within molecules as tiny spinning tops, their nuclear spins serving as our qubits (NMR stands for nuclear magnetic resonance). Using radio-frequency pulses, we can control the rotation and interaction of these nuclear spins to represent various quantum states and perform quantum logic [21, 23].
- **Topological qubits**: Envision a world where information is stored not in traditional qubits but in the knotty configurations of trajectories of quasiparticles often referred to as anyons. To manipulate these qubits, you'd move these anyons around each other. Temporally, the worldlines of the anyons form a braid.

Note that the first four platforms are primarily designed to implement the quantum circuit model. In contrast, the last platform, topological qubits, aims to bring the topological quantum computing model to life.

1.1.3 Why Do We Need Quantum Computers? Facing the Limits and Beyond

It's time to explore the compelling reasons that make quantum computers not just an intriguing academic endeavor but a necessary advancement in computational

---

[2] Coherence time will be explained in Chapter 6. Briefly, it is the time period during which a qubit state remains quantum.

technology. From the limitations of classical computing to the specific challenges that only quantum machines can efficiently tackle, we'll examine why there's so much investment in making quantum computing a reality.

The Impending Doom of Moore's Law

All good things must come to an end. The microprocessor's golden age is tapering off, and we're running out of ways to keep Moore's prediction alive. In plain speech, it's getting tougher to keep shrinking transistors and packing them into chips [11, 24].

Beyond the Limits: Problems Classical Computers Struggle With

Imagine a Sunday morning – there's a steaming cup of coffee on your table, and you're grappling with a Sudoku puzzle. Just as you're working through its logic, classical computers are tackling much larger puzzles, like keeping our digital world secure. One of the most significant of these challenges is the Rivest–Shamir–Adleman (RSA) cryptosystem [25, 26], the bedrock of our digital security protocols.

RSA is the linchpin of online banking, secure email services, and digital signatures [27, 28, 29]. Its computational robustness is anchored in the difficulty of prime factorizing large numbers – a problem so hard that even the best classical computers struggle with it. And that robustness has made RSA the gold standard in public-key cryptography.

But here comes the twist: Quantum computing promises to upend all this by tackling such problems exponentially faster [13]. Before we dig into this revolutionary transition, let's hop into a time machine and head back to the 1980s, where two brilliant minds – Yuri Manin and Richard Feynman – began contemplating the future of computing.

Manin and Feynman were deeply motivated by the inadequacies of classical computers in simulating complex quantum systems. You see, the quantum world obeys rules that are fundamentally different from our classical reality. Manin observed that "A classical computer is doomed to simulate quantum systems with an exponential slowdown" [30]. Feynman, meanwhile, found it intolerable that classical computers were inadequate for simulating the quantum behaviors seen in nature. "Nature isn't classical, dammit," he said, "and if you want to make a simulation of nature, you'd better make it quantum mechanical [31]."

Their vision was ahead of its time but deeply rooted in their frustration with classical computing's limitations. What they suggested was more than a shift in computational power; it was a shift in how we understand and interact with the universe itself.

So, get ready for a journey into the deepest puzzles of computation and reality. This isn't just an academic exercise; it's a real-world revolution. The world is already in the midst of a quantum upheaval, and this book is your passport to that brave new universe. Are you up for the challenge?

Table 1.1 AND gate truth table

a	b	a AND b
0	0	0
0	1	0
1	0	0
1	1	1

Table 1.2 OR gate truth table

a	b	a OR b
0	0	0
0	1	1
1	0	1
1	1	1

Saving Energy, One Qubit at a Time

You might not be aware of it, but classical computing can be quite the energy guzzler. Why? Say hello to Landauer's Principle [32]. According to this principle, any logically irreversible operation – such as erasing a bit of information – requires a minimum amount of energy to be dissipated as heat. To be exact, that's $kT \ln(2)$ joules for each bit erased, where k is the Boltzmann constant and T is the temperature of the system. We shall return to this principle later in Chapter 2.

To get a clearer picture, let's look at classical logic gates, like the AND and OR gates, which are, you guessed it, irreversible. Their truth tables are shown in Tables 1.1 and 1.2.

Notice anything peculiar? You can't backtrack from the output to the inputs in either the AND or OR gates. That's a one-way street. The reason? These gates take in two bits but spit out just one. So, they erase a bit of information, and according to Landauer, that's where the energy consumption comes in.

Quantum gates, on the other hand, are like the frugal cousins in a family: They're inherently reversible [13]. To be precise, a quantum gate is represented by a unitary matrix U; by definition it meets the condition $UU^{\dagger} = \mathbb{1}$, where $U^{\dagger}$ is the transpose of the complex conjugate of U, and $\mathbb{1}$ is the identity matrix of the commensurate size. As such, any quantum gate always has an inverse gate $U^{-1} \equiv U^{\dagger}$ and is thus reversible. Hold your horses! This doesn't mean quantum computers are some sort of free lunch; they too consume energy, but in a different manner [33].

Concluding Thoughts

As we hit the limits of Moore's Law, struggle with computationally insane problems, and wince at our escalating energy bills, the quantum computer emerges as a game-changing ally. Will it replace the classical computer? Unlikely, but they'll make an unbeatable tag team for tackling problems that are currently out of reach.

1.1.4   State of the Art in Quantum Computing: A Comprehensive Survey

Here, we will quickly explore the current frontiers in quantum computing, offering a survey that covers key milestones, technologies, and challenges that define the state of the art in this dynamic field. You will be left with two take-home messages about the status of quantum computation a well. So, buckle up.

A Historical Roadmap

The timeline for quantum computing has been a whirlwind, beginning with the conceptual foundations laid by Yuri Manin [30], Richard Feynman [31], and David Deutsch in the early 1980s [34]. The groundbreaking algorithms developed by Peter Shor and Lov Grover in the mid-1990s provided computational tasks where quantum methods could outperform classical ones [35, 36]. Since the early 2000s, we've seen small-scale implementations of quantum circuits and algorithms, largely due to advancements in quantum error correction methods [13, 37, 38]. Fast forward to the late 2010s, since when commercial companies including IBM, Google, and Rigetti Computing have been offering cloud access to their emerging quantum processors [39, 40, 41, 42, 43, 44, 45, 46, 47, 48]. Institutions in China have also launched cloud quantum platforms based on superconducting quantum chips, such as Zuchongzhi 2 developed by University of Science and Technology of China [49, 50], and Quafu developed by Beijing Academy of Quantum Information Sciences [51, 52, 53].

Current Records: A Platform-by-Platform Breakdown

- Superconducting circuits: Google's 72-qubit Bristlecone and the 53-qubit Sycamore processors have made headlines [40]. Notably, Sycamore was engineered for a specific task and was instrumental in Google's claim of "quantum supremacy" [39]. Scaling further, IBM's Condor chip debuted in 2023 as the world's first quantum processor with over 1,000 qubits [41]. Beyond raw scale, progress in error correction is critical; in 2025, Google's Willow chip showcased exponential error suppression on a 105-qubit array, marking a historic step toward fault-tolerant computation [42].
- Neutral atom arrays: A team led by Pan Jianwei at the University of Science and Technology of China used AI-guided optical tweezers to arrange 2,024 rubidium atoms into precise 2D and 3D arrays [43]. This breakthrough addresses a major scalability bottleneck in neutral-atom systems. More recently, a team in Harvard led by Mikhail D. Lukin managed to assemble and maintain an array of over 3,000 atom qubits for more than two hours [44]. Ion traps: Quantinuum's H2-1 system boasts the highest qubit count (56 qubits) for a commercial ion-trap quantum computer [45], renowned for its high-fidelity gates and advanced error correction capabilities.
- Semiconductor qubits: Intel is currently leading in silicon qubits, having developed the Tunnel Falls chip, a 12-qubit silicon-based processor provided to the research community [46]. Intel's partnership with QuTech demonstrated a two-qubit logic gate using silicon spin qubits, a step promising easier integration with existing semiconductor manufacturing [47].
- Topological qubits: Microsoft has pioneered this field with its Majorana 1 processor, the first to utilize a topological core architecture [48]. This chip leverages

a new class of materials called topological superconductors to create and control Majorana zero modes, which form the basis of topological qubits [54].

Keep in mind that the field of quantum computing is rapidly advancing; the records cited here may already be surpassed by the time you read this book. Nonetheless, this snapshot provides a historical perspective on the state of quantum computing at a certain point in time.

The Reality of Quantum Advantage

Google's claim of "quantum supremacy" in 2019 has stirred both excitement and controversy [39]. The term "quantum advantage" may be more appropriate, as quantum computing is not universally superior to classical computing. The existence of yet-to-be-discovered efficient classical algorithms cannot be ruled out [55, 56].

Take-Home Messages

In the fast-evolving field of quantum computing, it's easy to get caught up in the excitement and hyperbole. Media headlines often herald quantum computers as near-miraculous machines on the brink of revolutionizing everything from medicine to finance. While the potential is indeed staggering, it's crucial to temper this enthusiasm with a nuanced understanding of where the technology currently stands. The science is promising, but substantial engineering challenges remain. As we navigate through this transformative era, clear, accurate information is more valuable than ever. With this in mind, we offer two take-home messages to provide a balanced view of the state of the art in quantum computing.

- **The qubit quagmire**: Realizing a universal quantum computer, akin to the ENIAC in its era, remains a future milestone. The key bottleneck is the high number of physical qubits required for fault-tolerant computation. Current error-correction methods, such as the surface code, suggest around 1,000 physical qubits may be needed for a single logical qubit, although the number could vary depending on the specific implementation and error rates [57, 58, 59].
- **Engineering over science**: The focus has shifted from the question of "Can it be done?" to "How do we build it?" While the science is mostly settled, the engineering remains an immense challenge. Overcoming hurdles like decoherence and scaling up quantum systems are the next steps [60]. The rise of quantum software, middleware, and error-correction techniques will go hand in hand with hardware advancements.

Quantum Computation in Practice

While quantum computing is still in its infancy, several companies and institutions have already started tapping into its potential to revolutionize various sectors. The promise of unprecedented computational power has inspired pioneering practical applications in diverse fields, from automotives to high-energy physics:

- **Mercedes-Benz:** The automobile giant has been exploring quantum algorithms to optimize routes for its fleet of delivery vans. Traditional route optimization can be computationally challenging, especially as the number of vans and delivery points increases. Quantum algorithms offer a more efficient solution, potentially minimizing fuel consumption and reducing delivery times [61].
- **ExxonMobil:** The energy sector stands to gain immensely from quantum computing. ExxonMobil has been investigating the applications of quantum algorithms in energy sector challenges, such as carbon capture and storage or optimizing large-scale operations and manufacturing. These endeavors are not just about increasing efficiency but also about pioneering more sustainable energy solutions for the future [62].
- **CERN:** The European Organization for Nuclear Research (CERN) pushes the boundaries of what we know about the universe. Analyzing data from particle collisions in the Large Hadron Collider (LHC) requires immense computational resources. Quantum algorithms could drastically speed up this data analysis, helping scientists at CERN get answers faster about the fundamental structure of the universe [63].
- **Mitsubishi Chemical:** The chemical industry relies heavily on simulations to predict molecular properties and reactions. Nevertheless, simulating complex molecules and reactions can be beyond the reach of even the most powerful classical computers. Mitsubishi Chemical has been partnering with quantum computing companies to develop algorithms that could simulate chemical reactions more efficiently, paving the way for new material discoveries and more efficient production processes [64].

These applications underscore the real-world potential of quantum computing. While the technology has yet to fully mature, its implications in reshaping industries and advancing scientific frontiers are undeniably profound.

Concluding Thoughts

The current state of the art in quantum computing is awe-inspiring yet humbling. The major challenges ahead are more in the realm of engineering than basic science. As an evolving field, the state of the art is a moving target, poised for unexpected leaps as technologies mature.

1.2 Quantum Information: Beyond Computation

Quantum information refers to the information content within a quantum system. It is the lifeblood of quantum computing and extends to a broader field known as quantum information theory. This theory explores the mathematics and physics behind the properties, manipulation, and transmission of quantum information. Its applications form a vast landscape that goes beyond computation, revolutionizing cryptography, communication, and sensing. While we'll touch on the foundational elements of quantum information, the primary focus of this book remains on quantum computing, leaving other captivating aspects of quantum information for you to explore elsewhere.

In what follows, I'll present a roadmap of quantum communication – one of the most important applications of quantum information theory – and highlight the interplay between quantum information theory and high-energy physics.

The field of quantum communication has evolved substantially in recent years, especially in quantum key distribution (QKD), entanglement distribution, quantum teleportation, quantum memory, and quantum repeaters. Notable milestones are:

- **Long-distance QKD records**:
  - *Land-to-Land*: A 404 km terrestrial QKD was established over fiber-optic networks, marking the longest in this category [65]. The current long-distance Quantum Key Distribution (QKD) record is 12,900 km, achieved via a satellite-based quantum communication link between China and South Africa using the microsatellite Jinan-1 [66].
  - *Space-to-Land*: The Micius satellite achieved a record 1,200 km QKD link with a ground station [67].
- **Quantum entanglement distribution**:
  - *Land-to-Land*: A distance of 144 km was achieved between two of the Canary Islands [68].
  - *Space-to-Land*: Micius successfully distributed entangled photons over a distance of 1,200 km to separate ground stations [67].
- **Quantum teleportation**:
  - This is a process of transferring quantum states from one location to another without physical transmission of the underlying qubits [69].
  - The farthest distance quantum teleportation, over 1,200 km, was again carried out by Micius [67].
- **Quantum memory**:
  - This is a device that can store and retrieve quantum states on demand [70]. Quantum memory research has achieved extended storage times, improved coherence, and higher fidelity in various platforms. The most advanced approaches use atomic ensembles [71, 72, 73], rare-earth-ion-doped crystals at cryogenic temperatures [74], and optomechanical or superconducting resonators coupled to quantum bits [75]. Record coherence times exceed seconds and can reach hours in rare-earth-doped crystals, though operational complexity and requirements remain high. Integrated photonic architectures have improved compatibility and scalability, enabling compact designs and on-demand storage. Nonetheless, challenges persist in simultaneously achieving long storage times, high efficiency, and robust scalability. Hybrid systems

and novel materials may optimize memory performance and integration with quantum networks.

- **Quantum repeaters**:
  - These are crucial for extending quantum communication networks and are currently in the developmental phase [76]. They combine quantum memory and quantum teleportation to relay quantum information over long distances.
- **Commercialization and prototypes**:
  - Several companies are commercializing QKD systems, and some countries are planning or have deployed quantum-secure networks [77].
  - The commercialization of quantum computing has accelerated, with over 20 companies, such as IBM, Amazon, and Microsoft, now offering services that include hardware access, software tools, and industry-specific applications.

While the focus of this book is on quantum computation, it's worth mentioning that quantum information theory has found applications in high-energy physics. These include attempts to understand phenomena such as the black hole information paradox [78, 79], entanglement entropy in field theories [80, 81, 82, 83, 84], and the emergence of spacetime geometry from quantum correlations [85, 86, 87]. Recent studies have demonstrated how quantum entanglement can appear in collider experiments [88], shedding light on parton distributions and the internal structure of hadrons. Connections between quantum complexity measures and gravitational phenomena, as suggested by the AdS/CFT (anti-de Sitter/conformal field theory) correspondence, have also been explored [89, 90], providing a window into the geometric interpretation of complexity and the holographic duality. Quantum error correction has been linked to the robustness of holographic codes that describe bulk gravitational physics [91], bridging concepts from quantum information and quantum field theory. Although fascinating, these topics are beyond the scope of this book.

1.3 Outlook: The Quantum Future

Imagine a world where quantum computers are as ubiquitous as smartphones. In such a world, the barriers of classical computation would be shattered, opening new horizons for humanity. Medical research could leap forward, discovering cures for diseases that have stymied scientists for generations. Climate modeling would achieve unprecedented accuracy, equipping us to face the planet's most pressing challenges. And let's not forget about cryptography: In our quantum future, your online security could be practically unbreakable, thanks to quantum encryption techniques.

At the heart of this revolution would be our newfound ability to simulate quantum systems, an idea conceived by visionaries such as Yuri Manin and Richard Feynman. It would no longer take years to simulate complex molecular structures; pharmaceutical companies could do it over a cup of coffee. Feynman's dream of mimicking nature "as it really is – quantum mechanical" would finally come true.

But this future doesn't come without its challenges. Quantum computers could potentially crack encryption algorithms that secure our current digital world. Hence, quantum-safe encryption is not just an option; it's a necessity. Moreover, the ethical implications of such computational power must be addressed.

The roadmap to this quantum future is still under construction. But every milestone reached in quantum key distribution, entanglement distribution, and quantum error correction brings us closer to this new reality. Researchers and engineers are the architects of this future, and the next generation of students could well be the builders. The challenges are many, but the rewards – both intellectual and societal – are monumental.

And speaking of grand visions, let's not forget the realm of science fiction, where boundless imagination has frequently blazed the trail for real-world science and new technologies. In the blockbuster movie "The Wandering Earth," humanity faces an existential crisis as the Sun is about to expand into a red giant and devour the Earth. The solution? A project to propel the Earth itself to a safer orbit around another star, using gigantic "Earth Engines." And guess what's at the heart of these colossal engines? You guessed it – quantum computers! While we're not suggesting that quantum computers will literally move planets (at least not anytime soon), the film serves as an exciting testament to how deeply the idea of quantum computing has permeated our collective imagination.

As a student embarking on this quantum journey, you are not just a passive recipient of knowledge, but potentially an active contributor to a field that has the potential to redefine reality as we know it. Will you be one of the pioneers who conquers these quantum challenges and helps realize this extraordinary vision? The opportunities are boundless, and the stage is set for groundbreaking innovation. Welcome to the frontier of quantum computing; your imagination is the only limit.

Subsequent chapters will not only explore the intricate workings of quantum computers but also shed light on how they could reshape our understanding of computation and, by extension, our understanding of the universe.

1.4 Physical Limitations of Computation

You may be contemplating the limitless possibilities of ever faster and ever more capable computing architectures, whether classical or quantum. Nevertheless, it's essential to recognize that all forms of computation are fundamentally constrained by physical and mathematical principles. These limitations are not just engineering challenges but are rooted in the laws governing matter, energy, and the fabric of spacetime itself, as well as in the mathematical properties inherent to computation. This section provides an overview of some of the most salient constraints to bear in mind:

- The **clock frequency** is the rate at which a computer executes instructions. It is limited by the speed at which signals can propagate through the hardware components, such as wires, transistors, and capacitors. The speed of signal propagation is ultimately limited by the vacuum speed of light, which is about 3×10^8 meters per second. Therefore, the clock frequency cannot exceed the inverse of the time it takes for a signal to travel across the smallest component of the computer. For example, if the smallest component is one nanometer in size, the maximum clock frequency is about 300 GHz [92]. The clock frequency is also affected by the power consumption and heat dissipation of the computer. Higher clock frequencies require higher voltages and currents, which increase the power consumption and generate more heat. Excessive heat can damage the hardware components or reduce their performance. Therefore, cooling systems are needed to maintain a safe operating temperature. That said, cooling systems also consume power and space, and may introduce noise and vibrations. These factors impose practical limits on how much the clock frequency can be increased without compromising the reliability and efficiency of the computer.

- The **speed of light** limits not only the clock frequency but also communication between different parts of a computer or between different computers. The speed of light sets a lower bound on the latency, or delay, of any communication. For example, if two computers are separated by one meter, the minimum latency is about three nanoseconds. This means that any computation that requires data exchange between distant locations will incur some overhead due to the finite speed of light. This overhead can be significant for parallel or distributed computing, where many computers work together to solve a problem. The speed of light also limits the size and shape of a computer. A larger computer will have longer distances between its components, which will increase communication latency and limit synchronization. A spherical computer will have the minimum possible latency for a given volume, but it may not be the optimal shape for other aspects of computation, such as modularity, scalability, or fault tolerance. A computer shaped like a torus or a hypercube may have better properties for parallel or distributed computing, but it will also have longer communication paths and higher latency.

- The **Planck time** is the smallest unit of time that has any physical meaning. It is about 5.4×10^{-44} seconds. It is derived from the Planck constant, which is a fundamental constant that relates energy and frequency, the gravitational constant, and the speed of light in the vacuum. The Planck time sets a limit on how fast any physical process can occur, including computation. It is believed that quantum gravity effects become significant at this scale, and that the laws of physics as we know them break down [93]. Therefore, no computer can perform more than one operation per Planck time, and no computer can measure time intervals smaller than the Planck time. The Planck time is so small that it is unlikely that any physical process can reach it. Nonetheless, there may be other quantum limits on the speed of computation that are closer to reality. For example, some quantum algorithms require a certain amount of coherence time, which is the time during which the quantum system maintains its superposition state without being disturbed by external factors. The coherence time depends on the quality of the quantum hardware and the environment, and it sets a limit on how long a quantum computation can run before it becomes corrupted by noise or decoherence.

- The **memory density** is the amount of information that can be stored in a given volume of space. It is limited by the number and size of the physical units that store information, such as bits or qubits. The number of units is limited by the number of particles or states that can exist in a given volume, and the size of units is limited by the minimum distance at which they can be distinguished. The ultimate limit on memory density is given by the Bekenstein bound, which states that the entropy (or information) contained in a spherical region of space cannot exceed one quarter of its surface area in Planck units [94]. This means that there is a maximum amount of information that can be stored in any physical system. The memory density is also influenced by the error rate and access time of the memory units. A higher memory density may imply a higher error rate, as the units become more susceptible to interference or damage. Therefore, error correction codes are needed to ensure the reliability of the data storage. Nevertheless, error correction codes also consume space and time, and may reduce the effective memory density. Similarly, a higher memory density may imply a longer access time, as the units become more crowded or distant. Therefore, memory hierarchy and caching techniques are needed to optimize the performance of data retrieval. But these techniques also consume space and time, and may reduce the effective memory density.

- The **uncertainty principle** is a fundamental principle of quantum mechanics that states that certain pairs of physical quantities, such as position and momentum, cannot be simultaneously measured with arbitrary precision. This means that there is an inherent randomness and fuzziness in the behavior of quantum systems, which can affect computation in various ways. For example, quantum computers use qubits, which are quantum units of information that can exist in superpositions of two states (0 and 1). Qubits can perform parallel computations by exploiting quantum interference, but they are also prone to errors due to quantum decoherence. Therefore, quantum computers face trade-offs between speed and accuracy, and require error correction techniques to achieve reliable results. The uncertainty principle also limits the precision and accuracy of computation. A higher precision or accuracy may require a higher energy or time expenditure, which may not be feasible or desirable for some applications. For example, some physical and mathematical constants may have infinite decimal expansions, but only a finite number of digits can be computed or stored with finite resources. Therefore, approximation methods and numerical analysis techniques are needed to deal with irrational or transcendental numbers. Nevertheless, these methods also introduce errors and uncertainties, and may affect the validity or quality of the computation.

Be aware that any physical law has its applicable regime and is based on our accumulated yet limited understanding of nature. Thus, it is not entirely impossible that future observations or experiments may reveal a broader regime, or phenomena that contradict some of the known physical laws. And you, reader, may be the one who would breaks through the barriers.

Conclusion

In this chapter, we introduced the foundational concepts of classical and quantum computing, setting the stage for a deeper exploration of quantum computation. We began by revisiting the history of classical computers, from early mechanical machines to modern digital systems, examining their core architectures and limitations. We then explained why quantum computing is necessary and represents a revolutionary leap beyond classical computing, especially for tackling computational tasks such as simulating quantum systems and breaking classical encryption schemes. We also provided a survey of various quantum computing models, including the quantum circuit model and topological quantum computation, which will be explored in more detail later.

In the next chapter, we will review the key concepts of classical computation, including Turing machines, universal computation, and computational complexity. We'll also cover examples of problems in different complexity classes, examine energy considerations like Landauer's principle, and introduce reversible computation, which plays a crucial role in both classical and quantum computing. By strengthening your understanding of these classical fundamentals, you'll be well prepared to engage with the more advanced topics in quantum computing.

2   Classical Computation

In this chapter, we will explore some of the key aspects of classical computation. This will help you understand the basics of computation and prepare you for quantum computation. If you already have a solid background in classical computation, you can skip this chapter and go to the next one.

We will start by introducing the Turing machine [95, 96], which is the theoretical model of classical computation. Nevertheless, since Turing machines are not very practical to implement, we will also learn about the circuit model, which is the basis of modern computers. Next, we will discuss two important concepts: computational complexity and reversible computation. Computational complexity measures how hard it is to solve a problem using a computer. Reversible computation means that the input can be recovered from the output. Classical computers often struggle with complex problems and are usually irreversible. Irreversibility has a hidden cost of energy. These two limitations of classical computation motivate us to go for quantum computation.

2.1   Turing Machines

The story of computation began with a question posed by David Hilbert [97]: Is there an algorithm that can solve any mathematical problem? Before we can answer this question, we need to know what an algorithm is and what it means to solve a problem. We also need to consider the implications of Kurt Friedrich Gödel's incompleteness theorem [98], which states that for any consistent formal system capable of expressing elementary arithmetic, there are true propositions that cannot be proven or disproven within the system itself.[1] Therefore, a more realistic question is: What kinds of problems can be solved or what kinds of functions can be computed by an algorithm? To answer this question, we need rigorous definitions of an algorithm and of computability. This is where Alonzo Church and Alan Turing came in. They proposed the following thesis, which relates the notion of an algorithm to the concept of a Turing machine, invented by Turing in 1936.

---

[1] This theorem has a trivial exception: Presburger arithmetic (the first-order theory of the natural numbers with addition only, no multiplication).

The Church–Turing thesis [99, 100]: The class of all functions computable by a Turing machine is equivalent to the class of functions computable by an algorithm. In other words, if we can define a Turing machine precisely, we can also define an algorithm and computability.

In plain words, a Turing machine is a simple, abstract model of computation that consists of a tape, a head, and a set of rules. The tape is divided into cells that can store symbols. The head can read and write symbols on the tape and move left or right. The rules tell the head what to do based on the current symbol and state. A Turing machine can perform any computation that can be described by an algorithm, as long as it has enough time and space. Turing also proved that there are some problems that cannot be solved by any Turing machine, such as the halting problem, which asks whether a given Turing machine will ever stop. (We shall look at the halting problem later.)

Definition 2.1. *Formally, a Turing machine can be defined as a 7-tuple*

$$M = (Q, \Sigma, \Gamma, \delta, q_0, q_f, q_h), \tag{2.1}$$

where

- *Q is a finite set of states;*
- *Σ is a finite set of input symbols;*
- *Γ is a finite set of tape symbols, such that $\Sigma \subseteq \Gamma$ and the blank symbol $\sqcup \in \Gamma \setminus \Sigma$;*
- *$\delta \colon Q \times \Gamma \to Q \times \Gamma \times \{L, R, -\}$ is the transition function, which maps the current state and tape symbol to the next state, tape symbol, and head movement (indicated by L, moving one cell to the left, by R, moving one cell to the right, and by $-$, staying);*
- *$q_0 \in Q$ is the initial state;*
- *$q_f \in Q$ is the accept state, which is one of the possible halting states;*
- *$q_h \in Q$ is the reject state, which is the other possible halting state, such that $q_f \neq q_h$.*

A Turing machine also has an infinite tape comprising cells that can store symbols from Γ. The tape is initially filled with the blank symbol $\sqcup$, except for a finite segment that stores the input string $w = w_1 w_2 \ldots w_n$, where each w_i is a symbol from Σ. The input string occupies the leftmost cells of the tape, and the rest of the tape is blank. The machine also has a head that can read and write symbols on the tape and move left or right. The definition of a Turing machine requires that the sets Q, Σ, and Γ are finite and that the tape is infinite. These conditions are important because they ensure that the machine can be described and implemented in a precise and unambiguous way.

The finitenesses mean that the machine has a finite number of states, a finite number of input symbols, and a finite number of tape symbols. These limit what the machine can do and understand. The infiniteness means that the machine has an unlimited amount of memory to store information on the tape. This allows the machine to perform any computation that can be described by an algorithm, as long as it has enough time and space. You can think of these finitenesses and the infiniteness as analogous to the limitations of human brains. For example, human brains have a finite number of neurons, which can be seen as the states of the brain. Human brains can also process only a finite number of symbols, such as words or images, which can be seen as the input and tape symbols of the brain. These limitations constrain what human brains can do and understand. But a human brain can read an infinite amount of information, at least in principle.

A Turing machine can accept or reject an input string $w = w_1 w_2 \ldots w_n$ by starting from the initial state q_0 and scanning the leftmost symbol of the tape. Then, it applies the transition function δ to determine the next state, the symbol to write on the tape, and the direction to move the head. It repeats this process until it either reaches a halting state or loops forever. A halting state is a state that makes the machine stop. There are two possible halting states: the accept state q_f and the reject state q_h. The outcome of the computation depends on which halting state the machine reaches. If it reaches q_f, it halts and accepts the input. If it reaches q_h, it halts and rejects the input. If it never reaches either halting state, it never halts and loops forever. Note that often, a single halting state suffices, as can be seen in Exercise 2.4. To better understand Turing machines, let us study two examples.

Example 2.1

A simple example of a Turing machine is one that decides whether an input string is of the form $a^n b^m$: that is, whether it consists of n copies of a followed by m copies of b, where $n, m \geq 0$. For example, "", "a", "b", "ab", and "aabbb" are of this form, but "ba", "aba", and "aabab" are not. We can design, as follows, a Turing machine that accepts strings of this form and rejects otherwise.

- The input alphabet is $\Sigma = \{a, b\}$, which means that the machine can only accept strings that consist of a's and b's.
- The tape alphabet is $\Gamma = \{a, b, \sqcup\}$, which means that the machine can also use the blank symbol $\sqcup$ on the tape.
- The states are $Q = \{q_0, q_1, q_f, q_h\}$, where q_0 is the initial state, q_1 is an intermediate state, q_f is the accept state, and q_h is the reject state.
- The transition function δ is defined by Table 2.1.

Table 2.1 Transition function of the Turing machine in Example 2.1.

Current state	Current symbol	Next state	Next symbol	Head movement
q_0	a	q_0	a	R
q_0	b	q_1	b	R
q_0	$\sqcup$	q_f	$\sqcup$	–
q_1	b	q_1	b	R
q_1	$\sqcup$	q_f	$\sqcup$	–
q_1	a	q_h	a	–

To explain the machine step by step, let's take an example input string: "*aabbb*". We can represent the initial configuration of the machine as

$$aabbb \sqcup \ldots \tag{2.2}$$

$$\uparrow$$
$$\boxed{q_0}$$

Note that without any loss of generality, the read/write head always starts at the leftmost cell of the tape, where the leftmost symbol of the input string is *a*. The machine is in state q_0, which is the initial state. The rest of the tape is filled with blanks ($\sqcup$). The machine looks at the current state and symbol, and consults the transition function δ to determine what to do next. In this case, the current symbol is *a*. According to Table 2.1, if the current state is q_0 and the current symbol is *a*, then the machine should stay in state q_0, keep *a* on the tape, and move the head right. The new configuration of the machine is

$$aabbb \sqcup \ldots \tag{2.3}$$

$$\uparrow$$
$$\boxed{q_0}$$

The machine repeats this process until it reaches a *b* or a blank. In this case, it reaches a *b* after two steps. The configuration of the machine is

$$aabbb \sqcup \ldots \tag{2.4}$$

$$\uparrow$$
$$\boxed{q_0}$$

According to the table, if the current state is q_0 and the current symbol is *b*, then the machine should move to state q_1, keep symbol *b*, and move the head right. The new configuration of the machine is

$$aabbb \sqcup \ldots \tag{2.5}$$

$$\uparrow$$
$$\boxed{q_1}$$

The machine now enters state q_1, which is an intermediate state. According to the table, if the current state is q_1 and the current symbol is *b*, the machine should stay in state q_1, keep *b* on the tape, and move the head right. This means that the machine leaves every *b* it sees unchanged, and moves to the next symbol. The new configuration of the machine is

$$aabbb \sqcup \ldots \tag{2.6}$$

$$\uparrow$$
$$\boxed{q_1}$$

The machine repeats this process until it reaches a blank or an *a*. In this case, it reaches a blank after two steps. The configuration of the machine becomes

$$aabbb \sqcup \ldots \tag{2.7}$$

$$\uparrow$$
$$\boxed{q_1}$$

According to the table, if the current state is q_1 and the current symbol is $\sqcup$, then the machine should move to state q_f, write $\sqcup$ on the tape, and halt. The machine's final configuration reads

$$aabbb \sqcup \ldots \qquad (2.8)$$

$$\uparrow$$
$$\boxed{q_f}$$

The accept state q_f indicates that the machine accepts the input string "$aabbb$" as valid.

A Turing machine modifies its state when it needs to change its behavior based on the input or the tape content. For instance, in the Turing machine above, the machine changes its state from q_0 to q_1 when it sees a b after some a's. This is because the machine needs to switch from marking a's to leaving b's unchanged. The state change reflects this change of behavior. Similarly, the machine changes its state from q_1 to q_f when it sees a blank after some b's. This is because the machine needs to switch from scanning b's to accepting the input. The state change reflects this change of behavior as well.

> **Exercise 2.1** *Try the Turing machine in Example 2.1 with an input string that is to be rejected by the machine. Explain why it is rejected step by step.*
>
> **Exercise 2.2** *Design a Turing machine T that subtracts 1 from any given natural number $n \geq 1$, so that $T(n) = n - 1$. Hint: Take the unary representation of natural numbers[a].*
>
> ---
> [a] As a unary, a natural number n is expressed as $\underbrace{11 \cdots 1}_{n}$, and zero is simply the blank symbol $\sqcup$.

Example 2.2

A more interesting example of a Turing machine decides whether an input string takes the form $a^n b^n$, where $n \geq 0$. It is defined as follows.

- The input alphabet is $\Sigma = \{a, b\}$, which means that the machine can only accept strings comprising a's and b's.
- The tape alphabet is $\Gamma = \{a, b, \sqcup\}$, which means that the machine can also use the blank symbol $\sqcup$ on the tape.
- The states are $Q = \{q_0, q_1, q_2, q_3, q_4, q_5, q_f, q_h\}$, where q_0 is the initial state, q_1 and q_2 are intermediate states, q_3 and q_4 are looping states, q_f is the accept state, and q_h is the reject state.
- The transition function δ is defined by Table 2.2.

To understand the machine, let us try an example string "ab". The initial state is q_0 and the head points to the leftmost symbol

$$ab \sqcup \ldots \qquad (2.9)$$

$$\uparrow$$
$$\boxed{q_0}$$

Table 2.2 Transition function of the Turing machine deciding whether an input string takes the form $a^n b^n$, where $n \geq 0$.

Current state	Current symbol	Next state	Next symbol	Head movement
q_0	a	q_1	⊔	R
q_0	⊔	q_f	⊔	–
q_0	b	q_h	b	–
q_1	a	q_1	a	R
q_1	⊔	q_h	⊔	–
q_1	b	q_2	b	R
q_2	a	q_h	a	–
q_2	⊔	q_3	⊔	L
q_2	b	q_2	b	R
q_3	a	q_h	a	–
q_3	⊔	q_0	⊔	R
q_3	b	q_4	⊔	L
q_4	a	q_5	a	L
q_4	⊔	q_0	⊔	R
q_4	b	q_4	b	L
q_5	a	q_5	a	L
q_5	⊔	q_0	⊔	R
q_5	b	q_h	b	–

According to the table, the machine should switch to state q_1, erase the cell, and move the head right, arriving at the configuration

$$\underset{\underset{\boxed{q_1}}{\uparrow}}{⊔\,b\,⊔\,\ldots} \tag{2.10}$$

The transition function then instructs the machine to switch to state q_2, keep the cell symbol b, and move the head right, arriving at the configuration

$$\underset{\underset{\boxed{q_2}}{\uparrow}}{⊔\,b\,⊔\,\ldots} \tag{2.11}$$

The current configuration and the transition function tells the machine to switch to state q_3, retain the cell symbol, and move the head left, resulting in the configuration

$$\underset{\underset{\boxed{q_3}}{\uparrow}}{⊔\,b\,⊔\,\ldots} \tag{2.12}$$

Then, according to the transition function, the machine would switch to state q_4, erase the cell, and move the head left, arriving at

$$\sqcup\ \sqcup\ \sqcup\ldots \atop \uparrow \quad\quad (2.13)$$

Now, the state should switch back to state q_0, retain the cell symbol, and move the head right, resulting in

$$\sqcup\ \sqcup\ \sqcup\ldots \atop \uparrow \quad\quad (2.14)$$

Seeing an empty cell in state q_0, the machine should switch to halting state q_f, accept the input string, and halt. The final configuration is

$$\sqcup\ \sqcup\ \sqcup\ldots \atop \uparrow \quad\quad (2.15)$$

Can you figure out the problem-solving idea underpinning this example Turing machine? I hope you can. In any case, here you are: This Turing machine starts in state q_0, erases an a, and enters state q_1 to find a b. Once found, the machine transitions to state q_2 to traverse through remaining b's and then erases the rightmost b in state q_3. From there, state q_4 brings the head back to the leftmost position in the input to start the process again in q_0. In this way, the machine successfully erases pairs of a and b symbols from the string.

If it encounters an a in either q_2 or q_3, or a b in q_5, its indicates the presence of extra a's or b's, and hence the input is not of the form $a^n b^n$. The machine thus rejects the input. If upon returning to q_0 it finds a blank symbol, indicating the string is exhausted, the machine accepts the input, having found a matching b for every a from the initial sequence.

> **Exercise 2.3** *Try the Turing machine in Example 2.2 with an input string that is to be rejected by the machine. Explain step by step why the string is rejected.*
>
> **Exercise 2.4** *Table 2.3 is the transition function of a Turing machine. Is the transition function correctly defined? If true, what does this Turing machine do? Note that in the table, N means that the read/write head stays still, because − now represents minus.*

2.2 Universal and Probabilistic Turing Machines

There are alternative definitions of a Turing machine, as well as generalizations and variations of a Turing machine. We shall take a quick look at two of these.

2.2.1 The Universal Turing Machine

As we have seen in Section 2.1, one might think that each Turing machine can only realize one specific function. In fact, this is not the case. In fact, it is possible to

Table 2.3 Transition function for Exercise 2.4.

Current state	Current symbol	Next state	Next symbol	Head movement
q_0	$+$	q_1	$+$	R
q_0	$-$	q_2	$+$	R
q_1	1	q_1	1	R
q_1	$\sqcup$	q_3	$\sqcup$	R
q_1	$+$	q_{end}	$+$	N
q_1	$-$	q_{end}	$-$	N
q_3	1	q_3	0	R
q_3	$\sqcup$	q_{end}	1	N
q_3	$+$	q_{end}	$-$	N
q_3	$-$	q_{end}	$+$	N
q_2	1	q_2	$\sqcup$	L
q_2	$\sqcup$	q_2	$\sqcup$	L
q_2	$+$	q_2	$\sqcup$	L
q_2	$-$	q_{end}	$-/+$	R

define a Turing machine – the universal Turing machine – that can "do anything"; that is, it can realize the function of any specific device. To understand the universal Turing machine, we need the notion of Turing number: Each machine can be uniquely specified by a natural number, namely its Turing number. How is this done?

Recall that a Turing machine is defined by a finite number of symbols (including the head movements L, R, and –) and a transition function consisting of a finite number of instructions, each of which is a string of the symbols. It is then possible to map the symbols to natural numbers, such that each instruction becomes a string of natural numbers, and the transition function becomes a juxtaposition of the instruction strings. (Before moving forward, you might challenge yourself to invent such a mapping. The unary representation of numbers will be your friend.)

Consider the following convention for this mapping. For each device, do the mapping

- states: $q_i \mapsto 3\underbrace{111\cdots1}_{i}$;

- alphabets: $a_i \mapsto 3\underbrace{222\cdots2}_{i}$;

- head movements: $L \mapsto 4,\ R \mapsto 5,$ and $- \mapsto 6$;

- separation of instructions: $; \mapsto 7$.

Then, each instruction (Current state | Current symbol | Next state | Next symbol | Head movement) in the transition function of the device can be mapped to a natural number

$$\underbrace{3111\cdots1}_{q}\,\underbrace{3222\cdots2}_{a}\,\underbrace{3111\cdots1}_{q'}\,\underbrace{3222\cdots2}_{a'}\,d, \tag{2.16}$$

where $d \in \{4, 5, 6\}$ is the number corresponding to head movement. Now, if we separate the instructions in the transition function of a Turing machine by ;, mapped to 7 in our convention, the entire transition function is expressed as a natural number n_T:

$$\underbrace{\underbrace{3111\cdots13222\cdots23111\cdots13222\cdots2d}_{\text{row 1}}\ \underbrace{7}_{;}}{}$$
$$\underbrace{\underbrace{3111\cdots13222\cdots23111\cdots13222\cdots2d}_{\text{row 2}}\ \underbrace{7}_{;}\cdots. \tag{2.17}$$

This natural number n_T completely and uniquely defines a Turing machine and is called the **Turing number** of the Turing machine. A Turing number is definitely finite, albeit an astronomical number.

A Turing machine T specified by its Turing number n_T can take an input string x and output the result $T(x)$. To realize this Turing machine, the **universal Turing machine** T_U can take the input string $n_T \sqcup x$ and output the result $T(x)$. That is,

$$T_U(n_T \sqcup x) \equiv T(x). \tag{2.18}$$

In other words, while reading the input Turing number n_T, T_U has its own transition function that can interpret what T is supposed to do; T_U then computes the input x as if it were T. Let us consider an example.

Example 2.3

Consider the Turing machine T that can add 1 to any natural number. That is, for any input natural number n, $T(n) = n + 1$. If we represent natural numbers as unaries with digits 1, T would need only two alphabets, 1 and $\sqcup$. The transition function is shown in Table 2.4.

Table 2.4 Transition table for Example 2.3.

Current state	Current symbol	Next state	Next symbol	Head movement
q_1	1	q_1	1	R
q_1	$\sqcup$	q_2	1	–

In our convention, if we let $a_1 = 1$ and $a_2 = \sqcup$, the Turing number n_T of this machine should read

$$n_T = \underbrace{\overset{q_1}{31}\ \overset{1}{32}\ \overset{q_1}{31}\ \overset{1}{32}\ \overset{d}{5}}_{\text{row 1}}\ 7\underbrace{\overset{\sqcup}{31}\ 322\ \overset{q_2}{311}\ 32\ \overset{d}{6}}_{\text{row 2}}. \tag{2.19}$$

Note that here, q_2 plays the role of the accept state q_f. Now if we would like to use the universal Turing Machine U_T to realize the function of this T, the initial configuration of U_T should be

$$
\underset{\underset{\boxed{q_0}}{\uparrow}}{3}1323131325731322311326 \sqcup \underbrace{11\cdots 1}_{n} \sqcup \cdots. \tag{2.20}
$$

Here, q_0 is the initial state of T_U, and n is the natural number to be added by 1. Then, U_T will read in this input and operation under the instruction of its own transition function (not shown here), such that when it halts, the output is

$$
U_T(31323131325731322311326 \sqcup \underbrace{11\cdots 1}_{n}) = \underbrace{11\cdots 1}_{n+1} = n+1 = T(n). \tag{2.21}
$$

> **Exercise 2.5**[*] *Try to roughly describe the transition function of the universal Turing machine.*

2.2.2 The Probabilistic Turing Machine

An interesting generalization or variation of the original Turing machine is the probabilistic Turing machine. A probabilistic Turing machine is a type of non-deterministic Turing machine that can make random choices at each step of the computation. They can have different outcomes on the same input, depending on the random choices they make. Probabilistic Turing machines were motivated by several reasons. Some motivations are related to the limitations or drawbacks of standard Turing machines:

- Standard Turing machines are deterministic, meaning that they always have the same outcome on the same input. This means they are unable to model some natural phenomena or processes that are inherently random or stochastic, such as quantum mechanics, genetics, cryptography, and gambling. By using probabilistic Turing machines, we can simulate or analyze these phenomena or processes more realistically and accurately than using standard Turing machines [31, 101].
- Standard Turing machines are sequential, meaning that they perform one computation at a time. This makes them inefficient or impractical for some problems that require parallelism or exploration of different computational paths or possibilities. For example, some optimization or search problems may have multiple local optima, and standard Turing machines may get stuck in one of them without finding the global optimum or maximum. By using probabilistic Turing machines, we can avoid getting stuck in local optima or maxima by introducing random perturbations or variations in the computation. For example, simulated annealing is a probabilistic algorithm that uses random perturbations to escape from local optima and converge to a global optimum [102].

- Standard Turing machines are certain, meaning that they assume complete and accurate information about the input and the output. So they are unable to deal with uncertainty and learning in computation, which are essential for many real-world applications. For example, some data may be incomplete or noisy, some hypotheses may be uncertain or unknown, and some decisions may involve risks or rewards. Probabilistic Turing machines enable us to incorporate uncertainty and learning into computation, by using probability theory and statistics as tools for reasoning and decision making. For example, Bayesian inference is a probabilistic method that uses Bayes' theorem to calculate the posterior probability of a hypothesis given some data [103].

On the other hand, there are also some motivations that stem from the pros of probabilistic Turing machines themselves:

- Probabilistic Turing machines can perform some computations faster or easier than standard Turing machines.[2] For example, some problems that are hard for standard Turing machines (such as NP-complete problems) can be solved efficiently by probabilistic Turing machines with some error probability (these are called **BPP** problems).
- Probabilistic Turing machines can have different complexity classes than standard Turing machines, meaning that they can capture different notions of computational difficulty or feasibility. For example, some complexity classes are defined by using probabilistic Turing machines with different error probabilities (such as **BPP**, **RP**, **ZPP**) or different sources of randomness (such as **PP**, **BQP**). These classes can help us understand the power and limitations of probabilistic computation.

Computational complexity terminology such as P and BPP problems will be introduced in Section 2.4. Let us check out the formal definition of a probabilistic Turing machine and then unpack it with a simple example.

Definition 2.2. *A probabilistic Turing machine (PTM) is defined as a 7-tuple $M = (Q, \Sigma, \Gamma, \delta, q_0, q_{acc}, q_{rej})$, where*

- *Q is a finite set of states;*
- *Σ is the input alphabet, not containing the blank symbol $\sqcup$;*
- *Γ is the tape alphabet, where $\sqcup \in \Gamma$ and $\Sigma \subseteq \Gamma$;*
- *$q_0 \in Q$ is the initial state;*
- *$q_{acc} \in Q$ is the accepting state;*
- *$q_{rej} \in Q$ is the rejecting state;*
- *$\delta : Q \times \Gamma \rightarrow Q \times \Gamma \times \{L, R, -\} \times [0, 1]$ is the transition function.*

The transition function δ in a PTM includes a probability value in $[0, 1]$ along with the new state, new symbol, and direction. For each state-symbol pair, the sum of the probabilities for all possible transitions should be 1.

---

[2] Conversely, some problems that are easy for standard Turing machines (such as P problems) can be hard for probabilistic Turing machines with zero error probability (such as ZPP problems) [104, 105].

Operationally, when a PTM reads a symbol a in state q, it randomly chooses one of the transitions defined by $\delta(q, a)$, based on the associated probabilities, and acts accordingly.

The machine is said to "accept" an input string if the computation ends in state q_{acc} and to "reject" if it ends in state q_{rej}.

Example 2.4

Consider a probabilistic Turing machine designed to estimate whether a given binary string has an even number of 1's. The machine has four states: $q_{\text{even}}, q_{\text{odd}}, q_{\text{acc}}, q_{\text{rej}}$, where q_{acc} is the accepting state and q_{rej} is the rejecting state. The input alphabet and the tape alphabet are both $\{0, 1\}$. The transition functions are defined in Equation (2.2):

$$
\begin{aligned}
\delta(q_{\text{even}}, 0) &= \{(q_{\text{even}}, 0, R, 1)\}, \\
\delta(q_{\text{odd}}, 0) &= \{(q_{\text{odd}}, 0, R, 1)\}, \\
\delta(q_{\text{even}}, 1) &= \{(q_{\text{odd}}, 0, R, p), (q_{\text{even}}, 0, R, 1 - p)\}, \\
\delta(q_{\text{odd}}, 1) &= \{(q_{\text{odd}}, 0, R, 1 - p), (q_{\text{even}}, 0, R, p)\}, \\
\delta(q_{\text{even}}, \text{blank}) &= \{(q_{\text{acc}}, \text{blank}, -, 1)\}, \\
\delta(q_{\text{odd}}, \text{blank}) &= \{(q_{\text{rej}}, \text{blank}, -, 1)\}.
\end{aligned}
\tag{2.22}
$$

where $p \in [\frac{1}{2}, 1]$ is the probability that controls the likelihood of correctly updating parity when encountering a 1. The machine starts in the state q_{even} and reads the input from left to right. Every time it reads a 0, it remains in the current state; every time it reads a 1, it probabilistically decides whether to switch its counter state between q_{even} and q_{odd}. Finally, it deterministically moves to the accepting state q_{acc} if it ends in q_{even} upon reading the blank symbol at the end, and to the rejecting state q_{rej} if it ends in q_{odd}.

Note that this machine may not always return the correct result due to the probabilistic choices. The behavior of the machine can be analyzed in terms of probabilities. For each 1 in the input string, the machine will either stay in the same state with probability $1 - p$ or switch to the other state ($q_{\text{even}} \leftrightarrow q_{\text{odd}}$) with probability p.

Suppose the input has n occurrences of 1. Then after reading all these 1's, the probability that the machine ends up in state q_{acc} (indicating an even number of 1's) or in state q_{rej} (indicating an odd number of 1's) can be modeled by the binomial distribution. Specifically, if the machine starts in q_{even}, the probability of observing k successful transitions to q_{odd} and back, and $(n - k)$ failed transitions, is:

$$
\text{Prob}(k \text{ transitions}) = \binom{n}{k} p^k (1 - p)^{n-k}.
\tag{2.23}
$$

Assuming that the string contains an even number of 1's (n is even), the probability of a correct result can be approximated as

$$
\text{Prob}(q_{\text{acc}}) = \sum_{k=0}^{n/2} \text{Prob}(2k \text{ transitions}) = \frac{1 + (2p - 1)^n}{2}.
$$

Therefore,

- If $p = 1$, the probabilistic Turing machine reduces to a deterministic Turing machine, and one can immediately determine whether there is an even or odd number of 1's in the string after a single run.
- If $p > \frac{1}{2}$ and the input string contains an even number of 1's, the probability that the machine ends in the q_{acc} state is greater than $\frac{1}{2}$ but approaches $\frac{1}{2}$ at large n.
- If $p = \frac{1}{2}$, however, the transitions become completely random, and the probability that the machine ends in the q_{acc} state is always $\frac{1}{2}$, implying that it is impossible to determine whether the number of 1's is even or odd.

Therefore, in this case, the probabilistic Turing machine with low transition reliability p can perform much worse than a deterministic Turing machine. In realistic computational settings, though, even deterministic systems operate under physical constraints and noise, meaning that truly exact computation is an idealization. What we can do in practice is to enhance the reliability of computation, either by improving the probability parameter p (e.g., in Example 2.4) to be closer to certainty or by designing more efficient algorithms that tolerate or even exploit the inherent randomness to make robust and scalable decisions.

2.3 The Circuit Model

While Turing machines offer a theoretical framework for understanding the fundamentals of computation, they are far removed from the practical considerations of real-world computing systems. Turing machines operate in a serial manner, one instruction at a time, and on an infinitely long tape. Such characteristics make them ill-suited for direct implementation or for capturing the nuances of modern parallel and finite computing resources. Therefore, to bridge the gap between theoretical computation and practical implementation, computer scientists often turn to the circuit model of classical computation [103]. In this model, computation is represented as a network of simple logic gates that interact to perform complex tasks. This abstraction is much closer to the architecture of actual computers and serves as a critical stepping stone for understanding how we can build and analyze practical computing systems.

2.3.1 Binary Arithmetic

Binary Representation of Numbers

In classical computation, and in fact in all digital systems, the binary number system is foundational. Unlike the decimal number system, which is base 10, the binary system is base 2, meaning it employs only two symbols: 0 and 1.

A nonnegative integer N can be represented in binary as follows:

$$N = a_n 2^n + a_{n-1} 2^{n-1} + \ldots + a_2 2^2 + a_1 2^1 + a_0 2^0, \tag{2.24}$$

where each a_i is either 0 or 1, and n is the highest power of 2 in the series. This can also be written more compactly as

$$N = \sum_{i=0}^{n} a_i 2^i. \tag{2.25}$$

For example, the binary number 1101 corresponds to $1 \times 2^3 + 1 \times 2^2 + 0 \times 2^1 + 1 \times 2^0 = 8 + 4 + 0 + 1 = 13$ in decimal.

The representation can be extended to nonintegers by using negative powers of 2 for the terms after the binary point. A number x can be represented as

$$x = a_n 2^n + \ldots + a_0 2^0 + b_1 2^{-1} + b_2 2^{-2} + \ldots \tag{2.26}$$

where a_i and b_i are either 0 or 1. The binary number 101.01 can be converted to decimal as follows: $1 \times 2^2 + 0 \times 2^1 + 1 \times 2^0 + 0 \times 2^{-1} + 1 \times 2^{-2} = 4 + 0 + 1 + 0 + 0.25 = 5.25$.

Understanding this binary representation is crucial for understanding how arithmetic operations and logic gates function in the realm of classical computation.

Binary Operations

Binary arithmetic follows the same principles as arithmetic in any other number base. The most basic operations are addition and multiplication.

Binary addition is straightforward. Equation (2.27) shows the binary addition rule for each possible pair of binary digits, the sum, and the carry-over digit.

$$\begin{aligned}
0 \;+\; 0 \;&=\; 0, \quad \text{carry-over} = 0; \\
0 \;+\; 1 \;&=\; 1, \quad \text{carry-over} = 0; \\
1 \;+\; 0 \;&=\; 1, \quad \text{carry-over} = 0; \\
1 \;+\; 1 \;&=\; 0, \quad \text{carry-over} = 1.
\end{aligned} \tag{2.27}$$

For example, to add 1011 and 1101, we perform the addition as

$$\begin{array}{ll}
1\,0\,1\,1 & \text{(augend)} \\
\underline{+\,1\,1\,0\,1} & \text{(addend)} \\
1\,1\,0\,0\,0 & \text{Result: } 11000
\end{array} \tag{2.28}$$

Binary multiplication is also simple and follows the standard multiplication rule, which is $0 \times 0 = 0$, $1 \times 0 = 0$, $0 \times 1 = 0$, and $1 \times 1 = 1$. The example in Equation (2.29) would suffice to show you how such a multiplication is done. To multiply 1011 by 1101:

$$\begin{array}{ll}
1\,0\,1\,1 & \text{(multiplicand)} \\
\underline{\times\quad 1\,1\,0\,1} & \text{(multiplier)} \\
1\,0\,1\,1 & (1011 \times 1, \text{ shift 0 positions}) \\
0\,0\,0\,0 & (1011 \times 0, \text{ shift 1 position})
\end{array} \tag{2.29}$$

$$1\,0\,1\,1 \qquad (1011 \times 1, \text{ shift 2 positions})$$
$$\underline{1\,0\,1\,1} \qquad (1011 \times 1, \text{ shift 3 positions})$$
$$1\,0\,0\,0\,1\,1\,1\,1 \quad \text{Result: } 10001111$$

Understanding these basic operations is essential for comprehending more complex circuit operations and algorithms in classical computation.

> **Exercise 2.6** *Try to do some binary subtraction and division.*

2.3.2 Elementary Logic Gates and Universal Sets

To treat logic gates more rigorously, we should first bring up the notion of logical function. A **logical function** is a map $f\colon \{0,1\}^n \to \{0,1\}$. For example, a constant function can be defined as $f(a) = 0$ for all $a \in \{0,1\}$. Another example is the AND function $f(a,b) = a \wedge b$.

A **logic gate** is essentially a circuit implementation of a logical function. Figures 2.1 through 2.7 show the circuit diagrams, truth tables, and binary arithmetic functions of respectively some elementary logic gates:

Figure 2.1 (a) Circuit diagram and (b) truth table of the **trivial gate**. Binary arithmetic of the gate is $a = a$.

a	Output
0	0
1	1

Figure 2.2 (a) Circuit diagram and (b) truth table of the **NOT gate**. Binary arithmetic of the gate is $\bar{a} = 1 - a$.

a	Output
0	1
1	0

Figure 2.3 (a) Circuit diagram and (b) truth table of the **AND gate**. Binary arithmetic of the gate is $a \wedge b = a \cdot b$.

a	b	Output
0	0	0
0	1	0
1	0	0
1	1	1

Figure 2.4 (a) Circuit diagram and (b) truth table of the **OR gate**. Binary arithmetic of the gate is $a \vee b = a + b - a \cdot b$.

a	b	Output
0	0	0
0	1	1
1	0	1
1	1	1

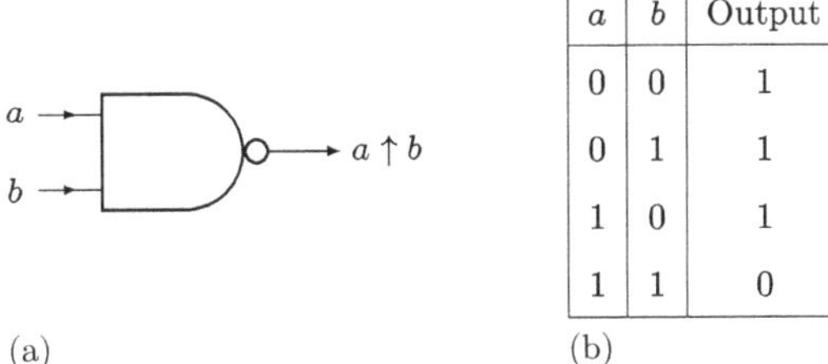

a	b	Output
0	0	1
0	1	1
1	0	1
1	1	0

(a)   (b)

Figure 2.5 (a) Circuit diagram and (b) truth table of the **NAND gate**. Binary arithmetic of the gate is $a \uparrow b = 1 - a \cdot b$.

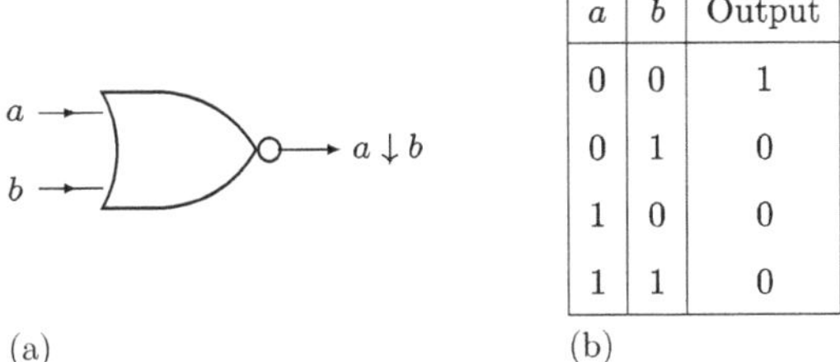

a	b	Output
0	0	1
0	1	0
1	0	0
1	1	0

(a)   (b)

Figure 2.6 (a) Circuit diagram and (b) truth table of the **NOR gate**. Binary arithmetic of the gate is $a \downarrow b = 1 - a - b + a \cdot b$.

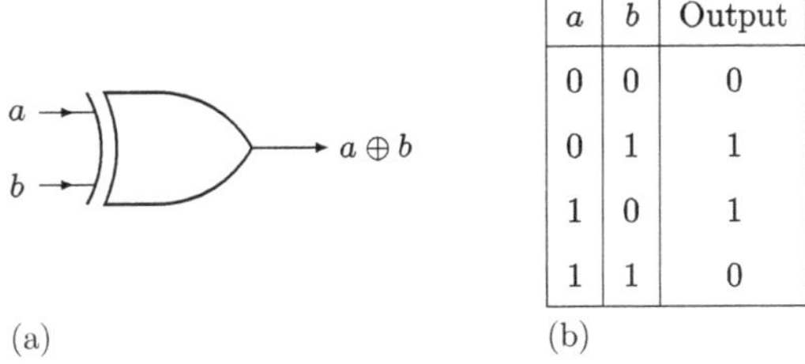

a	b	Output
0	0	0
0	1	1
1	0	1
1	1	0

(a)   (b)

Figure 2.7 (a) Circuit diagram and (b) truth table of the **XOR gate**. Binary arithmetic of the gate is $a \oplus b = a + b \mod 2$.

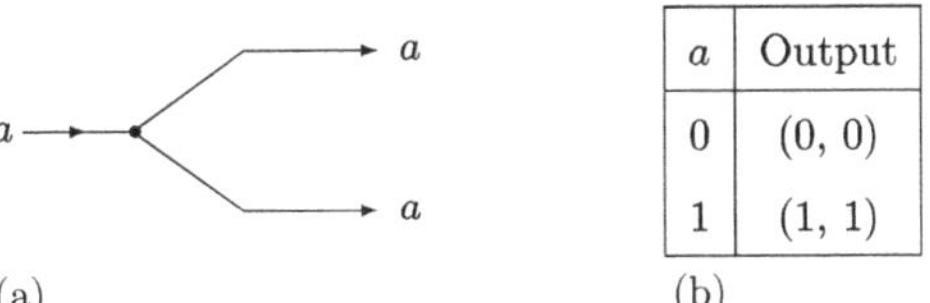

a	Output
0	(0, 0)
1	(1, 1)

(a)   (b)

Figure 2.8 (a) Circuit diagram and (b) truth table of the **FANOUT (COPY) gate**. Binary arithmetic of the gate is $a \rightarrow (a, a)$.

There are two more elementary gates that do not implement the usual logical functions, shown in Figures 2.8 and 2.9.

To better elucidate the circuit model of computation, let's explore the circuitry needed for integer addition, represented as $s = a + b$. Since this addition operation can be executed on a bit-by-bit basis, we design a specific circuit (refer to Figure 2.10) that takes care of adding the i-th bits of a and b. Importantly, this circuit

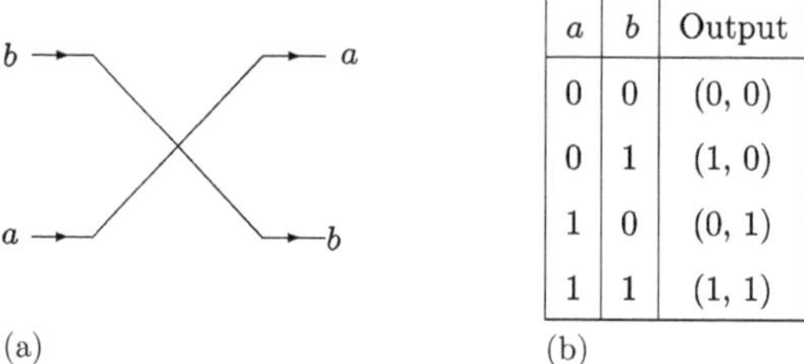

Figure 2.9 (a) Circuit diagram and (b) truth table of the **CROSSOVER (SWAP) gate**. Binary arithmetic of the gate is $(a, b) \rightarrow (b, a)$.

Figure 2.10 Circuit that adds the i-th bits of a and b.

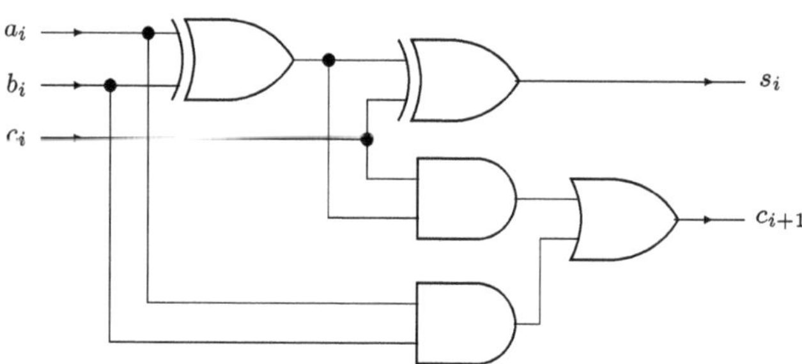

must also incorporate the carry-over bit c_i, which arises from adding the $(i - 1)$-th bits. The resultant sum and carry-over bits are denoted by s_i and c_{i+1}, respectively. By chaining together multiple instances of this circuit, we can successfully perform the complete addition of a and b.

> **Exercise 2.7** *Explain how the circuit in Figure 2.10 works.*

A set of gates is said to be *universal* if any logical function can be implemented using only the gates in that set. Importantly, the set of universal gates is not unique. Multiple combinations of gates can be sufficient for universal computation.

For instance, $\{\text{AND}, \text{OR}, \text{NOT}, \text{FANOUT}\}$ is a universal set of gates. Nonetheless, a minimal universal set could be $\{\text{NAND}, \text{FANOUT}\}$. Another example of a minimal universal set could be $\{\text{NOR}, \text{FANOUT}\}$. This nonuniqueness in the choice of universal set highlights the flexibility in designing computational circuits and the range of possible implementations for the same logical functions. To demonstrate the universality of a minimal set, consider the examples in Equation (2.30):

$$\bar{a} = 1 - a = 1 - a \cdot a = a \uparrow a;$$

$$a \wedge b = a \cdot b = a \cdot b - a \cdot b + a \cdot b = 1 - [1 - a \cdot b - a \cdot b + (a \cdot b)^2]$$
$$= 1 - (1 - a \cdot b)(1 - a \cdot b)$$
$$= (a \uparrow b) \uparrow (a \uparrow b). \tag{2.30}$$

In the derivations above, $a = a \cdot a = a^2$ is a property of bits. These show that NOT and AND can be re-expressed using only NAND and FANOUT gates, validating the universality of the set.

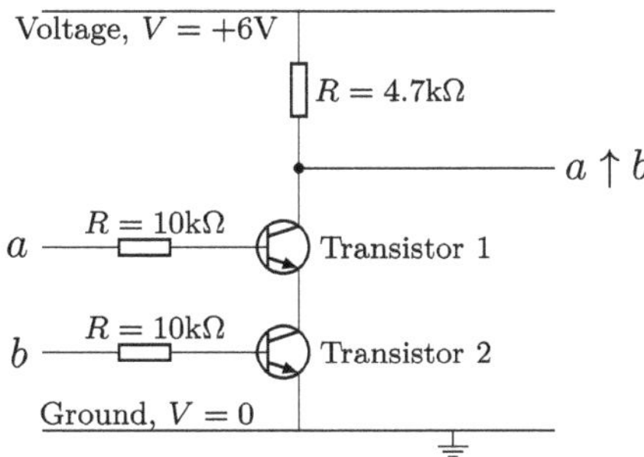

Figure 2.11 Sketch of the discrete circuit of an NAND gate. The output is connected to the ground, rendering $a \uparrow b = 0$ only when $a = b = 1$, causing both transistors to be in their conducting state. For all other combinations of a and b, $a \uparrow b = 1$.

Since NAND is a universal gate, you may wonder how it may be realized by a discrete circuit. Figure 2.11 gives a sketch.

> **Exercise 2.8** *Express $a \vee b$ and $a \oplus b$ in terms of NAND and FANOUT gates and draw the corresponding circuit diagrams.*

2.4 Computational Complexity

After discussing the circuit model of computation, it becomes essential to assess the efficiency of various computational models and algorithms. In any computational process, resources such as time, memory space, and energy are consumed. Understanding how these resources scale with the size of the input data is not just an academic concern, but is crucial for evaluating the feasibility and efficiency of real-world implementations. This scaling behavior is what we term as "computational complexity," a formal measure that quantifies the minimal resources required to solve a particular problem [106, 107].

For example, consider adding two n-bit numbers. In this case, the size of the input is n. The most efficient method to accomplish this is to add the numbers bit by bit. Therefore, the time t required to perform this operation is proportional to n, often expressed as $t = \mathcal{O}(n)$. This is read as "t is of the order of n."

> **Exercise 2.9** *How would the time consumption scale with the input size in the case of multiplying two n-bit numbers?*

In the ensuing sections, we will dive deeper into the nuances of computational complexity, exploring its various dimensions and implications. First, we'll distinguish between tractable and intractable problems, illustrating each with a set of characteristic examples. We will then move on to categorize problems based on their time and space complexities, introducing the notion of complexity classes. An important facet of computational complexity, often overlooked, is the intimate relationship between information and energy. We'll shed light on this relationship through the lens of Maxwell's Demon and Landauer's Principle. Finally, we will explore the fascinating concept of reversible computation, which promises to redefine the boundaries of what is computationally feasible.

2.4.1 Tractable versus Intractable Problems

Tractable, or feasible, problems are those that can be solved efficiently, requiring resources bounded by a polynomial function of the input size. In contrast, intractable or infeasible problems demand resources that grow superpolynomially with the input size, making them inefficient to solve. In other words, the resource requirements for solving intractable problems are not bounded by any polynomial function of the input size.

The Strong Church–Turing Thesis

You may wonder why polynomial and superpolynomial resource requirements serve as the dividing line between tractable and intractable problems. The explanation for this lies in the Strong Church–Turing Thesis, which posits:

A probabilistic Turing machine can simulate any realistic model of computation with at most a polynomial increase in the number of elementary operations required.

This thesis provides a foundational justification for why polynomial time serves as a useful demarcation between efficient and inefficient computation. If a problem can be solved efficiently (in polynomial time) by some computational model, then according to the Strong Church–Turing Thesis, it can also be solved efficiently by a probabilistic Turing machine. In this sense, polynomial-time solvability becomes a robust criterion for tractability, as it is model-independent and holds across a wide array of computational paradigms. Conversely, if a problem requires superpolynomial resources on a Turing machine, then it is considered intractable across all realistic models of computation, again due to the Strong Church–Turing Thesis.

Archetypal Problems

Next, let's examine a few archetypal problems that are either tractable or intractable, as well as one outlier – the halting problem – that defies categorization in either of these groups.

- **Integer factorization problem** (Intractable classically)
  This problem involves finding the prime factors of a given integer N. Classically, it is considered intractable for large N with the best-known algorithm, the General Number Field Sieve, having a time complexity of $\mathcal{O}\left(\exp\left(\frac{64}{9}^{1/3}(\ln N)^{1/3}(\ln \ln N)^{2/3}\right)\right)$ [108]. As of my last update in September 2021, the largest number that has been publicly factorized is RSA-240, a 795-bit

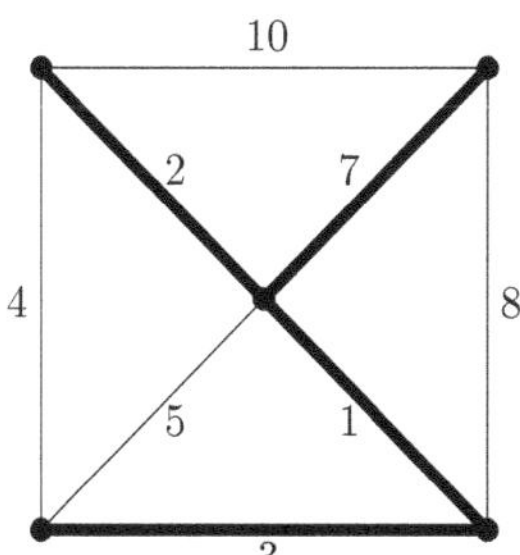

Figure 2.12 A weighted graph with a minimum spanning tree (bold edges).

number [109], which took 900 core-years (i.e., about a year) on a modern supercomputer to achieve. To put this into perspective, if one were to use the most advanced Intel CPU available as of that time, it would likely take a millennium. Remarkably, Shor's algorithm, a quantum computing algorithm, would make this problem tractable [35], shifting the landscape of cryptography and secure communication.

- **Minimum spanning tree problem** (Tractable)
  Given a weighted, connected graph $\mathcal{G}(V, E)$, where V is the set of vertices and E is the set of edges (see Figure 2.12 for an example), let $w : E \rightarrow \mathbb{N}$ be a weight function that assigns a positive integer weight to each edge. A subgraph T of $\mathcal{G}$ is a tree if it is free of any loop of edges. The goal is to find a tree T that spans all the vertices and minimizes the sum of the weights of its edges:

$$\min_{T \subseteq \mathcal{G}(V,E)} \sum_{e \in T} w(e). \tag{2.31}$$

Cayley's theorem [110] states that for a complete graph with n vertices, the number of distinct spanning trees is $n^{(n-2)}$.

The problem is known to be tractable, and there are polynomial-time algorithms to solve it, such as Kruskal's algorithm [111] and Prim's algorithm [112]. One intuitive way to see why the problem might be tractable is by considering a divide-and-conquer strategy: Divide the graph into two subgraphs recursively, find the minimum spanning tree for each, and combine them while minimizing the weight. Note that this is a simplification; the actual algorithms are more efficient.

- **Traveling salesman problem (TSP)** (Intractable)
  Given a weighted graph $\mathcal{G}(V, E)$, where V is the set of vertices (cities) and E is the set of edges (paths between cities), along with a weight function $w(e) \in \mathbb{N}$ (distances between cities) for each edge e, the objective is to find the shortest possible route that visits each vertex exactly once and returns to the original vertex. In graph theory terms, this is known as finding a **Hamiltonian cycle** (a cycle that visits each vertex exactly once) with minimum weight. See Figure 2.13 for an example.

The complexity of this problem becomes evident when considering that a graph with n vertices has $(n - 1)!$ possible Hamilton cycles, even greater than exponential complexity. The best-known classical algorithms for the TSP, such as the Held–Karp algorithm, have exponential time complexity $\mathcal{O}(2^n n^2)$

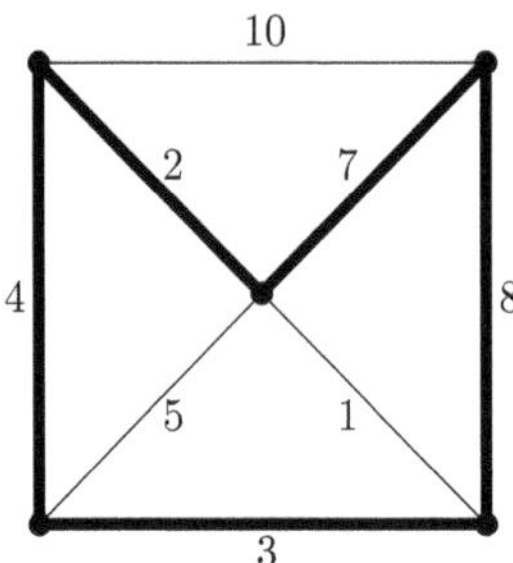

Figure 2.13 A weighted graph with a Hamiltonian cycle (bold edges).

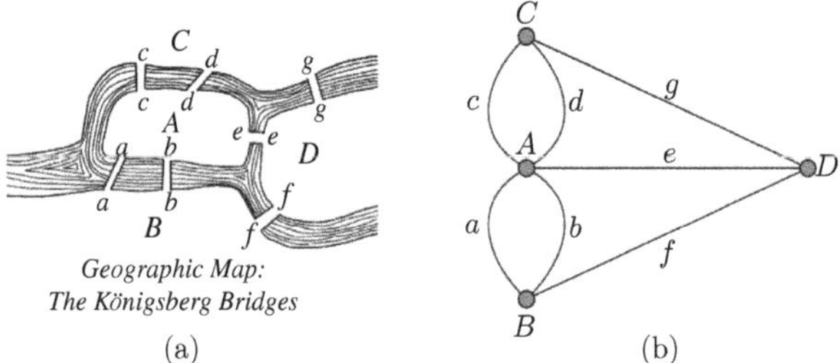

Figure 2.14 (a) Original graph of Königsburg's seven-bridge problem. (b) Euler's view of the problem.

[113]. Although there are heuristic and approximation methods designed to offer "good enough" solutions in a shorter time, the problem remains computationally intractable in general. It has important applications in logistics, route planning, and computational biology, to name but a few.

- **Euler cycle problem**

  Given an undirected graph $\mathcal{G}(V, E)$, where V is the set of vertices and E is the set of edges, the task is to determine whether the graph contains an Euler cycle – a cycle that visits every edge exactly once and returns to the starting vertex. The problem originated from the famous Seven Bridges of Königsberg problem, which asked if it was possible to walk through the city of Königsberg crossing each of its seven bridges exactly once. Figure 2.14 shows the original graph of Königsberg's seven bridges and Euler's view of the problem.

  Euler's theorem states that an Euler cycle exists if and only if the graph is connected and every vertex but an end vertex has an even degree [114], meaning an even number of incident edges. Algorithms to find an Euler cycle, such as Hierholzer's algorithm, run in linear time, $\mathcal{O}(|E|)$, where $|E|$ is the number of edges [115]. This makes the problem computationally tractable. Applications of Euler cycles can be found in network design [116, 117], DNA sequencing [118], and various optimization tasks [119, 120, 121, 122].

- **Satisfiability problem**

  To illustrate the concept of satisfiability problems, consider a scenario involving three individuals: Alice, Bob, and Charlie. Each makes a statement about the other two. The statements are as follows:
  - Alice: "If Bob is telling the truth, then Charlie is lying."
  - Bob: "Either Alice is lying or Charlie is lying."
  - Charlie: "Either Alice is telling the truth, or Bob is lying, but not both."

These statements can be initially translated into a Boolean formula as follows:

(A)lice's statement: $B \rightarrow \bar{C}$ becomes $\bar{B} \vee \bar{C}$

(B)ob's statement: $\bar{A} \vee \bar{C}$ stays the same $\bar{A} \vee \bar{C}$

(C)harlie's statement: $A \oplus \bar{B}$ becomes $(\bar{A} \vee \bar{B}) \wedge (A \vee B)$,

where $A, B, C \in \{0, 1\}$ are called literals or variables. Here, $A = 1$ (0) if A is lying (telling the truth). The naive Boolean formula representing all these statements combined is

$$F = (\bar{B} \vee \bar{C}) \wedge (\bar{A} \vee \bar{C}) \wedge ((\bar{A} \vee \bar{B}) \wedge (A \vee B)). \tag{2.32}$$

> **Exercise 2.10** *Describe the distributive and associative laws of the Boolean functions $\wedge$ and $\vee$.*

By applying distributive and associative laws, this formula can be simplified (as an exercise) to its **conjunctive normal form** (CNF):

$$F = (\bar{B} \vee \bar{C}) \wedge (\bar{A} \vee \bar{C}) \wedge (\bar{A} \vee \bar{B}) \wedge (A \vee B), \tag{2.33}$$

which can be easily verified to be true for the variable assignment: $A = 1, B = 1$, and C can be either 0 or 1.

In general, a Boolean formula in CNF is a conjunction (i.e., AND) of one or more clauses, where a clause is a disjunction (OR) of literals. Namely,

$$f(\mathbf{a}, \mathbf{b}, \cdots) = (a_1 \vee a_2 \vee \cdots \vee a_{n_a}) \wedge (b_1 \vee b_2 \vee \cdots \vee b_{n_b}) \wedge \cdots, \tag{2.34}$$

where $\mathbf{a}, \mathbf{b}, \ldots$ collectively denote the literals. The Satisfiability Problem (often referred to as SAT) asks whether there exists an assignment of truth values to the variables that makes the entire CNF formula true. Despite this simple definition, SAT problems are intractable, as no efficient solution has been found to solve them in general.

Interestingly, SAT problems have intriguing connections to statistical physics, particularly in the form of phase transitions. These phase transitions occur when the "order" parameter $\alpha = \frac{M}{N}$ crosses a specific critical value. Here, M is the number of clauses and N is the number of variables in a CNF Boolean formula.

When $M \ll N$, the formula is typically *underconstrained*, meaning there are fewer clauses to satisfy, making it easier to find a satisfying assignment. Conversely, when $M \gg N$, the formula becomes *overconstrained*, with so many conditions to meet that it is often impossible to find an assignment that satisfies all clauses.

As α varies, the problem undergoes a transition from being mostly satisfiable to mostly unsatisfiable, similar to phase transitions in physical systems. This is not just an academic curiosity but offers insights into the inherent complexity of these problems [123, 124].

- **Halting problem**

  The halting problem is one of the most well-known problems in the realm of theoretical computer science, particularly in the study of computability [96]. Formulated by Alan Turing in 1936, the problem asks whether, given a computer program P and an input I, it is possible to determine whether the program P will eventually halt (terminate) when run with input I.

  The halting problem is *undecidable*, meaning that there is no algorithm that can solve it for all possible program–input pairs. Undecidability is a category separate from tractable (polynomial-time solvable) and intractable (solvable, but not in polynomial time) problems. Essentially, an undecidable problem is one that cannot be solved by any algorithm.

  Turing's proof of the undecidability of the halting problem is a classic example of self-reference. Assume for contradiction that there exists an algorithm $H(P, I)$ that decides the halting problem. That is, H returns "yes" if P halts on I and "no" otherwise.

  Consider a new program Q defined as follows:

```
Program Q(x):
    if H(x, x) == ''no'':
        halt
    else:
        loop forever
```

  Now, what happens when we ask whether Q halts on its own source code – what does $H(Q, Q)$ return? If $H(Q, Q)$ returns "yes," then by definition of Q, it should loop forever, contradicting the "yes." If $H(Q, Q)$ returns "no," then Q should halt, again contradicting the "no." This contradiction proves that no such algorithm H can exist, establishing the undecidability of the halting problem.

> **Exercise 2.11** *Using an approach similar to Turing's proof of the undecidability of the halting problem, can you show why it is impossible to create a perfect algorithm that detects all possible bugs in any given program?*

2.4.2 Complexity Classes

While computational complexity theory boasts a wide array of classes [95], this subsection will select a few that are both fundamental and accessible. Figure 2.15 illustrates the relationships among a series of complexity classes to be introduced subsequently.

Time Complexity Classes

Time complexity classes are concerned with the computational time needed to solve problems.

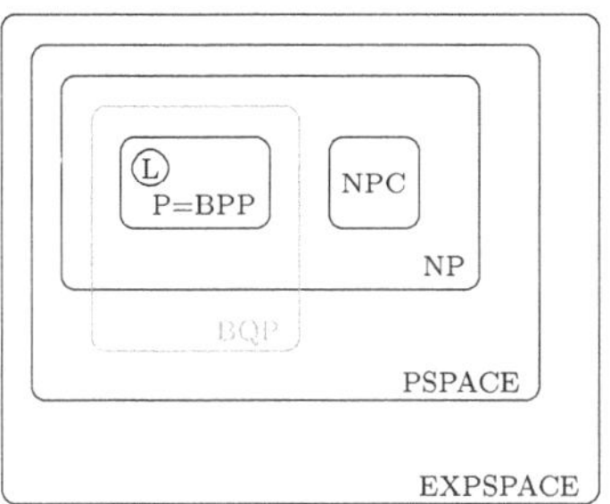

Figure 2.15 Inclusion relationships among various classes of complexity.

- **Class P (polynomial time)**: This class comprises decision problems that can be solved by a deterministic Turing machine in polynomial time. A problem is in Class P if there exists an algorithm that solves any instance of the problem in time that is polynomial in the size of the input.

  Examples include finding the greatest common divisor (GCD) via the Euclidean algorithm [125], the Shortest Path Problem solved by Dijkstra's [126] or Floyd–Warshall algorithms [127], and checking connectivity in a graph using Breadth-First Search (BFS) [128] or Depth-First Search (DFS). Additionally, problems we've previously discussed such as the Minimum Spanning Tree (MST) problem [111, 112], Eulerian Circuit (EC) problem [114], 1-SAT, and 2-SAT [129] also belong to Class P.

- **Class NP (Nondeterministic polynomial time)**: This class includes decision problems that, while potentially hard to solve, have solutions that can be verified in polynomial time on a deterministic Turing machine. Equivalently, these are problems that can be solved by a nondeterministic Turing machine in polynomial time (probabilistic Turing machines are a special case of nondeterministic Turing machines.).

  A key characteristic of NP problems is that the search space is often exponentially large, which is what makes them hard to solve in practice. Nonetheless, once a solution is found, it can be checked quickly.

  Examples include the *decision version* of the Travelling Salesman Problem (i.e., "Is there a tour of length at most k?"), Integer Factorization, and the Knapsack Problem [130]. Additionally, 3-SAT and problems like finding a Hamiltonian cycle in a graph (HC) belong to Class NP. These problems can have their proposed solutions verified in polynomial time, but finding the solution itself is generally not polynomial.

  Relation to Class P: One of the outstanding questions in computer science is whether $P = NP$ or $P \neq NP$. If $P = NP$, then every problem whose solution can be quickly verified can also be quickly solved. Yet, despite decades of research, no one has been able to prove either equality or inequality. If you manage to prove $P = NP$ or $P \neq NP$, not only will you revolutionize the field, but you'll also be almost certainly guaranteed a Turing Award. As the joke goes, the proof itself could very well be your ticket to computer science stardom.

- **Class NPC (NP-complete)**: A problem is considered NP-complete if it is in NP and every problem in NP is *polynomially reducible* to it. Mathematically, for a problem B to be in NPC, it must satisfy two conditions:

  1. $B \in$ NP
  2. For every problem A in NP, $A \leq_p B$

  The notation $A \leq_p B$ signifies that problem A is polynomially reducible to problem B; that is, instances of problem A can be transformed into instances of problem B in polynomial time. In other words, problem A is not harder than B.

  For instance, 3-SAT $\leq_p$ Knapsack and Knapsack $\leq_p$ TSP. HC $\leq_p$ TSP and vice versa. In fact, both HC and TSP are NPC problems; however, the integer factorization problem is believed not so.

  The crux of NP-complete problems lies in their universality. If a polynomial-time algorithm for any NP-complete problem is found, then $P = NP$, implying polynomial-time algorithms for all problems in NP. Conversely, if it's proven that even one NP-complete problem cannot be solved in polynomial time, then $P \neq NP$. Proving either would be groundbreaking and would virtually guarantee a Turing Award.

- **Class BPP (bounded-error probabilistic polynomial time)**: This class comprises decision problems that can be solved efficiently by a randomized algorithm. A decision problem is in BPP if there exists a polynomial-time probabilistic Turing machine that solves the problem with an error probability of less than $\frac{1}{3}$ for every input.

  The choice of $\frac{1}{3}$ is conventional and not strict; any constant less than $\frac{1}{2}$ would work. The key is that the error probability must be bounded away from $\frac{1}{2}$, which represents random guessing. To improve the algorithm's accuracy and reduce the error probability below a desired threshold ϵ, one common technique is to run the algorithm multiple times and take the majority vote. This is where the Chernoff bound [131] comes into play. The Chernoff bound is a probabilistic inequality that provides an upper bound on the tail distribution of the sum of independent random variables. In the context of BPP, the Chernoff bound states that the probability that the majority vote is incorrect decreases exponentially as the number of repetitions t increases. Specifically, the error probability can be bounded by $O(e^{-t})$. Thus, even a moderate number of repetitions can reduce the error probability to an arbitrarily low value, making the problem efficiently solvable.

Example 2.5

Consider the problem of determining whether an n-bit string is either constant (all zeros or all ones) or balanced (an equal number of zeros and ones). A simple randomized algorithm can randomly pick k bits and check if they are the same. While this algorithm can produce errors, repeating it multiple times and applying the Chernoff bound shows that the error probability can be made arbitrarily low, confirming that the problem is in BPP.

To explicitly calculate the number of repetitions t required to reduce the error probability below ϵ, we can use the Chernoff bound. In this example, the error probability δ for a majority vote over t runs is bounded as follows:

$$\delta \leq 2\exp\left(-\frac{t}{3 \times 2^{k-1}}\right). \tag{2.35}$$

To reduce this probability below ϵ, the number of repetitions t must satisfy

$$t \geq 3 \times 2^{k-1} \ln\left(\frac{2}{\epsilon}\right). \tag{2.36}$$

Thus, a moderate number of repetitions will reduce the error probability to an arbitrarily low level, confirming that the problem is in BPP.

While Class P represents problems that can be solved deterministically in polynomial time, Class BPP represents problems that can be solved probabilistically within a small, bounded error. BPP is often considered a more realistic representation of what can be efficiently solved by classical computers for several reasons:

1. **Randomization:** Randomized algorithms frequently exhibit superior performance compared to their deterministic counterparts. For example, the QuickSort [132] algorithm performs more efficiently when randomness is used for selecting the pivot.
2. **Practicality:** Computing systems in the real world are already designed to tolerate small probabilities of error, which could be due to hardware faults, external noise, or other stochastic factors. BPP algorithms naturally align with such practical tolerances.
3. **Resource utilization:** BPP algorithms often operate with reduced computational and memory resources compared to deterministic ones. For example, probabilistic algorithms for primality testing can achieve a high probability of correctness while using fewer resources.
4. **Cryptography:** Many cryptographic protocols and systems fundamentally rely on randomized algorithms for their security measures. Deterministic algorithms often cannot provide the level of security required, whereas BPP algorithms can.
5. **Algorithmic flexibility:** BPP includes a range of problems for which deterministic polynomial-time solutions are unknown, but for which probabilistic polynomial-time solutions are effective and efficient.
6. **Error tolerance:** The error probability in BPP algorithms can be adjusted to be arbitrarily small by increasing the number of algorithm repetitions. This offers a trade-off between computational time and reliability, making BPP algorithms both flexible and dependable.

- **Class BQP (bounded-error quantum polynomial time)**: This class contains decision problems that can be efficiently solved by a quantum computer, meaning the algorithm runs in polynomial time with respect to the size of the input. The term "bounded-error" indicates that the algorithm is probabilistic, with an error rate that can be made arbitrarily small but not zero.

A quintessential example of a problem in BQP is integer factorization, for which Shor's algorithm provides a polynomial-time quantum solution. This is in contrast to classical algorithms, which generally require exponential time for this problem. Shor's algorithm highlights the potential computational advantage of quantum algorithms over classical ones for certain types of problems.

Another example is the quantum version of the search problem, solved by Grover's algorithm. While classical search algorithms require $O(N)$ time to search through N unsorted items, Grover's algorithm can accomplish the task in $O(\sqrt{N})$ time.

It is currently an open question whether BQP $\subseteq$ P or BQP $\subseteq$ NP, or whether BQP has its own unique standing among complexity classes. What is clear, however, is that BQP contains problems that appear to be intractable for classical computers but are tractable for quantum ones.

Space Complexity Classes

Space complexity classes categorize problems based on the computational space required.

- **Class PSPACE (polynomial space)**: This class consists of decision problems that can be solved using a polynomial amount of memory with respect to the size of the input, regardless of the time it takes to solve the problem. Importantly, memory is reusable, which means that even NP problems, which may require exponential time but polynomial space, are also in PSPACE.

  In formal terms, a problem is in PSPACE if there exists a Turing machine that can solve the problem using polynomially bounded space. One example of a problem in PSPACE is the game of checkers, which can be formulated as a decision problem asking whether a particular player has a winning strategy. It's known that a polynomial amount of memory is sufficient to determine the outcome, even if the time to do so could be very long.

  Another example is the *quantified Boolean formula* (QBF) problem, which generalizes the SAT problem by allowing for existential and universal quantifiers over the variables. This problem is not just in PSPACE, but is also PSPACE-complete, meaning that it is as hard as the hardest problems in PSPACE.

  An interesting aspect of PSPACE is its relationship with other complexity classes. For instance, it is known that P $\subseteq$ NP $\subseteq$ PSPACE, yet whether these inclusions are proper remains unknown. Additionally, PSPACE $\subseteq$ EXPTIME means every PSPACE problem can be solved in exponential time. Nonetheless, we still do not know if this inclusion is strict.

  The study of PSPACE and its complete problems helps in understanding the limitations of what can be computed given polynomial memory constraints, regardless of computational time.

- **Class L (logarithmic space)**: This class consists of decision problems that can be solved with logarithmic space with respect to the size of the input. Logarithmic space is incredibly economical and often implies that the algorithm only needs to keep track of a constant number of pointers or counters, rather than storing substantial portions of the input itself.

  Formally, a problem is in L if it can be solved by a deterministic Turing machine using $O(\log n)$ space. A notable subclass is SL, which consists of problems solvable with logarithmic space by a symmetrical Turing machine (i.e., if a machine has a transition from state q_1 to q_2 under some condition, it must also have the reverse transition under the same condition).

  An example of a problem in L is *connectivity* in undirected graphs: Given an undirected graph and two nodes, determine if there exists a path connecting the two nodes. This problem can be solved using depth-first search with a stack that grows only logarithmically with the size of the graph.

  Interestingly, L is contained within P, which is itself contained in PSPACE, leading to $L \subseteq P \subseteq PSPACE$. It's however not yet proven whether these are proper subsets.

  The study of L and its relatives helps in understanding what can be computed when even polynomial space is too much, pushing the boundaries of algorithmic efficiency in terms of space.

- **Class EXPSPACE (exponential space)**: This class includes decision problems that can be solved using an exponential amount of memory with respect to the size of the input. Formally, a problem is in EXPSPACE if it can be solved by a deterministic Turing machine in $O(2^{p(n)})$ space, where $p(n)$ is a polynomial function of the input size n.

  An example of a problem in EXPSPACE is the generalization of chess or Go to an $n \times n$ board. Determining the winner from a given board configuration would require exponential space to store all possible game states.

  It is known that EXPSPACE contains PSPACE, which itself contains P, leading to $P \subseteq PSPACE \subset EXPSPACE$. Nonetheless, it is not yet proven whether the first inclusion is strict.

  Studying EXPSPACE problems provides insights into the boundaries of computational feasibility in terms of both time and space, showing the limits of what we can compute even with an exponentially growing amount of resources.

While this section provides an introduction to some of the key complexity classes, it is essential to understand that the landscape of computational complexity is far more vast and intricate. Numerous complexity classes, each with its unique characteristics and interrelations, are actively studied in theoretical computer science (see Figure 2.16). Unfortunately, the scope of this book does not permit a comprehensive exploration of all these classes. Nevertheless, understanding the classes discussed here should provide a solid foundation for further study. The reader is encouraged to dive deeper into this fascinating subject [103, 104, 107, 133, 134], as each class represents a unique aspect of what we understand, and don't understand, about computation.

Figure 2.16 A plethora of complexity classes.

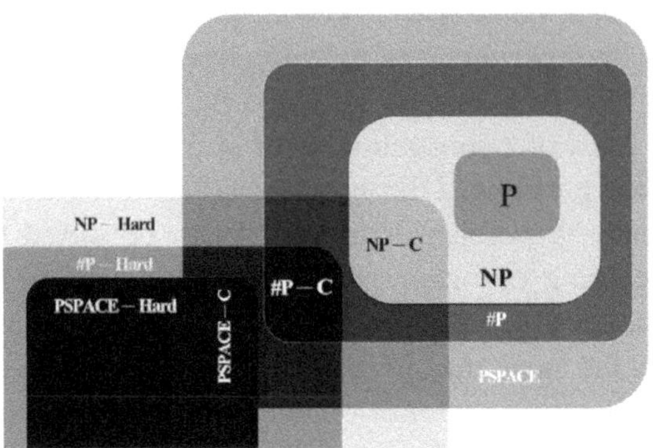

The Extended Chomsky Hierarchy Reloaded

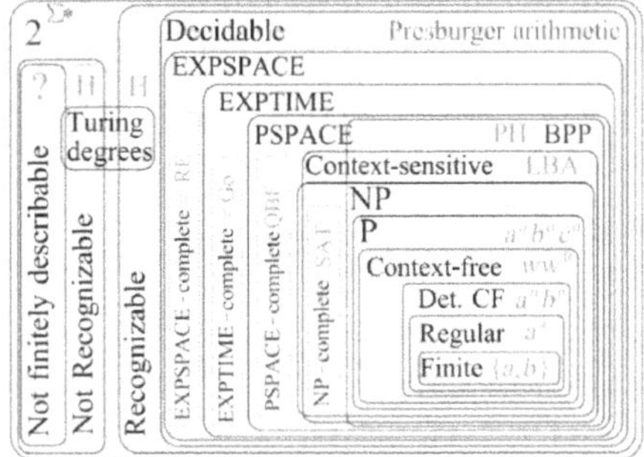

2.4.3 Information and Energy

In our universe, not only is matter equivalent to energy, as special relativity implies, but information too can be converted into and from energy. This surprising equivalence is deeply rooted in the principles of thermodynamics [135]. To appreciate this concept, we will expound on two fascinating key ideas: Maxwell's Demon [136, 137] and Landauer's Principle [32, 138].

Maxwell's Demon

James Clerk Maxwell's nineteenth-century thought experiment has long perplexed physicists and raised questions about the nature of thermodynamic laws. In this imaginary scenario, a "demon" oversees a container split into compartments A and B by an insulating wall with a tiny door. This demon has the ability to open and close the door to sort faster-than-average gas molecules into one compartment and slower-than-average molecules into the other, seemingly creating a temperature gradient and thus reducing the system's entropy, violating the Second Law of Thermodynamics. See Figure 2.17 for an illustration.

Various attempts have been made to resolve this paradox. Some have argued that the demon itself would gain entropy, offsetting the entropy reduction in the gas. Others have posited that the act of measurement would require an expenditure of energy, hence no real violation of the Second Law occurs. These explanations were not fully satisfactory for a few reasons. Some early solutions argued that the demon itself, being part of the system, would gain entropy or expend energy, thus balancing out the system's overall entropy. Unfortunately, this does not rigorously adhere

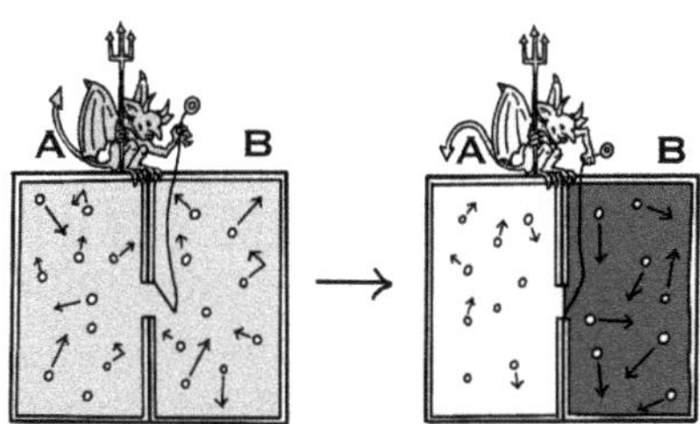

Figure 2.17 Maxwell's demon. The demon
selectively allows faster-than-average particles
to cross into the right box, while
slower-than-average particles are confined to
the left, posing a thought experiment that
challenges the Second Law of
Thermodynamics.

to the laws of thermodynamics, as it doesn't provide a clear mechanism or calculation to quantify this entropy gain or energy expenditure. In other proposals, the act of measurement and observation were said to consume energy, thus seemingly nullifying the paradox. Yet, these interpretations often resorted to extraphysical assumptions about the nature of measurement and observation. For instance, some early solutions posited that the very act of "observing" or "knowing" the position of a particle inherently required an expenditure of energy. While this idea touches on concepts familiar in quantum mechanics, it did not provide a universally applicable, classical thermodynamic explanation. These explanations sometimes ventured into philosophical territory, invoking concepts such as the "nature of knowledge" and "conscious observation," which are not quantifiable in the same way that thermodynamic variables are.

As we are going to see, the first adequate resolution came with Landauer's Principle, which explores the informational aspect of the thought experiment.

Landauer's Principle: The Resolution to Maxwell's Demon

Landauer's Principle [32] states that the erasure of one bit of information is accompanied by a minimum amount of energy dissipation, given by $kT \ln(2)$, where k is the Boltzmann constant and T is the temperature of the system in kelvins. Here is a simple proof of the principle.

Physically, consider a box divided into two equal parts, left and right. A single molecule is inside this box. If the molecule is on the left side, we say the system encodes the bit "0," and if it's on the right side, it encodes the bit "1." To "erase" this bit, we need to reset the system to a standard state, say "0." This would involve pushing the molecule from the right half to the left half if it's on the right, effectively "erasing" the information stored.

Initially, we might think that we can perform this erasure operation without any energy expenditure. Nevertheless, if we are to comply with the Second Law of Thermodynamics, we have to consider the entropy change of this operation. By confining the molecule to one-half of the box, we have effectively decreased the entropy of the system. According to the Second Law, this decrease in entropy has to be compensated by an increase in entropy elsewhere, resulting in energy dissipation.

The physical argument can be made mathematically precise as follows. The entropy S of a system in a given macrostate is defined by $S = k \ln(W)$, where W is the number of corresponding microstates. Initially, the molecule can be either on the left or the right, $W = 2$, and the entropy is $S_{\text{initial}} = k \ln(2)$.

After erasure, the molecule is confined to the left half, $W = 1$, and the entropy becomes $S_{\text{final}} = k \ln(1) = 0$. The change in entropy $\Delta S = S_{\text{final}} - S_{\text{initial}} = -k \ln(2)$. According to the Second Law, this entropy decrease must be compensated by dissipating heat Q into the surroundings. This is described by $\Delta S = \frac{Q}{T}$, leading to $Q = -kT \ln(2)$. Thus, the minimum energy that must be dissipated to reset the bit is $kT \ln(2)$.

Landauer's Principle elegantly resolves Maxwell's Demon paradox by quantifying the energy required for the demon's computations and reconciling it with the laws of thermodynamics.

The resolution of Maxwell's Demon has been a point of contention not just in classical thermodynamics but also in the broader realm of physics, including quantum mechanics. While early attempts to resolve the paradox sometimes invoked quantum concepts, these were often seen as "extraphysical" within the context of classical thermodynamics, which was the original framework where the paradox was posed. Landauer's Principle is often cited as the first "satisfactory" solution within this classical framework, providing a concrete, quantifiable relationship between information and energy. Nevertheless, it's worth noting that quantum interpretations offer their own set of intriguing solutions, often involving the fundamental nature of measurement and observation.

2.4.4 Reversible Computation

As we have seen from Landauer's Principle, the erasure of information inherently requires a minimum amount of energy to be dissipated. In standard models of classical computation, operations are generally irreversible; gates like the AND, OR, and NOT gates erase information, making them inherently dissipative according to Landauer's Principle. This naturally raises the question: Is there a form of computation that minimizes or even completely avoids such energy dissipation? Enter the intriguing realm of *reversible computation* [139].

Reversible computation offers a way to perform logical operations in a manner that is, in principle, fully reversible and hence could be exempt from the energy cost stipulated by Landauer's Principle. Given that quantum mechanics, a cornerstone of our understanding of nature, is fundamentally reversible, it seems natural to forecast that quantum computation would fall under this category of computation. Nevertheless, before we dive into that, it is essential to ask: What about classical computation?

It turns out that there is no fundamental obstruction to reversibility in classical physics. Newtonian mechanics, for example, is time-reversal invariant, allowing for the possibility of reversible operations even in classical systems. In this subsection, we shall explore the key features and limitations of reversible computation, both in classical and quantum contexts.

Embedding an Irreversible Function into a Reversible One

If a logic gate is irreversible, we are looking to replace it with a reversible one that does the same job. Since a logic gate implements a binary function, we can ask whether an irreversible binary function can be converted into a reversible one. Suppose we have an irreversible function $f: \{0, 1\}^n \to \{0, 1\}^m$, where n and m are natural numbers. We can convert this function into a reversible function $g: \{0, 1\}^{n+m} \to \{0, 1\}^{n+m}$ as follows:

$$g(x_1, x_2, \ldots, x_n, y_1, y_2, \ldots, y_m) = (x_1, x_2, \ldots, x_n, y_1 \oplus f_1(x), y_2 \oplus f_2(x), \ldots, y_m \oplus f_m(x)) \tag{2.37}$$

Here, $f_i(x)$ is the i-th bit of the output of the function $f(x)$, and $\oplus$ represents bitwise XOR. Essentially, the function g takes an n-bit input x and an m-bit "garbage" string y, and it maps them to a new $n + m$-bit string. The first n bits are unchanged, and the remaining m bits are the bitwise XOR of y and $f(x)$.

The function g is reversible because the original inputs x and y can be perfectly recovered from the output. To revert, one could apply f to the first n bits of the output and then XOR it with the last m bits. This makes g a reversible embedding of f.

> **Exercise 2.12** *The reversibility of the function g in (2.37) can also be easily seen as $g = g^{-1}$. Try applying g twice to recover the original inputs.*

This technique shows that any irreversible computation can be embedded into a reversible one without violating the laws of classical physics. Thus, it becomes plausible to construct reversible computers that align with Landauer's principle, minimizing the energy dissipation associated with information erasure.

Reversible Gates

All 1-bit gates are evidently reversible. Two-bit gates like AND, OR, and XOR are often not reversible. To see how we can make an irreversible function reversible, consider the XOR gate that implements the function $f(x, y) = x \oplus y$, where x and y are two bits. By Equation (2.37), this function can be embedded into a 3-bit function

$$\tilde{f}(x, y, z) = (x, y, z \oplus (x \oplus y)), \tag{2.38}$$

which is reversible. To see its reversibility, consider:

$$\tilde{f} \circ \tilde{f}(x, y, z) = \tilde{f}(x, y, z \oplus (x \oplus y)) = (x, y, z), \tag{2.39}$$

which shows that $\tilde{f}^{-1} = \tilde{f}$.

We can implement this function $\tilde{f}$ by a 3-bit reversible gate, whose third output bit is the output of the XOR gate in the case where the third input bit $z = 0$. If we need only to realize the XOR gate reversibly, we can simply set $z = 0$ in the 3-bit gate, reducing it to a 2-bit reversible gate – the CNOT gate, with the function becoming $\tilde{f}'(x, y) = (x, y, x \oplus y)$.

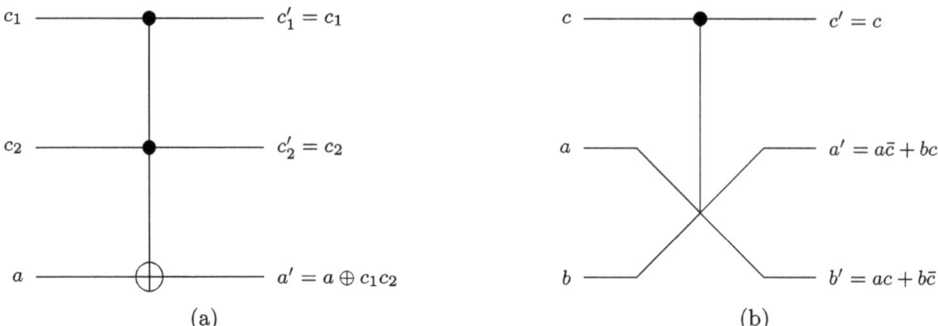

(a)　　　　　　　　　　　　　　　　　　　　　(b)

Figure 2.18 Circuit diagrams of the Toffoli gate (a) and the Fredkin gate (b).

> **Exercise 2.13** *How many 2-bit reversible gates are there? What about n-bit gates?*

More Reversible Gates: Toffoli and Fredkin

After discussing how to construct reversible versions of basic irreversible gates, it's worth mentioning a couple of universal reversible gates that play a foundational role in reversible computation.

The *Toffoli gate* is a 3-bit reversible gate defined by the function

$$\text{Toffoli}(x, y, z) = (x, y, z \oplus (x \wedge y)),$$

whose circuit diagram and truth table are depicted in Figure 2.18(a) and Table 2.5(a) respectively. From the circuit diagram, you can see why the Toffoli gate is also known as the controlled-controlled NOT (CCNOT or C_2NOT) gate. Interestingly, by setting $z = 1$ in the Toffoli gate, the third output bit mimics the NAND operation on the first two input bits. So, the Toffoli gate embeds the NAND gate, which belongs to the minimal universal set of gates. It also embeds the FANOUT function when $y = 1$ and $z = 0$, as the third output then becomes a copy of the first input x.

The *Fredkin gate* is another universal reversible gate, defined as

$$\text{Fredkin}(x, y, z) = \begin{cases} (x, y, z), & \text{if } x = 0, \\ (x, z, y), & \text{if } x = 1. \end{cases} \tag{2.40}$$

Its circuit diagram and truth table are shown in Figure 2.18(b) and Table 2.5(b).

> **Exercise 2.14** *Can you use the Toffoli gate to realize a reversible XOR gate? Can you realize the NAND and COPY gates using the Fredkin gate?*

Unlike the Toffoli gate, the Fredkin gate is also conservative, meaning it preserves the number of 0s and 1s in the input string. Apart from being reversible, *conservative gates* have several attractive features that make them of particular interest both theoretically and practically:

Table 2.5 Truth tables of the Toffoli gate (a) and the Fredkin gate (b).

(a)

c_1	c_2	a	c_1'	c_2'	a'
0	0	0	0	0	0
0	0	1	0	0	1
0	1	0	0	1	0
0	1	1	0	1	1
1	0	0	1	0	0
1	0	1	1	0	1
1	1	0	1	1	1
1	1	1	1	1	0

(b)

c	a	b	c'	a'	b'
0	0	0	0	0	0
0	0	1	0	0	1
0	1	0	0	1	0
0	1	1	0	1	1
1	0	0	1	0	0
1	0	1	1	1	0
1	1	0	1	0	1
1	1	1	1	1	1

- **Energy efficiency**: Ideal conservative gates could operate without dissipating energy, aligning well with low-power computational needs.
- **Signal integrity**: The conservation laws in the gates could reduce errors due to signal degradation.
- **Fault tolerance**: The reversibility of conservative gates makes error detection and correction more straightforward.
- **Simpler physical implementation**: Conservation laws in the gate could align well with physical laws, making hardware implementation easier.
- **Compatibility with quantum computing**: Reversible and conservative gates are naturally suited for quantum mechanical operations.
- **Theoretical insights**: Studying these gates can provide valuable insights into the fundamental limits of computation and its relation to physical laws [140].

Both Toffoli and Fredkin gates are reversible and universal, showing the variety and richness in the world of reversible computation.

Models of Reversible Classical Computation

While the concept of reversible computing has been highly influenced by the advent of quantum computing, it also has various classical models that are worth noting. These classical models for reversible computation offer alternative frameworks for understanding how reversible logic can be practically implemented. Here we discuss some of the most notable ones.

- **Reversible circuit model:** A straightforward way to realize reversible computing is to build circuits composed of reversible gates like the Toffoli and Fredkin gates. These reversible circuits can implement any Boolean function without dissipating energy on information erasure.
- **Billiard-ball model:** Proposed by Edward Fredkin and Tommaso Toffoli, the billiard-ball model is a mechanical model for reversible computation [141]. In

this model, billiard balls collide elastically in a frictionless environment, emulating the behavior of reversible logic gates. This model serves as an interesting proof of concept that classical mechanics can perform reversible computation.

- **Reversible cellular automata:** This model extends the idea of cellular automata [142] to a reversible framework [143]. In a reversible cellular automaton, each state has a unique predecessor, thereby ensuring reversibility. The rules governing the automaton are constructed so that the system can evolve forwards and backwards deterministically.
- **Adiabatic computation:** Though often associated with quantum systems [15], adiabatic processes can also be classical. In adiabatic computation, the system evolves slowly enough to stay in its ground state, ensuring that the computation is reversible. This is particularly useful for computations that require low energy dissipation.
- **Reversible turing machines:** These are Turing machines designed to be reversible. At each step, they retain enough information to revert to the previous state, offering a more theoretical angle to reversible computing.
- **Reversible stack machines:** Analogous to regular stack machines, reversible stack machines are designed to operate without information loss [144]. They offer an alternative, more abstract model for reversible computation.

In summary, these models showcase the versatility and feasibility of reversible computation in a classical setting. The study of these classical models is crucial for understanding the fundamental limits and capabilities of reversible computing.

Conclusion

In this chapter, we introduced foundational concepts in classical computation, starting with Turing machines, which laid the groundwork for understanding algorithmic processes and computational limits. We examined the universal Turing machine, capable of simulating any other Turing machine, and explored the probabilistic variant that broadens our view on computation under uncertainty. Shifting from theory to practical application, we explored the circuit model, a framework central to modern computing systems, which helps bridge the gap between theoretical and real-world computation.

We then discussed computational complexity, differentiating between tractable and intractable problems, and mapped these through various complexity classes. This section highlighted how classical resources like time and space scale in relation to input size, a key factor for evaluating computational feasibility. Additionally, we discussed the interplay between information and energy, especially through Landauer's Principle, which underscores the energy cost of information erasure in any computation. Finally, we examined reversible computation, an area that hints at energy-efficient, potentially quantum-compatible future computational methods.

In the next chapter, we will review the basics of quantum mechanics that are relevant to our expedition in quantum computation in the rest of the book.

3   A Crash Course On Quantum Mechanics

In the journey to understand the deepest recesses of our universe, classical physics – despite its elegance and precision – proves to be insufficient. The need for a more comprehensive framework brings us to quantum mechanics, a theory born from the limitations of classical descriptions when dealing with the subatomic world. In this chapter, we will explore the foundational postulates of quantum mechanics, then proceed to unravel the nuances of mixed states, density matrices, and quantum coherence. As we navigate through the enigmatic phenomena of superposition and entanglement, we'll introduce key concepts like separable states, entangled states, and the related notion of entanglement entropy. The versatile Schmidt decomposition provides us with a valuable tool for examining composite systems. To round off our exploration, we'll enter the realm of generalized measurements, which allows for a more complete understanding of quantum systems. This chapter aims to equip you with the essential mathematical and conceptual tools for a deeper journey into the quantum realm, setting the stage for forthcoming discussions on quantum computation.

3.1   Why Quantum Mechanics?

While an array of compelling arguments and phenomena make the necessity of quantum mechanics self-evident, this section will zoom in on a singularly illuminating experiment that is closely related to quantum computation: the Stern–Gerlach experiment. This seminal work didn't merely throw a wrench in the gears of classical intuition, it also laid the groundwork for the fundamental concept of quantum bits, or qubits, upon which the field of quantum computing is constructed.

3.1.1   The Stern–Gerlach Experiment

Conceived in the early years of the twentieth century, the Stern–Gerlach experiment [145] was initially intended to validate the quantization of angular momentum. Otto Stern and Walther Gerlach planned to direct a stream of silver atoms through a spatially varying magnetic field and record the angular distribution of these atoms. According to classical expectations, one would anticipate the silver atoms to scatter across a continuous array of angles, painting a continuous distribution

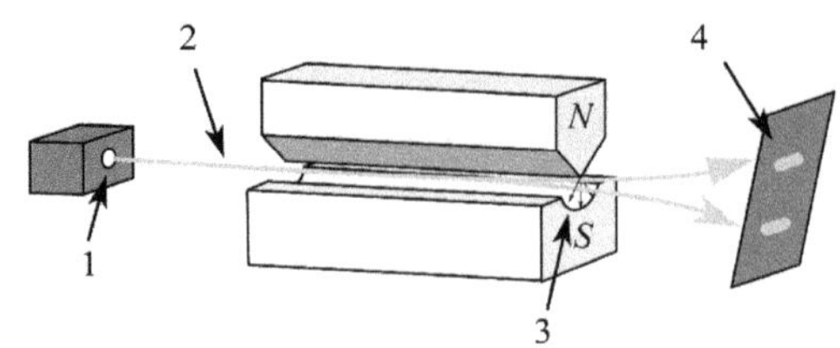

Figure 3.1 Stern–Gerlach experiment. Silver atoms travelling through an inhomogeneous magnetic field, and being deflected up or down depending on their spin; (1) furnace, (2) beam of silver atoms, (3) inhomogeneous magnetic field, (4) observed result.

Figure 3.2 State collapse and superposition in the Stern–Gerlach experiment.

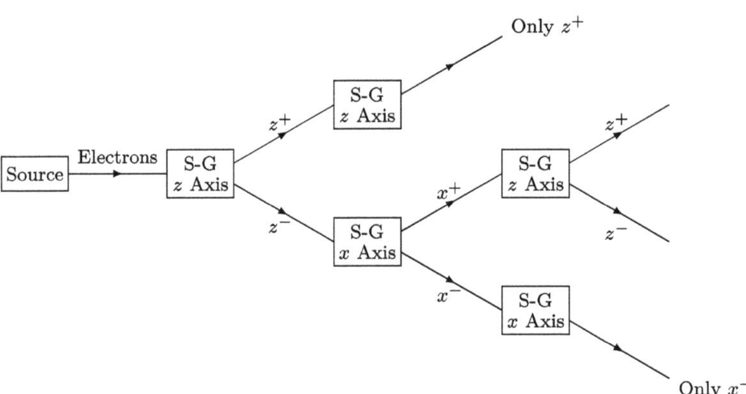

on a detection screen. The outcome, however, deviated dramatically from these predictions.

The experiment was set up as follows: An oven emitted a beam of thermally excited silver atoms, which then traversed a magnetic field with a gradient along the vertical direction (see Figure 3.1 for the experimental setup). The magnetic field was constructed to interact specifically with the atoms' intrinsic angular momentum or "spin." Following this region, a detection screen was placed to capture the landing points of the deflected atoms.

What Stern and Gerlach observed was revelatory: Rather than appearing in a continuum as classically expected, the atoms were detected at two vertically discrete points on the screen. Specifically, these points corresponded to atom states of "spin up" and "spin down." This quantized behavior was at odds with classical physics and stood as a robust validation of quantum mechanics.

Among the experiment's far-reaching implications is the concept of state collapse and superposition. As in Figure 3.2, suppose that after moving through the magnetic field aligned along the z-axis, a silver atom experiences a "measurement" of its spin and collapses into a state of either "spin up" or "spin down" along that axis. Now, if we change the orientation of the magnetic field to align along the x-axis and pass the atom through it, the atom will exhibit a superposition of states in this new basis. Specifically, its state will be a quantum mixture of "spin left" and "spin right" along the x-axis.

Even more intriguingly, as also seen in Figure 3.2, if we were to "measure" the state in this x-axis basis, causing it to collapse into one of these x-axis states, and then realign our apparatus to measure along the z-axis once more, the atom would again be in a superposition of "spin up" and "spin down" states along the z-axis. This showcases the fundamentally probabilistic and nondeterministic nature of quantum systems, where measurements in one basis affect the state's behavior in other bases.

As we venture into the realm of quantum computing, it is crucial to emphasize that the two states of the silver atom ("spin up" and "spin down") serve as a prime example of a qubit. More broadly, a qubit can be any quantum system with two discernible, mutually reachable states. Thus, the Stern–Gerlach experiment not only provided a foundational pillar for quantum mechanics but also sowed the seeds for the qubits that are essential in quantum computers.

To conclude, it's worth noting that other seminal experiments such as the double-slit experiment, the EPR paradox, and the photoelectric effect have been instrumental in shaping our understanding of quantum mechanics. Nevertheless, their immediate relevance to quantum computation is not as pronounced as that of the Stern–Gerlach experiment. For this reason, we've centered our discussion around the Stern–Gerlach experiment, which provides a direct introduction to the concept of a qubit – the cornerstone of quantum computation.

3.2   The Postulates of Quantum Mechanics

Traditionally, quantum mechanics has been framed through a set of fundamental postulates that outline the basic principles of the theory. Our modern understanding offers a more unified perspective:

Quantum mechanics can be fundamentally described within the formalism of linear algebra.[1]

Consequently, many of what were once considered to be separate, foundational postulates can now be understood as natural outcomes of a modern, unified viewpoint. Moreover, certain aspects of quantum mechanics, such as the interpretation of the wavefunction and the role of measurement, extend beyond mathematical questions to touch upon philosophical or interpretational issues. Nonetheless, for practical calculations and specifically for applications in quantum computing, this unified approach remains both effective and streamlined. In line with the focus of this course on quantum computation, we will introduce the subject through three key postulates. These postulates emerge from the broader principle of quantum mechanics as a mathematical framework rooted in linear algebra and hold particular importance for the realm of quantum computation.

---

[1] In simpler terms, quantum mechanics differs from Newtonian mechanics by being linear. The ultimate reason for this inherent linearity remains elusive.

3.2.1 Postulate 1

A quantum system S is described by a state space, which is a Hilbert space $\mathcal{H}_S$ tailored to the system. The vectors $|\psi\rangle$ in $\mathcal{H}_S$ serve as possible states of the system. For practical purposes, especially for probability calculations, we typically work with normalized (unit) vectors in $\mathcal{H}_S$. The time evolution of a state $|\psi(t)\rangle$ is governed by the Schrödinger equation

$$i\hbar\frac{d}{dt}|\psi(t)\rangle = H(t)\,|\psi(t)\rangle\,, \tag{3.1}$$

where $H(t)$ is the Hamiltonian of the system, and $\hbar = \frac{h}{2\pi}$ is the reduced Planck constant.

A few remarks are warranted to clarify the concepts encompassed and implied by this postulate.

A **Hilbert space** is a vector space over the complex field $\mathbb{C}$ equipped with an inner product that induces a complete metric. Specifically, a **metric space** in this context is a vector space endowed with a nondegenerate,[2] positive-definite[3] norm that satisfies the triangle inequality. The space is complete if it contains the limit points of all **Cauchy sequences** within it.

States are represented in **Dirac's bra–ket notation** as $|\psi\rangle$, often referred to as a **Dirac ket** or simply a **ket**. Any state $|\psi\rangle \in \mathcal{H}_S$ has an associated dual state $\langle\varphi|$, read as **Dirac bra** and defined as a linear functional mapping $|\psi\rangle$ to a complex number:

$$\langle\varphi|: |\psi\rangle \mapsto \langle\varphi|\psi\rangle \in \mathbb{C}. \tag{3.2}$$

Here, $\langle\varphi|\psi\rangle$ can be taken as the *inner product* of $|\varphi\rangle$ and $|\psi\rangle$. Usually, $|\psi\rangle$ can be treated as a column vector, whereas $\langle\varphi|$ is a row vector.

The set of all such dual states $\{\langle\varphi|\}$ forms the dual space $\mathcal{H}_S^*$ of $\mathcal{H}_S$. This allows us to define the notion of a wavefunction: $\psi(x) = \langle x|\psi\rangle$ represents the **wavefunction** corresponding to the state $|\psi\rangle$ in position space, where x denotes the spatial coordinate.[4]

As a vector space, $\mathcal{H}_S$ can have an infinite number of bases. Physicists often work with a specific complete, orthonormal basis $\{|i\rangle\,|\,i = 1,\cdots,N = \dim\mathcal{H}_S\}$, where N can be finite or infinite. Orthonormality means that $\langle i|j\rangle = \delta_{ij} = 1$ if $i = j$ and otherwise 0. The *completeness* of the basis can be expressed as the **resolution of identity**:

$$\mathbb{1} = \sum_{i=1}^{\dim\mathcal{H}_S} |i\rangle\langle i|, \tag{3.3}$$

---

[2] A norm $||\cdot||$ on a vector space V is nondegenerate if $||v|| = 0 \Leftrightarrow v = 0$ for $v \in V$.

[3] A norm $||\cdot||$ on a vector space V is positive definite if $||v|| \geq 0$ for all $v \in V$, and $||v|| = 0$ if and only if $v = 0$.

[4] The concept of the dual state $\langle x|$ is subtle and its rigorous definition is beyond the scope of this book. Interested readers may explore the notion of a *rigged Hilbert space* for a deeper understanding. While physicists might overlook these mathematical nuances, quantum computer scientists are often even less concerned with them.

where $\mathbb{1}$ is the identity operator (matrix)[5] acting on $\mathcal{H}_S$. In this basis, any state $|\psi\rangle \in \mathcal{H}_S$ can be expanded as

$$|\psi\rangle = \sum_{i=1}^{N} \psi_i \, |i\rangle, \tag{3.4}$$

where $\psi_i := \langle i|\psi\rangle$ and $\sum_{i=1}^{N} |\psi_i|^2 = 1$ for $|\psi\rangle$ to be normalized.

> **Exercise 3.1** *If you are not familiar with linear algebra in Dirac's bra–ket notation, practice using this notation with some simple examples to become acquainted with it. Start by expressing basic vectors and operators, and then move on to more complex expressions involving inner products and outer products.*

According to the widely accepted **Copenhagen interpretation**, $|\langle i|\psi\rangle|^2$ is the probability of finding the basis state $|i\rangle$ in $|\psi\rangle$, and $\langle i|\psi\rangle$ is the corresponding **probability amplitude**. This elucidates the normalization condition $\sum_{i=1}^{N} |\psi_i|^2 = 1$, ensuring the total probability sums to unity. More generally, $\langle\varphi|\psi\rangle$ is the probability amplitude for finding state $|\varphi\rangle$ in state $|\psi\rangle$. This also implies that $|\langle\psi|\psi\rangle|^2 = 1$, as one would expect, since the probability of finding $|\psi\rangle$ in $|\psi\rangle$ should be unity. In this framework, the wavefunction $\langle x|\psi\rangle$ can be interpreted as the probability amplitude for finding a particle at position x when it is in state $|\psi\rangle$.

That a generic quantum state can be expressed as a linear superposition, coupled with the Copenhagen interpretation, underscores an essential distinction between quantum and classical physics – namely, the phenomenon of quantum coherence. Consider two states: $|\psi\rangle$ given by Equation (3.4), and another state $|\varphi\rangle = \sum_{i=1}^{N} \varphi_i \, |i\rangle$. The probability amplitude of finding $|\varphi\rangle$ in $|\psi\rangle$ or vice versa can be written as

$$\langle\varphi|\psi\rangle = \sum_{i=1}^{N} \varphi_i^* \psi_i, \tag{3.5}$$

which simplifies due to the orthonormality of the basis states, $\langle i|j\rangle = \delta_{ij}$. Consequently, the associated probability is

$$|\langle\varphi|\psi\rangle|^2 = \sum_{i=1}^{N} |\varphi_i^* \psi_i|^2 + \underbrace{\sum_{i \neq j} \varphi_i^* \psi_i \varphi_j \psi_j^*}_{\text{coherence terms}}. \tag{3.6}$$

The terms above the brace are known as **coherence terms**. Originating from the interference between different components in $|\psi\rangle$ and $|\varphi\rangle$, they give rise to a unique quantum phenomenon – **quantum coherence**. Unlike classical probabilities, which

---

[5] $|\varphi\rangle\langle\psi|$ is the outer product of two vectors $|\varphi\rangle$ and $|\psi\rangle$: multiplying an N-dimensional row vector with a column vector as a multiplication of two matrices results in an $N \times N$ matrix. In continuum, Equation (3.3) is understood as $\mathbb{1} = \int |x\rangle\langle x|\mathrm{d}x$.

simply sum up individual likelihoods, quantum coherence introduces interference effects that can significantly influence the system's behavior. This capacity for interference not only makes quantum coherence an intrinsic feature of quantum systems but also underpins its critical role in quantum computing, quantum cryptography, and quantum sensing. It is precisely these coherence properties that enable quantum systems to perform tasks and computations that are practically impossible for classical systems to achieve. Therefore, quantum coherence is not just a mathematical quirk; It's an indispensable physical resource with far-reaching implications. We will later explore how the influence of an external environment can diminish these coherence terms, giving rise to **quantum decoherence**, which is the primary adversary in the realization of quantum computing.

The last remark concerns the time evolution of a quantum state. The formal solution to the Schrödinger equation (3.1) can be written as

$$|\psi(t)\rangle = \underbrace{\mathcal{T} \exp\left[-\frac{i}{\hbar} \int_{t_0}^{t} H(t')dt'\right]}_{=:U(t,t_0)} |\psi(t_0)\rangle. \tag{3.7}$$

Here, $\mathcal{T}$ is the time-ordering operator, which takes care of the potential noncommutativity of the Hamiltonian at different times. It is defined as

$$\mathcal{T}[A(t_1)B(t_2)] = \begin{cases} A(t_1)B(t_2) & t_1 > t_2, \\ B(t_2)A(t_1) & t_1 < t_2. \end{cases} \tag{3.8}$$

The term above the brace is defined to be the time evolution operator $U(t, t_0)$, which evolves the state from time t_0 to t.

For a time-independent Hamiltonian H, the situation simplifies considerably. The time-ordering operator $\mathcal{T}$ becomes unnecessary, and the time evolution operator $U(t, t_0)$ takes on the much simpler form:

$$U(t, t_0) = \exp\left[-\frac{i}{\hbar}H(t - t_0)\right]. \tag{3.9}$$

This simplification arises because the Hamiltonian commutes with itself at all times, rendering the time-ordering irrelevant. The expression makes it evident how a quantum state $|\psi(t)\rangle$ evolves from its initial state $|\psi(t_0)\rangle$ in a straightforward manner. It's a commonly encountered case in many quantum systems and serves as a useful point of departure for understanding more complex, time-dependent situations.

So concludes our discussion of Postulate 1, which lays the foundation for understanding quantum states and their time evolution. While we've touched on the Hamiltonian's role in determining a system's energy, it's worth noting that energy alone doesn't capture all the nuances of a quantum system. What other information can we glean from a state, and how do we go about extracting it? These questions will be the focus of Postulates 2 and 3.

3.2.2 Postulate 2

*Any **physical observable** O of a quantum system S is represented by a **self-adjoint** operator[6] on the Hilbert space $\mathcal{H}_S$. Self-adjoint operators have real eigenvalues, ensuring that the outcomes of physical measurements are real numbers. A complete set of information for the system S can be extracted from a maximal set of commuting observables,[7] often denoted as a **complete set of commuting operators (CSCO)**. In an appropriate basis $\{|i\rangle\}$, these observables are diagonalized, simplifying the interpretation and calculation of physical quantities.*

The eigenvalues of an observable O correspond to the possible outcomes of a measurement of O on any state of the system. If the eigenvalue equation of O is $O\,|i\rangle = o_i\,|i\rangle$, and a generic state is expanded as $|\psi(t)\rangle = \sum_i \psi_i(t)\,|i\rangle$, then the probability of obtaining the outcome o_i upon measuring O at time t in the state $|\psi(t)\rangle$ is given by

$$p_i(t) = |\,\langle i|\psi(t)\rangle\,|^2 = |\psi_i(t)|^2. \tag{3.10}$$

Three remarks are in order.

Hermiticity

Quantum mechanics is inherently probabilistic. When one measures an observable O on a given quantum state $|\psi\rangle$, the outcome can vary from one instance to another, each with a certain probability of occurrence. Therefore, what can be reliably measured for an observable O in the state $|\psi\rangle$ is its expectation value, defined by

$$\langle O \rangle := \langle \psi|O\psi \rangle. \tag{3.11}$$

The adjoint $O^\dagger$ of O is defined by

$$\langle O^\dagger \varphi|\psi \rangle = \langle \varphi|O\psi \rangle, \tag{3.12}$$

where $|\varphi\rangle$ is an arbitrary state in the Hilbert space. To understand the relationship between the adjoint $O^\dagger$ and the original operator O, let us consider their matrix elements in a certain basis $\{|i\rangle\}$. For O, its matrix elements O_{ij} are given by

$$O_{ij} = \langle i|Oj \rangle. \tag{3.13}$$

Similarly, for $O^\dagger$, its matrix elements $O^\dagger_{ij}$ can be defined as

$$O^\dagger_{ij} = \langle i|O^\dagger j \rangle. \tag{3.14}$$

Letting $|\varphi\rangle = |j\rangle$ and $|\psi\rangle = |i\rangle$ in the definition (3.12) of the adjoint and taking complex conjugation of both sides of the equation, we have

$$\langle i|O^\dagger j \rangle = \langle j|Oi \rangle^*. \tag{3.15}$$

Equation (3.15) implies that

$$O^\dagger_{ij} = O^*_{ji}, \tag{3.16}$$

---

[6] Usually, the hatted notation $\hat{O}$ is used to emphasize that it is a quantum operator acting on the states of the corresponding Hilbert space. But since we deal with quantum operators exclusively in this book, we shall simply use O instead of $\hat{O}$.

[7] Two observables O and O' commute if their commutator $[O, O']$ vanishes:
$[O, O'] := OO' - O'O = 0.$

and thus when we write down $O^\dagger$ as a matrix, it corresponds to taking the complex conjugate of the transpose of O, yielding

$$O^\dagger = \left(O^T\right)^*. \tag{3.17}$$

Hence, the adjoint $O^\dagger$ of O is also called the **Hermitian conjugate** of O.[8]

Physical observables must yield real numbers in measurements. Thus, the expectation value must satisfy

$$\langle O \rangle = \langle \psi | O\psi \rangle = \langle O\psi | \psi \rangle = \langle O \rangle^*, \; \forall \, |\psi\rangle, \tag{3.18}$$

which, along with Equation (3.12), leads to

$$\langle O^\dagger \psi | \psi \rangle = \langle O\psi | \psi \rangle, \; \forall \, |\psi\rangle, \tag{3.19}$$

from which it follows that $O^\dagger = O$. This property is known as the hermiticity of O, indicating that O is a Hermitian operator.

Hermitian operators possess real eigenvalues, and this property extends to each physical observable in a CSCO. Consequently, the eigenvalues of the physical observables of a quantum system are termed the **good quantum numbers**, as they represent the only values that can be experimentally derived from system measurements. By measuring all the observables in the CSCO of a quantum system, one can determine a corresponding set of good quantum numbers for a given quantum state, thus fully characterizing it.

It is important to recognize that good quantum numbers do not always assume discrete values. For example, the energy levels of a quantum system and the mass of a quantum particle are not exclusively discrete.

> **Exercise 3.2** *What is the role of the dual operator A' of an operator A in quantum mechanics? In matrix form, how is A' related to A?*

Distinguishing Two States

One of the most intriguing and critical questions in both quantum mechanics and quantum information theory concerns how to distinguish quantum states. This question has pivotal implications in areas such as quantum computing, quantum cryptography, and quantum error correction.

First of all, any state $|\psi\rangle$ cannot be distinguished from $e^{i\theta} |\psi\rangle$ with arbitrary $\theta \in \mathbb{R}$, because they have the same inner product $\langle \psi | \psi \rangle$. Such a phase factor $e^{i\theta}$ is immaterial in any measurable probabilities.

---

[8] Nuances arise in the case of infinite dimensions; however, in quantum computation, you will almost always deal with finite, albeit possibly large, dimensions. Thus, Hermitian conjugation and adjoint can be used interchangeably. In fact, physicists usually do not concern themselves with these nuances. For those who are particularly interested in the details, Chapter 2 of the book "What Is Quantum Field Theory" [146] is a good reference and requires little prior knowledge.

For two orthogonal states, $|i\rangle$ and $|j\rangle$ with $\langle i|j\rangle = \delta_{ij}$, we can distinguish these states with absolute certainty by constructing an observable $O_i = |i\rangle\langle i|$. The expectation values of O_i in these states yield

$$\langle O_i\rangle_i = \langle i|O_i|i\rangle = \langle i|i\rangle\langle i|i\rangle = 1,$$
$$\langle O_i\rangle_j = \langle j|O_i|j\rangle = \langle j|i\rangle\langle i|j\rangle = 0. \tag{3.20}$$

Thus, orthogonal states are perfectly distinguishable.

In the case of nonorthogonal states $|\varphi\rangle$ and $|\psi\rangle$ where $\langle\varphi|\psi\rangle \neq 0$, these can be connected as

$$|\psi\rangle = \alpha|\varphi\rangle + \beta|\psi'\rangle, \tag{3.21}$$

with $\alpha, \beta \in \mathbb{C}$ and $\langle\varphi|\psi'\rangle = 0$. Given that $|\psi\rangle$ has a component parallel to $|\varphi\rangle$, there exists a nonzero probability $|\alpha|^2/(|\alpha|^2 + |\beta|^2)$ of obtaining an outcome in any measurement that misidentifies $|\psi\rangle$ as $|\varphi\rangle$.

This nonzero overlap has significant practical implications. For instance, in quantum key distribution, the inability to perfectly distinguish nonorthogonal states provides a mechanism to detect eavesdropping [147]. In quantum computing, this limitation prompts the use of quantum error correction methods to correct for the inevitable errors caused by state overlap. Therefore, this issue serves as a cornerstone that influences various branches of quantum mechanics and quantum information theory.

Conservative Systems, Stationary States, and Distinguishability

When dealing with a time-independent Hamiltonian H, the system allows for special states known as *stationary states*. These states are eigenstates of H and can be represented as

$$H|n\rangle = E_n|n\rangle. \tag{3.22}$$

A stationary state remains the same up to a phase factor during its time evolution, thereby making it "stationary" in a probabilistic sense. This property is an extreme case of the distinguishability problem discussed a moment ago. In a stationary state, the system becomes physically indistinguishable from itself at different times.

Any time-dependent state $|\psi(t)\rangle$ can be expanded in terms of these stationary states as

$$|\psi(t)\rangle = \sum_n c_n(t)|n\rangle, \tag{3.23}$$

where the time-dependence is carried by the coefficients $c_n(t)$. Inserting this into the Schrödinger equation (3.1) gives us the time-evolved state as

$$|\psi(t)\rangle = \sum_n c_n(0)\exp\left(-\frac{i}{\hbar}E_n t\right)|n\rangle. \tag{3.24}$$

If the system starts in a stationary state $|\psi(0)\rangle = |n\rangle$, the state at any later time t will be

$$|\psi(t)\rangle = \exp\left(-\frac{i}{\hbar}E_n t\right)|n\rangle \propto |n\rangle. \tag{3.25}$$

Physically, states that differ by a phase factor are indistinguishable. Therefore, the system remains in the same stationary state $|n\rangle$ and conserves energy for all time.

3.2.3 Degeneracy and Symmetry

In the preceding discussion, it was implicitly assumed that each energy eigenvalue of the system's Hamiltonian corresponds to a unique eigenstate, modulo phase factors.[9] Specifically, if $H|\psi\rangle = E|\psi\rangle$ and $H|\psi'\rangle = E|\psi'\rangle$, then it follows that $|\psi\rangle = e^{i\theta}|\psi'\rangle$ for some $\theta \in \mathbb{R}$. Nevertheless, it is more typical that an eigenvalue E of a Hamiltonian has multiple eigenstates that are not proportional to each other. These eigenstates form a basis for a degenerate **eigenspace**, a subspace of the system's total Hilbert space. The dimension of this eigenspace is defined as the **degeneracy** of E.

The presence of degeneracy in a quantum system implies the existence of a **global symmetry**. To illustrate, consider a scenario where the energy eigenvalue E is two-fold degenerate with two orthogonal states $|\psi_1\rangle$ and $|\psi_2\rangle$ such that $H|\psi_1\rangle = E|\psi_1\rangle$ and $H|\psi_2\rangle = E|\psi_2\rangle$. These states are distinct, being orthogonal, but are indistinguishable through energy measurements alone. Therefore, there must exist another physical observable O that yields a distinct good quantum number for each state. This is expressed as $O|\psi_1\rangle = o_1|\psi_1\rangle$ and $O|\psi_2\rangle = o_2|\psi_2\rangle$, with $o_1, o_2 \in \mathbb{R}$ and $o_1 \neq o_2$. The requirement for O and H to commute, $[O,H] = 0$, facilitates simultaneous measurements (to be explained shortly in Section 3.2.4) of O and H, distinguishing the states. A concrete example would help you understand the relation between symmetry and degeneracy. Consider Example 3.1.

Example 3.1

Consider the Hamiltonian in Equation (3.26):

$$H = -J \begin{pmatrix} 1 & 0 & 0 & 1 \\ 0 & -1 & 1 & 0 \\ 0 & 1 & -1 & 0 \\ 1 & 0 & 0 & 1 \end{pmatrix}, \tag{3.26}$$

where J is a positive real constant. This Hamiltonian has the four eigenstates

$$|G_1\rangle = \frac{1}{\sqrt{2}}\begin{pmatrix} 1 \\ 0 \\ 0 \\ 1 \end{pmatrix}, \quad |G_2\rangle = \frac{1}{\sqrt{2}}\begin{pmatrix} 1 \\ 0 \\ 0 \\ -1 \end{pmatrix}, \quad |E_1\rangle = \frac{1}{\sqrt{2}}\begin{pmatrix} 0 \\ 1 \\ 1 \\ 0 \end{pmatrix}, \quad |E_2\rangle = \frac{1}{\sqrt{2}}\begin{pmatrix} 0 \\ 1 \\ -1 \\ 0 \end{pmatrix}, \tag{3.27}$$

---

[9] Phase factors are the only variations permitted as the states are normalized.

where $|G_1\rangle$ and $|G_2\rangle$ are the two degenerate ground states because $H|G_1\rangle = -2J|G_1\rangle$ and $H|G_2\rangle = -2J|G_2\rangle$, while $|E_1\rangle$ and $|E_2\rangle$ are the two nondegenerate excited states because $H|E_1\rangle = 0$ and $H|E_2\rangle = 2J|E_2\rangle$.

Exercise 3.3 *Verify that the states in Equation (3.27) are indeed the four eigenstates of the Hamiltonian (3.26) with the corresponding eigenvalues.*

One can construct a symmetry operator S as

$$S = \begin{pmatrix} 0 & 0 & 0 & 1 \\ 0 & 0 & 1 & 0 \\ 0 & 1 & 0 & 0 \\ 1 & 0 & 0 & 0 \end{pmatrix}. \tag{3.28}$$

Clearly, $[H, S] = 0$, $S|G_1\rangle = |G_1\rangle$, and $S|G_2\rangle = -|G_2\rangle$. So, indeed, the operator S is a symmetry operator of the Hamiltonian H and can distinguish the two degenerate ground states $|G_1\rangle$ and $|G_2\rangle$ of H by the quantum numbers 1 and -1. Since $S^2 = \mathbb{1}_4$, the 4×4 identity matrix, the symmetry generated by S is described by the group $\mathbb{Z}_2 = \{\mathbb{1}_4, S\}$.[10] But this doesn't mean that $\mathbb{Z}_2$ is the entire symmetry of the Hamiltonian (3.26) (see Exercise 3.4). Nevertheless, as long as there is a two-fold degeneracy in a quantum system, no matter what the total symmetry is, it must contain a $\mathbb{Z}_2$ symmetry.

Exercise 3.4 *Try to determine the total symmetry (as large as possible) of the Hamiltonian (3.26).*

Exercise 3.5 *Apply these symmetry considerations to classical systems, such as the harmonic system depicted in Figure 3.3. Identify the system's symmetry and possible energy degeneracies without using the Hamiltonian. Aided solely by the degeneracies and symmetry you found, can you find the system's motion modes with zero frequency? Such modes are termed* **zero modes**.

---

[10] For those who don't know anything about group theory: A group is a discrete or continuous set $\{g\}$ of elements g, equipped with a **"multiplication"** $\cdot: G \times G \to G$ satisfying the following properties: 1) G is closed under the multiplication, so that $g \cdot g' \in G$, $\forall g, g' \in G$. 2) The multiplication is associative, $(g \cdot g') \cdot g'' = g \cdot (g' \cdot g'')$, $\forall g, g', g'' \in G$. 3) There exists an **identity element** $e \in G$ such that $e \cdot g = g \cdot e = g$, $\forall g \in G$. 4) Any $g \in G$ has an **inverse element** $g' \in G$ such that $g \cdot g' = g' \cdot g = e$, so g' is conveniently denoted as g^{-1}. Often, the multiplication symbol is omitted: $gg' := g \cdot g'$. The multiplication is just a name for the binary operation that maps two group elements to a third, so it may not be arithmetical multiplication. For example, in the group $\mathbb{Z}$ of all integers (as an exercise, show that $\mathbb{Z}$ is indeed a group), the multiplication is in fact the addition: $a + b \in \mathbb{Z}$, $\forall a, b \in \mathbb{Z}$.

Figure 3.3 A harmonic system of three equal-mass balls connected by equal-length springs with the same spring constant, on a frictionless plane.

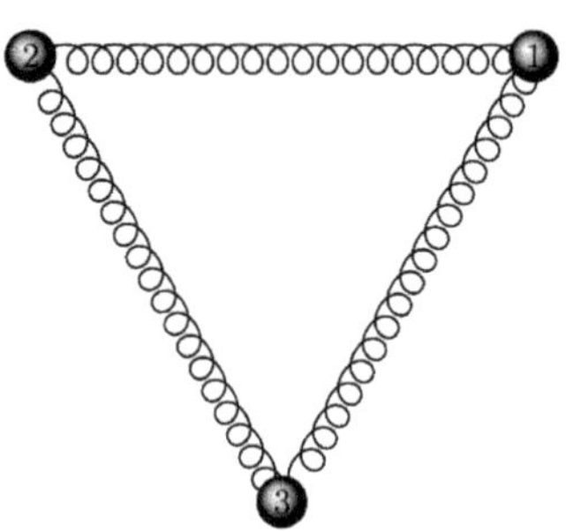

Two remarks warrant mention. First, quantum systems may exhibit accidental degeneracies, which seem coincidental and not related to any observable symmetry. Nonetheless, recent research [148] indicates that such two-fold accidental degeneracies stem from a *hidden symmetry* represented by an *anti-unitary operator*, which squares to $-\mathbb{1}$. Thus, accidental degeneracies are not purely accidental. Second, the study of physical system symmetries has evolved beyond simple group theories in recent years [149]. Concepts such as generalized [150] and categorical symmetries [151, 152] are now vibrant research areas, albeit too advanced for this text but pertinent to quantum computation [153].

In upcoming Chapters 7 and 8, we will further explore the implications of degeneracy and symmetry.

3.2.4 Postulate 3

Suppose a quantum system is in state $|\psi\rangle \in \mathcal{H}_S$, and suppose that the measurement of an observable O in this state returns the outcome o_i. Then the state of the system immediately after the measurement becomes

$$\frac{P_i \, |\psi\rangle}{\sqrt{\langle\psi|P_i \, |\psi\rangle}}, \tag{3.29}$$

where P_i is the projector onto the Hilbert subspace associated with the outcome o_i.

Remarks

- **Projectors and their properties:** The operator P_i in the postulate is a projector mapping the state $|\psi\rangle$ into the Hilbert subspace associated with the eigenvalue o_i of O. The action and form of P_i can differ depending on whether o_i is a degenerate or nondegenerate eigenvalue:
  - *Nondegenerate case:* When o_i is nondegenerate, $P_i = |i\rangle \langle i|$, where $O \, |i\rangle = o_i \, |i\rangle$. The action of P_i leaves $|i\rangle$ invariant and annihilates vectors orthogonal to $|i\rangle$.
  - *Degenerate case:* When o_i is degenerate, P_i is formed by summing up the projectors corresponding to the degenerate eigenstates $|i_k\rangle$, so that $P_i = \sum_k |i_k\rangle \langle i_k|$, where $O \, |i_k\rangle = o_i \, |i_k\rangle$ for all k.

Additional properties of projectors are (show them as an exercise):

$$P_i P_j = \delta_{ij} P_i,$$
$$\sum_i P_i = \mathbb{1}, \tag{3.30}$$

which is precisely the resolution of identity (3.3). These properties allow us to express the observable O in terms of projectors as:

$$O = \sum_i o_i P_i. \tag{3.31}$$

- **Normalization:** The denominator $\sqrt{\langle \psi | P_i | \psi \rangle}$ serves as a normalization factor. After the measurement, the resulting state should be a normalized state in the Hilbert space $\mathcal{H}_S$. This ensures that the state remains normalized even after it has been projected onto a subspace.

- **Nondeterministic nature:** The postulate reflects the inherently probabilistic nature of quantum mechanics. Even if the system is initially in state $|\psi\rangle$, we cannot deterministically predict which outcome o_i will be observed upon measurement. The quantum system "chooses" one of the possible eigenstates corresponding to o_i in a probabilistic manner, as dictated by Postulate 2.

- **Post-measurement state and state collapse:** The post-measurement state is determined by the action of the projector P_i on the pre-measurement state $|\psi\rangle$, followed by normalization. This transition from $|\psi\rangle$ to the post-measurement state can be interpreted as the "collapse" of the wavefunction, a concept that has been much discussed in the context of the philosophy of quantum mechanics. The measurements considered in this postulate are commonly referred to as *projective measurements* for an obvious reason. In Section 3.7, however, we will discuss more general quantum measurements.

  Nevertheless, if one simply adheres to the linear algebraic structure of quantum mechanics, this "collapse" is nothing but a mathematical formalism: After the measurement, the system is prepared into a specific state corresponding to the observed outcome. The action of the projector P_i ensures this, annihilating components not relevant to the outcome o_i and leaving the rest intact. Therefore, from this viewpoint, state collapse is simply an outcome of the linear algebra that underpins quantum mechanics.

- **Uncertainty principle:** A question arises: Can we measure multiple observables, say A and B, simultaneously with arbitrary precision? The uncertainty principle, in its generalized form, provides an answer. It states that for any two observables A and B,

$$\Delta A \Delta B \geq \frac{1}{2} |\langle [A, B] \rangle|, \tag{3.32}$$

where ΔA and ΔB are the *variances* of A and B in the state $|\psi\rangle$, defined as

$$\Delta A := \langle (A - \langle A \rangle)^2 \rangle = \langle A^2 \rangle - \langle A \rangle^2, \tag{3.33}$$

and $[A, B] = AB - BA$ is the commutator of A and B. By the uncertainty principle (3.32), two noncommuting observables cannot be measured with arbitrary precision simultaneously; the more precisely one of them is measured, the less the other.

The principle arises naturally in the framework of linear algebra: Observables that do not commute ($AB \neq BA$) cannot be diagonalized simultaneously. That is, they do not share a complete set of eigenvectors: If the state is an eigenstate of, say, A, the it must be a linear superposition of the eigenstates of B.

In cases where observables are connected by a Fourier transform, such as position and momentum, the principle takes the familiar form $\Delta x \Delta p \geq \hbar/2$.

This limitation is not an additional rule but a consequence of the basic mathematical structure underlying quantum mechanics.

Exercise 3.6 *Can classical probability appear in a quantum state?*

3.3 Mixed States and Density Matrices

If you completed Exercise 3.6, you would be well-prepared for this section that explores mixed states and density matrices. The quantum states we have encountered so far are known as pure states. These are states that can be expressed as superpositions of the basis ket states of a specific quantum system. Nevertheless, the majority of states in a quantum system are not pure states. As we will discuss in Chapter 4, the pure states represent a measure-zero subspace within the total Hilbert space of a given quantum system.

3.3.1 Mixed States

Quantum states can be combined to form what is known as a classical statistical mixture – or a mixed state. To grasp this notion, consider a set of states $|\psi_1\rangle$, $|\psi_2\rangle, \ldots, |\psi_m\rangle$ in an N-dimensional Hilbert space $\mathcal{H}_S$ of a quantum system S. These m states may not necessarily be orthogonal to one another. Assume that the system is in state $|\psi_k\rangle$ with probability p_k, such that

$$\sum_{i=1}^{m} p_i = 1. \tag{3.34}$$

In this context, the system is described as being in a **mixed state**.

Before introducing formal notation to describe mixed states, let's examine the expectation value of an observable in such a state. Consider an observable O with eigenvalues o_i, where $i = 1, \ldots, N$. The probability of measuring O in a mixed state and obtaining an outcome o_i is

$$p(o_i) = \sum_{k=1}^{m} p_k \langle \psi_k | P_i | \psi_k \rangle, \tag{3.35}$$

where P_i is the projector onto the i-th eigenspace of O. Given that state $|\psi_k\rangle$ may have a component in the i-th eigenspace of O and that the system has a probability p_k of being in $|\psi_k\rangle$, the expectation value of O in this mixed state is

$$\langle O \rangle = \sum_{k=1}^{m} p_k \underbrace{\sum_{i=1}^{N}}_{\substack{\uparrow \\ \text{classical} \\ \text{probability}}} o_i \underbrace{\langle \psi_k | P_i | \psi_k \rangle}_{\substack{\text{quantum} \\ \text{probability}}}, \tag{3.36}$$

where one should differentiate the two probabilities present. Because $O = \sum_{i=1}^{N} o_i P_i$ as in Equation (3.31), the equation for the expectation value becomes

$$\langle O \rangle = \sum_{k=1}^{m} p_k \langle \psi_k | O | \psi_k \rangle. \tag{3.37}$$

It is crucial to underscore that a mixed state cannot be represented as a state vector. Complete Exercise 3.7 to validate this point.

> **Exercise 3.7** *Attempt to express the mixed state as a superposition of $|\psi_k\rangle$, where $k = 1, \ldots, m$. Identify the issues that arise in this attempt. Remember that the system has a probability p_k of being in $|\psi_k\rangle$.*

3.3.2 Density Matrices

Instead of directly presenting the definition of density matrices, let's discover how this concept naturally emerges by revisiting the expectation value $\langle O \rangle$ given in Equation (3.37). Starting with the expression for $\langle O \rangle$:

$$
\begin{aligned}
\langle O \rangle &= \sum_{k=1}^{m} p_k \langle \psi_k | O | \psi_k \rangle \\
&= \sum_{k=1}^{m} p_k \langle \psi_k | O \underbrace{\sum_{i=1}^{N} |i\rangle \langle i|}_{\mathbb{1}} |\psi_k\rangle \\
&= \sum_{k=1}^{m} \sum_{i=1}^{N} p_k \underbrace{\langle \psi_k | O |i\rangle}_{\in \mathbb{C}} \underbrace{\langle i|\psi_k\rangle}_{\in \mathbb{C}} \\
&= \sum_{k=1}^{m} \sum_{i=1}^{N} p_k \langle i|\psi_k\rangle \langle \psi_k | O |i\rangle \\
&= \sum_{i=1}^{N} \langle i| \underbrace{\sum_{k=1}^{m} p_k |\psi_k\rangle \langle \psi_k|}_{=:\rho} O |i\rangle \\
&= \mathrm{Tr}(\rho O),
\end{aligned}
\tag{3.38}
$$

where in the second step, a resolution of identity is inserted. In the final step, we introduce the **density operator** ρ, defined as

$$\rho := \sum_{k=1}^{m} p_k \, |\psi_k\rangle \langle \psi_k|. \tag{3.39}$$

When expressed in the eigenbasis of the observable O, the matrix elements of ρ are given by

$$\rho_{ij} = \langle i|\rho\,|j\rangle. \tag{3.40}$$

Although the matrix representation of the density operator ρ is often referred to as the **density matrix**, it is more common to simply call ρ a density matrix. Hereafter, we shall use the terminologies density operator and density matrix interchangeably.

Note that a pure state $|\psi\rangle$ can be expressed as a density operator: By Equation (3.39), in this case the density operator is simply $\rho = |\psi\rangle \langle \psi|$.

> **Exercise 3.8** *Is a density operator (matrix) a projector? Provide your reasoning.*

To deepen our understanding of mixed states, it is essential to discuss the concept of "lack of knowledge": A mixed state of a quantum system is said to contain only incomplete information of the system. This term is not indicative of a property inherent to the quantum system itself but rather serves as a reflection of the observer's limitations in accurately describing the system. Consider, for example, an ensemble of two-level[11] quantum systems, some of which are prepared in the state $|0\rangle$ and others in $|1\rangle$. If you were to randomly select a system from this ensemble, your best possible description of its state would be the mixed state $\rho = 0.5\,|0\rangle \langle 0| + 0.5\,|1\rangle \langle 1|$, under the assumption of equal probabilities for each state. This depiction as a mixed state arises not from an intrinsic property of the individual systems but rather from your lack of complete information about which specific system you have selected.

Contrast this with a pure state, such as the superposition $\frac{1}{\sqrt{2}}(|0\rangle+|1\rangle)$. In this case, we know the system's exact quantum state, and the probabilities of measuring it to be in either $|0\rangle$ or $|1\rangle$ are simply a consequence of its quantum nature. In essence, the "lack of knowledge" associated with a mixed state is linked to our uncertainty about which of the many potential pure states the system actually occupies. This is fundamentally different from the inherent quantum uncertainty that exists even when we possess complete information, as is the case with pure states.

In Section 3.4.3 discussing quantum entanglement, we shall soon encounter another related facet of the "lack of knowledge" in quantum systems. Upon exploring the geometry of mixed states, Section 4.1.2 will also further address "information loss" in mixed states.

---

[11] A two-level quantum system is one whose Hilbert space is two-dimensional.

Properties of ρ

Given an operator ρ, it may not be readily recognizable as a density operator in the form described by Equation (3.39). You might naturally ask: How can we identify an operator as a density operator? In the context of a quantum system S, an operator qualifies as a density operator if it is self-adjoint and meets the following three criteria:

- The operator ρ is nonnegative: $\langle \varphi | \rho | \varphi \rangle \geq 0$ for every $| \varphi \rangle \in \mathcal{H}_S$.
- The trace of ρ is unity: $\mathrm{Tr}\rho = 1$.
- The square of the trace satisfies $\mathrm{Tr}\rho^2 \leq 1$. Equality is attained only when ρ represents a pure state.

Note that the latter two conditions are basis-independent, as the trace operation does not depend on the chosen basis.

Furthermore, by invoking the Schrödinger equation (3.1), one can readily show that the time evolution of a density operator adheres to the **Liouville–von Neumann equation**:

$$\frac{\mathrm{d}\rho(t)}{\mathrm{d}t} = \frac{1}{i\hbar}[H, \rho(t)], \tag{3.41}$$

where H is the pertinent Hamiltonian. The formal solution to this equation is given by

$$\rho(t) = U(t, t_0)\rho(t_0)U^\dagger(t, t_0), \tag{3.42}$$

with $U(t, t_0)$ being the time evolution operator defined in Equation (3.7).

> **Exercise 3.9** *Demonstrate the validity of Equation* (3.41).

Fidelity

Imagine a scenario where Alice has a state of a quantum system, and Bob asserts that he possesses an identical state. The critical question is: How can Alice ascertain the veracity of Bob's claim? Alternatively, she might wish to determine how closely Bob's state approximates hers. In the realms of quantum computation and quantum information, Alice can employ a metric known as *fidelity* to gauge the closeness between the two states.

For two pure states $| \psi_A \rangle$ and $| \psi_B \rangle$, the fidelity is straightforwardly given by the modulus of the inner product between them, denoted as $| \langle \psi_A | \psi_B \rangle |$. On the other hand, for mixed states ρ_A and ρ_B, the fidelity is formally defined as [154]

$$F(\rho_A, \rho_B) := \mathrm{Tr}\left(\sqrt{\sqrt{\rho_A}\rho_B\sqrt{\rho_A}} \right). \tag{3.43}$$

This metric possesses three essential properties:

- $F(\rho, \rho) \equiv 1$;
- $F \geqslant 0$, by its definition;
- $F(\rho_A, \rho_B) < 1$ if $\rho_A \neq \rho_B$.

It is worth emphasizing that a smaller fidelity implies a greater divergence between the two states. For pure states represented as $\rho_A = |\psi_A\rangle\langle\psi_A|$ and $\rho_B = |\psi_B\rangle\langle\psi_B|$, we have

$$
\begin{aligned}
F(\rho_A, \rho_B) &= \mathrm{Tr}\sqrt{\langle\psi_A|\,|\psi_A\rangle\langle\psi_B|\,|\psi_B\rangle\langle\psi_A|\,|\psi_A\rangle} \\
&= |\langle\psi_A|\psi_B\rangle|\,\mathrm{Tr}\langle\psi_A|\,|\psi_A\rangle \\
&= |\langle\psi_A|\psi_B\rangle|,
\end{aligned}
\tag{3.44}
$$

consistent with the fidelity measure for pure states mentioned earlier.

Exercise 3.10 *Consider two quantum states represented by density matrices ρ_A and ρ_B. Using the definition of fidelity in Equation (3.43), show that fidelity is symmetric; that is, $F(\rho_A, \rho_B) = F(\rho_B, \rho_A)$. Hint: Use the cyclic property of the trace operation to rearrange terms.*

Post-Measurement States

Postulate 3 addresses the post-measurement state for pure states after measuring an observable O and obtaining an outcome o_i. This concept naturally extends to mixed states. In the mixed-state context, suppose that measuring the observable O on state ρ yields the outcome o_i. The state ρ' after the measurement is given by

$$
\rho' = \frac{P_i \rho P_i}{\mathrm{Tr}(P_i \rho)}.
\tag{3.45}
$$

In this equation, the numerator $P_i \rho P_i$ projects[12] the state ρ onto the i-th eigenspace of O. The denominator $\mathrm{Tr}(P_i \rho)$ represents the probability of observing o_i, consistent with Equation (3.38). Thus, Equation (3.45) serves as the mixed-state counterpart to Equation (3.29).

3.3.3 Quantum Coherence in Mixed States

We have discussed the phenomenon of quantum coherence and have seen the coherence terms in Equation (3.6) for pure states. Quantum coherence also exists in mixed states. Consider a generic mixed state described by density operator (3.39), and expand the pure states $|\psi_k\rangle$ involved in the mixed state as

$$
|\psi_k\rangle = \sum_{n=1}^{N} \psi_n^{(k)} |n\rangle.
\tag{3.46}
$$

in the eigenbasis $\{|n\rangle\}$ of certain observable O. Then, the corresponding density matrix elements are

$$
\begin{aligned}
\rho_{mn} = \langle m|\rho|n\rangle &= \sum_{k} p_k \sum_{n'}\sum_{m'} \langle m|m'\rangle\langle n'|n\rangle\, \psi_{m'}^{(k)} \psi_{n'}^{(k)*} \\
&= \sum_{k} p_k \psi_m^{(k)} \psi_n^{(k)*}.
\end{aligned}
\tag{3.47}
$$

---

[12] Projecting a density matrix needs the projector twice, while in Equation (3.29), projecting a pure-state vector needs it only once.

The diagonal elements are

$$\rho_{nn} = \sum_k p_k \left|\psi_n^{(k)}\right|^2 = \text{Tr}\,(\rho P_n) = \text{Tr}\,(\rho |n\rangle\langle n|), \qquad (3.48)$$

which is precisely the probability that the system is in pure state $|n\rangle$ after measuring O and obtaining outcome o_n. These correspond to the diagonal terms in Equation (3.6) for pure states, so the diagonal matrix element ρ_{nn} are termed the population number of $|n\rangle$ in the mixed state ρ. Now, from Equation (3.47), we can extract the off-diagonal terms in the density matrix:

$$\rho_{mn} = \sum_k p_k \psi_m^{(k)} \psi_n^{(k)*} \quad m \neq n. \qquad (3.49)$$

These are the counterparts to the coherence terms in Equation (3.6). Importantly, quantum coherence is basis-dependent. Since ρ is a nonnegative self-adjoint operator, a basis $\{|i\rangle\}$ always exists to diagonalize ρ, yielding

$$\rho = \sum_{i=1}^{N} \omega_i \, |i\rangle \, \langle i|, \qquad (3.50)$$

where $\omega_i \in \mathbb{R}$ and $\sum_i \omega_i = 1$. In this basis, coherence terms are absent. This might lead to the misconception that quantum coherence is easily removable by basis changing. This is incorrect. Diagonalizing a density operator is not equivalent to quantum decoherence, which involves an interaction with an external environment. Furthermore, a quantum system is specified by a maximal set of commuting observables, and it is generally not possible for a density operator to commute with all of them. For instance, if the density operator ρ commutes with the Hamiltonian, which usually belongs to the set of commuting observables, it will not evolve in time according to Equation (3.41), rendering the state essentially classical. Therefore, a generic quantum state will exhibit quantum coherence for specific observables, despite the existence of a basis that apparently eliminates it.

3.4 Composite Systems

While the preceding discussions have focused on individual quantum systems, many real-world scenarios and fundamental experiments demand an understanding of composite systems comprised of multiple interacting subsystems. Insights gained from the study of these subsystems, and particularly their mutual relationships, offer a richer and finer understanding of the complete system as a whole.[13] In this section, we will discourse on the formalism for describing composite systems and grapple with one of the most enigmatic and essential phenomena in quantum

---

[13] Even in cases where a system initially appears to be singular, partitioning it into multiple subsystems often yields significant and valuable insights.

theory – entanglement. The groundwork laid in our exploration of mixed states will prove invaluable in navigating the complexities of composite quantum systems.

3.4.1 Separable versus Entangled

A composite system is comprised of multiple subsystems, each with its own Hilbert space, denoted as $\mathcal{H}_1$, $\mathcal{H}_2$, and so on. The Hilbert space $\mathcal{H}$ of the total system is then the tensor product of these individual spaces:

$$\mathcal{H} = \mathcal{H}_1 \otimes \mathcal{H}_2 \otimes \cdots. \tag{3.51}$$

To make the forthcoming discussion more concrete, we will narrow our focus to bipartite systems, which involve just two subsystems, $\mathcal{H} = \mathcal{H}_1 \otimes \mathcal{H}_2$. These simplified systems are sufficient for capturing the essential physics of composite quantum systems. Initially, we will focus exclusively on pure states in bipartite systems. You will find, as we dive deeper, that mixed states will emerge even within this restricted scope, most notably in our discussion on entanglement [155].

Hilbert Space Bases

Let's consider the Hilbert spaces $\mathcal{H}_1$ and $\mathcal{H}_2$ of the two subsystems to be

$$\begin{aligned}
\mathcal{H}_1 &= \mathrm{span}\{|n_1\rangle \,|\, n_1 = 0, 1, \ldots, N_1 - 1\}, \quad N_1 = \dim \mathcal{H}_1, \\
\mathcal{H}_2 &= \mathrm{span}\{|n_2\rangle \,|\, n_2 = 0, 1, \ldots, N_2 - 1\}, \quad N_2 = \dim \mathcal{H}_2.
\end{aligned} \tag{3.52}$$

The total Hilbert space $\mathcal{H}$ of the composite system is then $N_1 N_2$-dimensional and given by

$$\mathcal{H} = \mathrm{span}\,\{|n_1\rangle \otimes |n_2\rangle \,|\, n_1 = 0, 1, \ldots, N_1 - 1; n_2 = 0, 1, \ldots, N_2 - 1\},$$

allowing a pure state $|\psi\rangle$ in $\mathcal{H}$ to be expanded as

$$\begin{aligned}
|\psi\rangle &= \sum_{n_1, n_2} \psi_{n_1 n_2} |n_1\rangle \otimes |n_2\rangle \\
&= \sum_{n_1, n_2} \psi_{n_1 n_2} |n_1 n_2\rangle,
\end{aligned} \tag{3.53}$$

where $|n_1 n_2\rangle := |n_1\rangle \otimes |n_2\rangle$ serves as a shorthand notation for the basis states of $\mathcal{H}$. This setup for the Hilbert space basis will be the assumption for our subsequent discussions. Example 3.2 teaches you how to perform the tensor product of two vectors if you didn't know this before.

Example 3.2

Consider two vectors

$$|\psi\rangle = \begin{pmatrix} a \\ b \end{pmatrix} \quad \text{and} \quad |\varphi\rangle = \begin{pmatrix} c \\ d \\ e \end{pmatrix}. \tag{3.54}$$

Then

$$|\psi\rangle \otimes |\varphi\rangle = \begin{pmatrix} a\,|\varphi\rangle \\ b\,|\varphi\rangle \end{pmatrix} = \begin{pmatrix} ac & ad & ae & bc & bd & be \end{pmatrix}^T, \tag{3.55}$$

where the superscript T means transpose.

Exercise 3.11 *The tensor product of two matrices is similar. Perform the tensor product of any 2×2 and 3×3 matrices.*

Separable States

A state $|\psi\rangle \in \mathcal{H}$ is termed **separable** if there exist states $|\varphi_1\rangle \in \mathcal{H}_1$ and $|\varphi_2\rangle \in \mathcal{H}_2$, such that

$$|\psi\rangle = |\varphi_1\rangle \otimes |\varphi_2\rangle. \tag{3.56}$$

To illustrate, consider a case where both subsystems are two-dimensional ($N_1 = N_2 = 2$). The state

$$|\psi\rangle = \frac{1}{\sqrt{2}}\left(|01\rangle + |11\rangle\right) \tag{3.57}$$

is separable, as it can be recast in the form

$$|\psi\rangle = \frac{1}{\sqrt{2}}\left(|0\rangle + |1\rangle\right) \otimes |1\rangle. \tag{3.58}$$

Entangled States

Unlike the constrained set of separable states, a large portion of states in the Hilbert space of a bipartite system are inseparable, and these are known as *entangled states*. To illustrate, consider our example where $N_1 = N_2 = 2$. The state

$$|\psi\rangle = \frac{1}{\sqrt{2}}\left(|01\rangle - |10\rangle\right) \tag{3.59}$$

is inseparable because it cannot be decomposed into a tensor product state as in Equation (3.56); therefore, it is an entangled state. This state is commonly known as a *Bell state*.

To dive deeper into the nature of entanglement, consider the two subsystems as spin-$\frac{1}{2}$ systems. Here, $|0\rangle = |\uparrow\rangle$ represents a spin-up state, while $|1\rangle = |\downarrow\rangle$ denotes a spin-down state. We can then write Equation (3.59) as

$$|\psi\rangle = \frac{1}{\sqrt{2}}\left(|\uparrow\downarrow\rangle - |\downarrow\uparrow\rangle\right). \tag{3.60}$$

Even if Alice and Bob, who are the observers for subsystems 1 and 2 respectively, are separated by any distance, Alice's and Bob's measurement outcomes are entangled.

Specifically, if Alice measures her spin in the $|\uparrow\rangle$ state, then, by the postulates of quantum mechanics, Bob will find his spin in the $|\downarrow\rangle$ state, and vice versa.

Some might mistake this quantum phenomenon as having a classical analog. Consider a pair of black and white balls placed into two boxes, which are then randomly given to Alice and Bob. No matter their distance apart, if Alice finds a black ball in her box, Bob's must contain a white one, and vice versa. Nevertheless, this classical scenario fails to capture the true essence of quantum entanglement.

To differentiate further, consider a basis transformation for the Bell state (3.60):

$$|\pm\rangle = \frac{1}{\sqrt{2}}(|\uparrow\rangle \pm |\downarrow\rangle), \tag{3.61}$$

which changes the Bell state (3.60) to

$$|\psi\rangle = \frac{1}{\sqrt{2}}(|+-\rangle - |-+\rangle). \tag{3.62}$$

In this new basis, the entanglement remains. Alice and Bob are not only free to choose their basis for measurements, but also find that the entanglement persists regardless of the basis used. This freedom is unimaginable in the classical world.

As depicted in Figure 3.4, consider a photograph divided vertically and placed into two sealed envelopes, which are then given to Alice and Bob. While this setup might seem analogous to quantum entanglement, it crucially lacks the ability to change the "basis," or in this case, the orientation of the cut after the envelopes have been sealed and separated. Such an alteration is not just ineffective in changing the predetermined nature of classical entanglement; it's outright impossible. On the other hand, the very essence of quantum entanglement allows for such freedom.

You might be tempted to think that quantum entanglement implies instantaneous communication between distant parties. This is however not true. Although Alice's and Bob's measurement outcomes are correlated due to the entangled state, neither can immediately know the result of the other's measurement without some form of communication. One might wonder if they could utilize quantum communication to bypass the speed-of-light constraint. In fact, this is a misunderstanding. Quantum communication still requires the transmission of quantum states between the parties, which also cannot exceed the speed of light, as per the principles of special relativity. Moreover, to fully decipher the correlation and make use of the entanglement, classical communication is essential for sharing the measurement outcomes and bases used. Thus, regardless of whether classical or quantum communication is

Figure 3.4 A photo cut vertically into two halves.

used, the speed-of-light constraint remains intact. Special relativity holds in both the classical and quantum realms, as substantiated by extensive experimental evidence to date.

3.4.2 EPR Paradox and Bell's Theorem

The discovery of quantum entanglement sparked a series of debates and questions concerning the foundational aspects of quantum mechanics. The EPR paradox, posed by Einstein, Podolsky, and Rosen [156], stands as a significant milestone in these debates, questioning the very principles upon which quantum mechanics is built.

Historical Background and Einstein–Bohr Debate

In the early days of quantum mechanics, the philosophical implications of the theory were a subject of intense debate. Perhaps the most famous of these debates was between Albert Einstein and Niels Bohr [156, 157]. While Bohr was a proponent of the Copenhagen interpretation, which embraces the probabilistic nature of quantum mechanics, Einstein was uncomfortable with the idea of a theory that lacked deterministic predictability.

EPR Paradox and Hidden Variables

In their landmark 1935 paper [156], Einstein, Podolsky, and Rosen (**EPR**) tackled foundational questions about quantum mechanics by formulating a thought experiment. This experiment was crafted to scrutinize the completeness of quantum mechanics while focusing on two principles: the *reality principle* and the *locality principle*. The reality principle contends that physical quantities possess definite values even in the absence of observation, thereby asserting that physical reality exists independent of our measurements. The locality principle, on the other hand, maintains that physical interactions at one location cannot instantaneously influence outcomes at a separate, distant location.

Rather than outright stating that quantum mechanics would violate these principles if it were complete, EPR argued more subtly. They posited that if quantum mechanics aspires to provide a complete description of physical reality, then it must force us to relinquish one of these deeply ingrained principles – either reality, locality, or both. According to EPR, if quantum mechanics is complete and there are no "local hidden variables" to account for these principles, then quantum mechanics itself must be nonlocal and/or nonrealistic. This would mean that elements of physical reality (like the position or momentum of a particle) are not part of the complete description provided by quantum mechanics alone.

The concept of hidden variables is not introduced frivolously or capriciously; it is motivated by the desire to reconcile quantum mechanics with the principles of locality and realism. These hidden variables act as stand-ins for yet-to-be-discovered elements of physical reality, which are "hidden" in the sense that they are not encompassed by the quantum mechanical framework.

Bell's Theorem and Proof with Expectation Values

In 1964, physicist John Bell presented a theorem [158] that provided a concrete way to test the principles of the **EPR paradox**. He developed inequalities based on measurements taken from entangled states of two spin-$\frac{1}{2}$ particles. These inequalities, known as **Bell inequalities**, are derived assuming the principles of locality and realism. Importantly, these inequalities must be verified through repeated experiments to ensure statistical validity.

Suppose Alice and Bob measure the spins of entangled particles along directions $\vec{a}$ and $\vec{b}$, respectively. After repeating the experiments many times, they can calculate the expectation value $E(\vec{a}, \vec{b})$, defined as the average value of the product of their measurement outcomes:

$$E(\vec{a}, \vec{b}) = \langle AB \rangle, \tag{3.63}$$

where A and B are the outcomes of Alice's and Bob's measurements, being either $+1$ or -1. This expectation value characterizes the correlation between Alice's and Bob's measurement outcomes.

For a hidden-variable theory respecting both locality and realism, among all Bell inequalities, the most common one is known as the **Clauser–Horne–Shimony–Holt (CHSH) inequality** [159]:

$$\left| E(\vec{a}, \vec{b}) - E(\vec{a}, \vec{b}') \right| + \left| E(\vec{a}', \vec{b}) + E(\vec{a}', \vec{b}') \right| \le 2, \tag{3.64}$$

where $\vec{a}'$ and $\vec{b}'$ are two other directions of measurement. This inequality sets an upper bound that the expectation values must satisfy if the principles of locality and realism hold.

To derive the CHSH inequality, consider a local hidden variable λ that determines the outcomes for both Alice and Bob.[14] The outcomes $A(\vec{a}, \lambda)$ and $B(\vec{b}, \lambda)$ can either be $+1$ or -1. The expectation value $E(\vec{a}, \vec{b})$ is then

$$E(\vec{a}, \vec{b}) = \int d\lambda \rho(\lambda) A(\vec{a}, \lambda) B(\vec{b}, \lambda), \tag{3.65}$$

where $\rho(\lambda)$ is the probability density of λ.

Next, compute the difference in expectation values as follows:

$$E(\vec{a}, \vec{b}) - E(\vec{a}, \vec{b}') = \int d\lambda \rho(\lambda) \left[A(\vec{a}, \lambda) B(\vec{b}, \lambda) - A(\vec{a}, \lambda) B(\vec{b}', \lambda) \right]. \tag{3.66}$$

---

[14] In the context of local hidden variable theories, "locality" means that the measurement outcome on one particle is independent of the measurement setting or outcome on another, spatially separated particle. This principle is known as the independence of distant events. The hidden variables λ are shared by both particles and determine their outcomes based on the local measurement settings $\vec{a}$ and $\vec{b}$, respectively. This ensures that each result $A(\vec{a}, \lambda)$ and $B(\vec{b}, \lambda)$ is separately dependent only on its own measurement setting and the shared hidden variable, not on the setting or outcome of the other distant particle. These assumptions uphold the constraints imposed by relativity, ensuring no superluminal communication or influence.

Introduce additional terms that cancel out to rewrite the equation:

$$E(\vec{a}, \vec{b}) - E(\vec{a}, \vec{b}') = \int d\lambda \rho(\lambda) A(\vec{a}, \lambda) B(\vec{b}, \lambda) \left[1 \pm A(\vec{a}', \lambda) B(\vec{b}', \lambda) \right]$$
$$- \int d\lambda \rho(\lambda) A(\vec{a}, \lambda) B(\vec{b}', \lambda) \left[1 \pm A(\vec{a}', \lambda) B(\vec{b}, \lambda) \right]. \tag{3.67}$$

Since $|A(\vec{a}, \lambda)| = |B(\vec{b}, \lambda)| = 1$, and both $1 \pm A(\vec{a}', \lambda) B(\vec{b}', \lambda)$ and $1 \pm A(\vec{a}', \lambda) B(\vec{b}, \lambda)$ are nonnegative, taking the absolute value of both sides and invoking the triangle inequality gives

$$\left| E(\vec{a}, \vec{b}) - E(\vec{a}, \vec{b}') \right| \le 2 \pm \left[E(\vec{a}', \vec{b}') + E(\vec{a}', \vec{b}) \right]. \tag{3.68}$$

The RHS can be further bounded by the minimum possible value $2 - \left| E(\vec{a}', \vec{b}') + E(\vec{a}', \vec{b}) \right|$. Transferring this absolute value term to the LHS completes the derivation of the CHSH inequality (3.64).

Following this inequality, let us consider Example 3.3 where quantum mechanics blatantly violates the Bell inequality, thus refuting classical ideas of locality and realism.

Example 3.3

Assume Alice and Bob are measuring particles in the Bell state (3.60):

$$|\psi\rangle = \frac{1}{\sqrt{2}} (|\uparrow\downarrow\rangle - |\downarrow\uparrow\rangle), \tag{3.69}$$

The quantum mechanical prediction for $E(\vec{a}, \vec{b})$ is given by

$$E(\vec{a}, \vec{b}) = -\vec{a} \cdot \vec{b}. \tag{3.70}$$

In this case, we choose vectors as follows:

$$\vec{a} = (1, 0), \quad \vec{a}' = (0, 1), \tag{3.71}$$

$$\vec{b} = \left(\frac{1}{\sqrt{2}}, \frac{1}{\sqrt{2}} \right), \quad \vec{b}' = \left(-\frac{1}{\sqrt{2}}, \frac{1}{\sqrt{2}} \right). \tag{3.72}$$

The CHSH inequality becomes

$$S = \left| -\frac{1}{\sqrt{2}} - \frac{1}{\sqrt{2}} \right| + \left| -\frac{1}{\sqrt{2}} + (-\frac{1}{\sqrt{2}}) \right| = \sqrt{2} + \sqrt{2} = 2\sqrt{2} > 2, \tag{3.73}$$

clearly violating the Bell inequality (3.64) and thereby challenging classical notions of locality and realism.

Exercise 3.12 *Come up with another quantum-mechanical example that violates the Bell inequality (3.64).*

Resolution of the Paradox

The EPR paradox initially questioned the completeness of quantum mechanics by suggesting that if quantum mechanics is a complete theory, it must violate either the locality principle, the reality principle, or both. Einstein, Podolsky, and Rosen invoked the concept of hidden variables as a possible mechanism to explain quantum correlations while preserving both locality and realism. In this view, quantum mechanics would be incomplete as it would not account for these hidden variables.

Bell's theorem, however, radically changed the context of this debate. It showed that any theory based on local hidden variables would make certain statistical predictions (captured by Bell's inequalities) that are violated by the predictions of quantum mechanics. Experimental violations of Bell's inequalities, therefore, refute the very existence of local hidden variables as a viable explanation for quantum correlations.

So where does this leave us? It confirms that quantum mechanics is both special and complete in the sense that no local hidden variables can explain the observed quantum phenomena. By eliminating local hidden variables as a possible explanation, we are left with the implication that quantum mechanics indeed violates either the principle of locality, the principle of realism, or both.

Given these findings, there is little point in assuming a theory with nonlocal hidden variables that reproduces the predictions of quantum mechanics. The nature of quantum mechanics itself could be nonlocal, or it might require us to abandon traditional ideas of realism. In any case, the violation of Bell's inequalities obliges us to adapt our classical intuitions to the counterintuitive, yet experimentally confirmed, nature of quantum phenomena.

Before concluding this subsection, it's important to dispel a common misconception: Einstein's role in the EPR paradox was not that of a critic trying to disprove quantum mechanics. Rather, his contribution should be viewed as a deep probe into the philosophical and physical underpinnings of the theory. Not only was Einstein one of the founding figures of quantum theory, but his EPR paper also posed a question so fundamental that it pushed the scientific community to explore the very limits of our understanding of nature.

3.4.3 Entanglement Entropy: A Measure of Entanglement

Having understood the concept of entanglement and its physical implications, a natural question arises: Given two entangled states, can we tell which one is more entangled than the other? To answer this question, we would need to define a reasonable measure of the entanglement in a quantum state. This measure is termed entanglement entropy, which is one of the most important concepts in any physics area that involves quantum information.

To see how entanglement entropy surfaces in gauging the amount of entanglement in a quantum state, let us recast the generic pure state (3.53) of a composite system we began with in this section in the form of a density matrix. Namely, the state can be written as

$$\rho = |\psi\rangle\langle\psi| = \sum_{n_1 n_2}\sum_{n_1' n_2'} \underbrace{\psi_{n_1 n_2}\psi^*_{n_1' n_2'}}_{=:\rho_{n_1 n_2; n_1' n_2'}} |n_1\rangle\otimes|n_2\rangle\langle n_1'|\otimes\langle n_2'|. \tag{3.74}$$

We are now ready to introduce a stepping-stone notion – the reduced density matrix – before we define entanglement entropy.

The reduced Density Matrix

Consider an observable on subsystem 1 only, denoted by O_1. To compute its expectation value in state ρ, we can extend it trivially to an observable on the composite system: $O_1 \otimes \mathbb{1}$, where $\mathbb{1}$ is the identity operator on subsystem 2. So, this extended observable does nothing to subsystem 2 at all, as can be verified:

$$\langle O_1 \rangle = \langle O_1 \otimes \mathbb{1}_2 \rangle = \mathrm{Tr}\left(\rho O_1 \otimes \mathbb{1}_2\right) = \sum_{m_1, m_2} \langle m_1| \otimes \langle m_2|\rho O_1 \otimes \mathbb{1}_2 |m_1\rangle \otimes |m_2\rangle$$

$$\overset{(3.74)}{=\!=\!=} \sum_{m_1, m_2} \sum_{n_1', n_2'} \rho_{m_1 m_2; n_1' n_2'} \langle n_1'| \otimes \langle n_2' |O_1 \otimes \mathbb{1}_2| m_1\rangle \otimes| m_2\rangle$$

$$= \sum_{m_1, m_2} \sum_{n_1'} \rho_{m_1 m_2; n_1' m_2} \langle n_1' |O_1| m_1\rangle$$

$$= \sum_{m_1} \sum_{n_1'} \underbrace{\sum_{m_2} \rho_{m_1 m_2; n_1' m_2}}_{=: \rho_{1 m_1 n_1'}} (O_1)_{n_1' m_1}$$

$$= \mathrm{Tr}\left(\rho_1 O_1\right). \tag{3.75}$$

In the derivation above, we define the **reduced density matrix** ρ_1 as the **partial trace** of ρ over subsystem 2:

$$\rho_1 = \mathrm{Tr}_2(\rho) := \sum_{m_2} \langle m_2|\rho |m_2\rangle. \tag{3.76}$$

> **Exercise 3.13** *Consider an observable O_2 on subsystem 2 only. Compute $\langle O_2 \rangle$ similarly.*
>
> **Exercise 3.14** *Show that partial trace is a matrix operation invariant under basis transformation.*

One might wonder why the concept of a reduced density matrix is significant. The reduced density matrix, ρ_1, encapsulates all the information required for making predictions about measurements on subsystem 1, irrespective of what occurs in subsystem 2. In simple terms, if Alice, an experimenter, only has access to subsystem 1, then ρ_1 is her go-to guide for all possible measurement outcomes on that subsystem. This becomes particularly pertinent when discussing entangled states, where the individual subsystems are deeply correlated. Each subsystem might appear to be in a mixed state when looked at in isolation, even if the entire system is in a pure state. This "mixing" provides a quantitative measure of the entanglement, leading us naturally to the notion of entanglement entropy.

Information Loss: Separable versus Entangled

In the act of tracing out the degrees of freedom from subsystem 2 to obtain ρ_1, it seems like we lose some information about the original, composite system. An observer (Alice) in subsystem 1 would find that any observable behaves as if her subsystem is solely in the state described by ρ_1. Subsystem 2, having been traced out, becomes inaccessible. Now, you might ask: Is this loss of information the same for all kinds of quantum states? Importantly, as we'll soon see, the implications of information loss differ dramatically between separable and entangled states.

To illustrate this, let's first consider a separable state. Suppose our system begins in a state $|\psi\rangle = |\psi_1\rangle \otimes |\psi_2\rangle$. The density matrix then becomes

$$\rho = |\psi_1\rangle \otimes |\psi_2\rangle \langle\psi_1| \otimes \langle\psi_2|. \tag{3.77}$$

Since trace or partial trace is an operation invariant under basis transformation, as you showed in Exercise 3.14, to partial-trace subsystem 2 we can choose the basis of $\mathcal{H}_2$ where $|\psi_2\rangle$ is a basis state. Hence,

$$\rho_1 = \mathrm{Tr}_2\rho = |\psi_1\rangle \langle\psi_1|, \tag{3.78}$$

which remains a pure state. Alice, observing only subsystem 1, will find in the long run that the outcomes agree with the expectation value of the observable in pure state $|\psi_1\rangle$, so she never even knows about the existence of subsystem 2. Consequently, there's no real loss of information for her.

On the other hand, if we start with an entangled state, the consequence changes dramatically. Consider, for instance, the Bell state

$$|\psi\rangle = \frac{1}{\sqrt{2}}\left(|01\rangle - |10\rangle\right). \tag{3.79}$$

The corresponding density matrix is

$$\rho = \frac{1}{2}\left(|01\rangle\langle01| + |10\rangle\langle10| - |01\rangle\langle10| - |10\rangle\langle01|\right). \tag{3.80}$$

Taking the partial trace, we get

$$\rho_1 = \frac{1}{2}\left(|0\rangle\langle0| + |1\rangle\langle1|\right), \tag{3.81}$$

a mixed state. Alice, upon performing measurements, realizes that her subsystem appears to be in a mixed state, indicating a lack of complete information. (Remember the discussion about the lack of information in mixed states in Section 3.3.2?) Here, the information loss is not merely a mathematical artifact from the tracing operation but stems from the physical entanglement between the subsystems: The entanglement renders the state observable to Alice a mixed state. Bob, who observes subsystem 2, is similarly limited. A god-like observer who can access the entire composite system certainly sees no loss of information.

The importance of entanglement in modern physics can never be exaggerated. For example, entanglement has also been applied, and is perhaps a key, to resolving the **black hole information loss paradox** [160].

> **Exercise 3.15** *Show that for any entangled bipartite pure state, partial trace would result in a mixed state.*
>
> **Exercise 3.16** *What would be the condition on the density matrix of a bipartite system for it to describe a separable state, possibly mixed?*

Entanglement Entropy

To advance our exploration of entanglement, it is imperative to introduce a quantifiable measure for it. As we observed, the reduced density matrices of entangled states yield mixed states, whereas those for separable states yield pure states. This motivates us to identify a basis-independent attribute from the reduced density matrix to evaluate the degree of entanglement between two subsystems. This attribute is what we call entanglement entropy. Specifically, for a bipartite system in a pure state ρ, the entanglement entropy is defined as the **von Neumann entropy** of its reduced density matrix $\rho_1 = \mathrm{Tr}_2 \rho$:

$$S = -\mathrm{Tr}\,[\rho_1 \ln \rho_1]. \tag{3.82}$$

Example 3.4

Consider once more the Bell state (3.59). The reduced density matrix for this state is given by

$$\rho_1 = \begin{pmatrix} \frac{1}{2} & 0 \\ 0 & \frac{1}{2} \end{pmatrix}. \tag{3.83}$$

Using Equation (3.82), we find that the entanglement entropy for this state is

$$S = -\mathrm{Tr}\,[\rho_1 \ln \rho_1] = -\mathrm{Tr}\left[\begin{pmatrix} \frac{1}{2} & 0 \\ 0 & \frac{1}{2} \end{pmatrix} \ln \begin{pmatrix} \frac{1}{2} & 0 \\ 0 & \frac{1}{2} \end{pmatrix} \right]$$

$$= -\ln \frac{1}{2} = \ln 2. \tag{3.84}$$

This value is the maximal entanglement entropy achievable in a bipartite two-level system. Skeptical readers are invited to prove this claim through Exercise 3.17.

> **Exercise 3.17** *Consider a generic bipartite pure state*
>
> $$|\psi\rangle = \alpha\,|00\rangle + \beta\,|01\rangle + \gamma\,|10\rangle + \delta\,|11\rangle, \quad |\alpha|^2 + |\beta|^2 + |\gamma|^2 + |\delta|^2 = 1. \tag{3.85}$$
>
> *Show that the entanglement entropy of this state is bounded from above by $\ln 2$. What is the entanglement entropy of a separable state?*

> **Exercise 3.18** *What is the maximal entanglement entropy of a bipartite three-level system?*
>
> **Exercise 3.19** *Is the entanglement entropy for a bipartite pure state different from that in Equation (3.82) if it is defined using the von Neumann entropy of the reduced density matrix of subsystem 2? The answer is "No." Show that*
>
> $$\mathrm{Tr}\left[\rho_1 \ln \rho_1\right] = \mathrm{Tr}\left[\rho_2 \ln \rho_2\right], \tag{3.86}$$
>
> *where $\rho_2 = \mathrm{Tr}_1 \rho$.*

Entanglement entropy serves as a monotone for entanglement,[15] unambiguously gauging the extent of entanglement in bipartite pure states. While other entanglement monotones exist [161, 162], especially for multipartite systems [163, 164, 165], defining a clear-cut measure for such complex systems remains a challenging task. Indeed, quantifying multipartite entanglement is an active field of research and lies beyond the purview of this book.

Exercise 3.20 *With the preceding discussion in mind, compute the entanglement entropies for the tripartite states in Equations (3.87) and (3.88) after tracing out one or two subsystems.*

$$|GHZ\rangle_3 = \frac{1}{\sqrt{2}}(|000\rangle + |111\rangle), \tag{3.87}$$

$$|W\rangle_3 = \frac{1}{\sqrt{3}}(|001\rangle + |010\rangle + |100\rangle). \tag{3.88}$$

Although it may seem counterintuitive to compare entanglement entropy with *thermal entropy*, the two are essentially two sides of the same coin. You may remember that thermal entropy measures the disorder of a system. A more disordered system engenders greater ignorance, harmonizing with the role of entanglement entropy as a measure of the ignorance of local observers in composite systems. Recent research has highlighted the equivalences between thermal and entanglement entropies in a range of contexts [166, 167, 168, 169, 170].

As a final point, it's essential to recognize that the consideration of mixed states can complicate defining a measure of entanglement. Even though our focus here has primarily been on pure states, extending these concepts to mixed states necessitates a more comprehensive set of tools, which are beyond the scope of the book. On the other hand, given a mixed state of a system, as will be seen in Section 3.6, one can always describe the mixed state using the reduced density matrix of a pure state of a larger system. Moreover, physicists often idealize the situation, such that only pure states are relevant.

---

[15] A measure of entanglement that increases with the amount of entanglement.

> **Exercise 3.21** *Prove the **subadditivity** of entanglement entropy:*
>
> $$S(\rho) \le S(\rho_1) + S(\rho_2).$$ (3.89)

3.5 Schmidt Decomposition

In the realm of quantum systems, understanding the underlying structure of entangled states is pivotal for both theoretical insights and practical applications. A powerful tool to unravel this structure is the Schmidt decomposition [171], a mathematical technique that provides a unique and optimal way to represent bipartite quantum states. The Schmidt decomposition not only simplifies the mathematical form of an entangled state but also directly exposes key features such as the entanglement spectrum and the number of degrees of freedom involved in the entanglement. As we proceed in this section, we will explore the mathematical formulation of the Schmidt decomposition and explore its profound implications for quantum information theory.

Theorem 3.1. *Schmidt decomposition theorem:* *Any pure state $|\psi\rangle$ of a bipartite system $\mathcal{H}_1 \otimes \mathcal{H}_2$ admits the decomposition*

$$|\psi\rangle = \sum_n s_n |n\rangle_1 |n\rangle_2,$$ (3.90)

where $\{|n\rangle_1\}$ and $\{|n\rangle_2\}$ are orthonormal states in $\mathcal{H}_1$ and $\mathcal{H}_2$, $s_i \ge 0$, and

$$\sum_n s_n^2 = 1.$$ (3.91)

*The numbers s_i are the **Schmidt coefficients** of $|\psi\rangle$. The number of nonzero Schmidt coefficients is the **Schmidt number** of $|\psi\rangle$.*

Proof The theorem can be easily proven by means of singular value decomposition, which is a linear algebra technique not covered in this book. We first express a generic pure state of a bipartite system as

$$|\psi\rangle = \sum_{i_1,j_2} A_{i_1 j_2} |i_1\rangle |j_2\rangle$$ (3.92)

in terms of the basis states of the total Hilbert space. The superposition coefficients form a matrix A, which is not necessarily a square matrix because the two subsystems may have unequal dimensions N_1 and N_2. Any such matrix A admits a *singular value decomposition* [171]:

$$A = UDV^\dagger,$$ (3.93)

where U and V are $N_1 \times N_1$ and $N_2 \times N_2$ unitary matrices, while D is an $N_1 \times N_2$ matrix with all matrix entries being zero except the entries D_{ii}, i ranges from 0 to $\min(N_1, N_2) - 1$. Precisely speaking, we have

$$|\psi\rangle = \sum_{i_1, j_2, n} U_{i_1 n} D_{nn} V^*_{n j_2} \underbrace{\phantom{U_{i_1 n} D_{nn} V^*_{n j_2}}}_{A_{i_1 j_2}} |i_1\rangle |j_2\rangle \tag{3.94}$$

Now, if we define

$$|n\rangle_1 := \sum_{i_1} U_{i_1 n} |i_1\rangle, \tag{3.95}$$

$$|n\rangle_2 := \sum_{j_2} V^*_{j_2 n} |j_2\rangle, \tag{3.96}$$

$$s_n := D_{nn} \xrightarrow{\langle\psi|\psi\rangle=1} \sum_n s_n^2 = 1, \tag{3.97}$$

we obtain the Schmidt decomposition

$$|\psi\rangle = \sum_n s_n |n_1\rangle |n_2\rangle, \tag{3.98}$$

which is often written as

$$|\psi\rangle = \sum_n \sqrt{p_n} |n_1\rangle |n_2\rangle, \tag{3.99}$$

because $s_n \geq 0$. If you sense that the p_n here are certain probabilities, hooray for you. $\qquad\square$

An immediate corollary of the Schmidt decomposition theorem states that a bipartite state $|\psi\rangle$ is separable if and only if its Schmidt number is 1. Note however that Schmidt decomposition exists only for bipartite systems.

In summary, the Schmidt decomposition serves as an invaluable mathematical tool for dissecting the intricacies of bipartite quantum systems. For instance, in quantum optics, the decomposition is often used to simplify the analysis of entangled photon pairs [172]. Another application can be found in condensed matter physics, where Schmidt decomposition can be instrumental for understanding quantum phase transitions in spin chains [173]. By presenting a composite quantum state in its Schmidt basis, we not only make calculations more tractable but also obtain a more nuanced view of the entanglement between subsystems. Although the focus of this section has been primarily on the theorem and its formal proof, this important concept will reappear in our forthcoming discussion on purification. The scope of Schmidt decomposition's utility is extensive, affecting not only the bedrock principles of quantum mechanics but also practical applications ranging from quantum computing to quantum cryptography. While this chapter won't look further into the details of Schmidt decomposition, its pervasive influence in the broader landscape of quantum theory should not go unremarked.

Exercise 3.22 *Find the Schmidt decomposition of the state:*

$$|\psi\rangle = \frac{1}{2}(|00\rangle + |11\rangle + i\,|20\rangle + i\,|21\rangle). \tag{3.100}$$

> **Exercise 3.23** *Show that, for example, the Schmidt decomposition doesn't exist for tripartite systems.*

3.6 Purification

As opposed to partial trace, which reduces a pure state of a composite system to a mixed state of one of its subsystems, purification serves a crucial role in allowing us to extend our understanding of mixed states by embedding them in a larger Hilbert space [174]. Specifically, given a mixed state ρ_1 of a system $\mathcal{H}_1$, purification identifies a composite system $\mathcal{H}_1 \otimes \mathcal{H}_2$, such that the composite system is in a pure state $|\psi\rangle$ that satisfies $\rho_1 = \mathrm{Tr}_2 |\psi\rangle \langle\psi|$. While the added system 2 may not have any immediate physical significance, the concept of purification is indispensable for tasks such as quantum error correction, quantum cryptography, and the study of quantum correlations. It provides a powerful framework for understanding how local and global properties of quantum systems are interrelated.

Recall that a generic bipartite pure state takes the form of (3.53) and the corresponding density matrix takes the form (3.74). A generic mixed state of a single system, say, system 1, reads

$$\rho_1 = \sum_{n_1, n_1'} (\rho_1)_{n_1, n_1'} |n_1\rangle \langle n_1'|, \tag{3.101}$$

where $\{|n_1\rangle\}$ is an orthonormal basis of $\mathcal{H}_1$. We wish to enlarge system 1 to a bipartite system in a pure state ρ, such that

$$
\begin{aligned}
\rho_1 &= \mathrm{Tr}_2 \rho \\
&= \sum_{m_2} \langle m_2| \sum_{n_1 n_2 \, n_1' n_2'} \psi_{n_1 n_2} \psi^*_{n_1' n_2'} |n_1\rangle \otimes |n_2\rangle \langle n_1'| \otimes \langle n_2'|m_2\rangle \\
&= \sum_{m_2} \sum_{n_1 n_1'} \psi_{n_1 m_2} \psi^*_{n_1' m_2} |n_1\rangle \langle n_1'|,
\end{aligned}
\tag{3.102}
$$

with the identification

$$(\rho_1)_{n_1 n_1'} = \sum_{m_2} \psi_{n_1 m_2} \psi^*_{n_1' m_2}. \tag{3.103}$$

This relation is a generic result. For your sake, let's examine Example 3.5.

Example 3.5

Consider both systems 1 and 2 being two-level. Equation (3.103) leads to three equations:

$$(\rho_1)_{00} = \psi_{00}\psi^*_{00} + \psi_{01}\psi^*_{01}, \tag{3.104}$$

$$(\rho_1)_{01} = \psi_{00}\psi^*_{10} + \psi_{01}\psi^*_{11} = (\rho_1)^*_{10}, \tag{3.105}$$

$$(\rho_1)_{11} = \psi_{10}\psi_{10}^* + \psi_{11}\psi_{11}^*. \tag{3.106}$$

Clearly, the two diagonal matrix elements of ρ_1 are real. Instead of finding the general solution to these equations, a special solution is easily found by setting $\psi_{10} = 0$. We can also take $\psi_{00}, \psi_{11} \in \mathbb{R}$. Then, we have

$$\psi_{11} = \sqrt{(\rho_1)_{11}}, \quad \psi_{01} = (\rho_1)_{10}^* / \sqrt{(\rho_1)_{11}},$$
$$\psi_{00} = \sqrt{(\rho_1)_{00} - |(\rho_1)_{10}|^2 / (\rho_1)_{11}}. \tag{3.107}$$

Therefore, the purified state reads

$$|\psi\rangle = \sqrt{\frac{(\rho_1)_{00}\,(\rho_1)_{11} - |(\rho_1)_{10}|^2}{(\rho_1)_{11}}}\,|00\rangle + \frac{(\rho_1)_{10}^*}{\sqrt{(\rho_1)_{11}}}\,|01\rangle + \sqrt{(\rho_1)_{11}}\,|11\rangle. \tag{3.108}$$

Alternatively, recall that a density matrix can always be diagonalized by a certain basis; namely, we can always write

$$\rho_1 = \sum_n p_n |n_1\rangle \langle n_1|. \tag{3.109}$$

To purify ρ_1, we can simply append a system 2 of the same dimension as that of system 1, such that the purified state reads

$$|\psi\rangle = \sum_n \sqrt{p_n}\,|n_1\rangle\,|n_2\rangle, \tag{3.110}$$

which is precisely the Schmidt decomposition of $|\psi\rangle$. This result exemplifies the utility of Schmidt decomposition in understanding the structure of entangled systems. It serves as a practical example that fortifies the conceptual bridge between purification and Schmidt decomposition, and simplifies our understanding of entangled systems.

3.7   Generalized Measurement

In our discussion thus far, we have primarily focused on projective measurements, the most basic and well-understood type of quantum measurements. In the realm of composite quantum systems, however, the landscape of measurements broadens significantly. A projective measurement applied to an entire composite system may translate into a nonprojective, or generalized, measurement when observed at the level of its constituent subsystems. This observation leads us to venture beyond the confines of projective measurements to explore the wider spectrum of quantum measurements.

But what exactly defines a quantum measurement? Given the inherently probabilistic nature of quantum mechanics, the total probability of all possible outcomes of a quantum measurement must equal one. This requirement is akin to measuring the identity operator in the quantum realm. Thus, *a quantum measurement can be conceptualized as a specific resolution of the identity operator.* While projective measurements, characterized by a set of projectors that map onto the eigenspaces of particular observables, represent the most straightforward form of this resolution, they are not the only possibility.

In this section, we aim to unravel how projective measurements on a composite system can influence its subsystems in a nontrivial manner, thereby leading us to the more expansive concept of generalized measurements, which encompass a broader class of resolutions of identity beyond the scope of projective measurements. Let's first consider Example 3.6.

Example 3.6

Assume a bipartite pure state

$$|\Phi\rangle = \sum_{k_2} M_{k_2} |\psi_1\rangle |k_2\rangle, \tag{3.111}$$

where $\mathcal{H}_2 = \text{span}\{|k_2\rangle\}$, $|\psi_1\rangle$ is a certain pure state of subsystem 1, and each M_{k_2} is a $\dim \mathcal{H}_1 \times \dim \mathcal{H}_1$ $\mathbb{C}$-matrix (not self-adjoint in general) acting on $|\psi_1\rangle$ but indexed by the corresponding basis state $|k_2\rangle$ of $\mathcal{H}_2$. These matrices satisfy

$$\sum_k M_k^\dagger M_k = \mathbb{1}_1. \tag{3.112}$$

Now let's perform a projective measurement on subsystem 2 only, so the projector is

$$P_i = \mathbb{1}_1 \otimes |i_2\rangle \langle i_2|, \tag{3.113}$$

where $|i_2\rangle$ is a basis state of $\mathcal{H}_2$. Note that $\sum_i P_i = \mathbb{1}_1 \otimes \mathbb{1}_2$. We derive the probability of obtaining outcome i (bearing in mind that M_k acts on $|\psi_1\rangle$ only):

$$
\begin{aligned}
p_i = \text{Tr}\,(\rho P_i) &= \text{Tr}\left(\sum_{kk'} M_{k_2} |\psi_1\rangle |k_2\rangle \langle\psi_1|M_{k_2'}^\dagger \langle k_2'|\mathbb{1}_1 \otimes |i_2\rangle \langle i_2|\right) \\
&= \text{Tr}\left(\sum_k M_{k_2} |\psi_1\rangle |k_2\rangle \langle\psi_1|M_{i_2}^\dagger \langle i_2|\right) \\
&= \sum_n \sum_j \langle n_1|\langle j_2| \left(\sum_k M_{k_2} |\psi_1\rangle |k_2\rangle \langle\psi_1|M_{i_2}^\dagger \langle i_2|\right) |n_1\rangle |j_2\rangle \\
&= \sum_n \langle n_1|M_{i_2} |\psi_1\rangle \langle\psi_1|M_{i_2}^\dagger |n_1\rangle \\
&= \sum_n \langle\psi_1|M_{i_2}^\dagger |n_1\rangle \langle n_1|M_{i_2} |\psi_1\rangle \\
&= \langle\psi_1|M_{i_2}^\dagger M_{i_2} |\psi_1\rangle, \tag{3.114}
\end{aligned}
$$

where $M_{i_2}^\dagger M_{i_2}$ is clearly self-adjoint but in general not a projector! This operator is an example of a **Kraus operator**, to be introduced in Chapter 6. The gist of the derivation above is that a projective measurement over the bipartite system (over subsystem 2 in particular) turns out to be a non-projective measurement over subsystem 1. This motivates us to forget about composite systems and define in what follows for any single system a generalized form of measurement.

3.7.1 Definition of Generalized Measurement

In light of the subtleties revealed in Example 3.6, we find it essential to establish a formal definition of generalized measurements. A **generalized measurement** is described by a set $\{M_i\}$ of **measurement operators**, which are not necessarily self-adjoint or positive but satisfy the completeness condition

$$\sum_i M_i^\dagger M_i = \mathbb{1}, \tag{3.115}$$

which is precisely a resolution of identity. Note that the product $M_i^\dagger M_i$ is automatically self-adjoint and positive. This is crucial for ensuring well-defined probabilities, since $\langle \psi | M_i^\dagger M_i | \psi \rangle$ yields the probability of obtaining outcome i, as illustrated in Equation (3.114). Thus, the completeness condition (3.115) signifies that the probabilities must sum to unity. Analogous to a projective measurement, the post-measurement state of the system after measuring M_i is given by

$$|\psi_i'\rangle = \frac{M_i |\psi\rangle}{\sqrt{\langle \psi | M_i^\dagger M_i | \psi \rangle}}. \tag{3.116}$$

This operational framework equips us with the mathematical machinery to analyze a broad range of physical scenarios.

In retrospect, Example 3.6 provides a stepping stone towards appreciating **Neumark's theorem** [175], which asserts that a generalized measurement over a single system can equivalently be described by a projective measurement over a larger system. This intriguing equivalence intertwines the notion of generalized measurements with topics like purification and Schmidt decomposition, revealing an intricate web of relationships that form the backbone of modern quantum theory.

3.7.2 Positive Operator-Valued Measures

In various applications, especially those related to quantum information processing, it is often the measurement outcomes themselves that are of primary interest, rather than the post-measurement states. Such scenarios find relevance in quantum communication and cryptography where a detector is often configured to gather specific information, sidelining the need to consider the post-measurement state. This is where **positive operator-valued measures** (**POVMs**) come into play, offering a generalized form of quantum measurement suited to a wide range of physical situations.

A POVM is characterized by a set $\{F_i\}$ of positive operators, known as POVM elements, that comply with the completeness condition

$$\sum_i F_i = \mathbb{1}, \tag{3.117}$$

which is again a resolution of identity. For a state $|\psi\rangle$, the probability of obtaining a particular outcome i is

$$p_i = \langle\psi|F_i|\psi\rangle. \tag{3.118}$$

The completeness condition (3.117) guarantees that these probabilities sum to one, ensuring a consistent statistical interpretation of the measure.

Both projective and generalized measures naturally fall under the umbrella of POVMs. For generalized measures, we have $F_i = M_i^\dagger M_i$. To appreciate the utility and adaptability of POVMs, let's study a concrete example.

Example 3.7

We revisit the inquiry from Section 3.2.2 about the distinguishability of two quantum states. The lesson was that perfect discrimination is impossible in general. What if our goals are less ambitious? Suppose Alice is to send Bob one of two nonorthogonal states. Can Bob at least refrain from mistaking one state for the other?

To address this, let's look at the two states

$$\begin{aligned}
|\psi_1\rangle &= \sin\theta\,|0\rangle + \cos\theta\,|1\rangle, \\
|\psi_2\rangle &= \sin\theta\,|0\rangle - \cos\theta\,|1\rangle,
\end{aligned} \qquad 0 < \theta < \frac{\pi}{4}. \tag{3.119}$$

Notably, in this case, $\langle\psi_1|\psi_2\rangle \neq 0$. Alice is poised to send one of these states to Bob. Proficient in quantum information theory, Bob opts for a POVM consisting of three elements:

$$F_0 = \frac{1}{2}\begin{bmatrix} 1 & r \\ r & r^2 \end{bmatrix}, \quad F_1 = \frac{1}{2}\begin{bmatrix} 1 & -r \\ -r & r^2 \end{bmatrix}, \quad F_2 = \begin{bmatrix} 0 & 0 \\ 0 & 1-r^2 \end{bmatrix}, \quad r = \tan\theta. \tag{3.120}$$

He then determines the conditional probabilities $p(i|k) = \langle\psi_k|F_i|\psi_k\rangle$, namely the probabilities of finding each outcome i in state $|\psi_k\rangle$, $k = 1, 2$.

After performing the calculations, Bob discovers that

$$p(1\mid 1) = p(0\mid 2) = 0. \tag{3.121}$$

This informs Bob that if he registers outcome 1, the state must not be $|\psi_1\rangle$, and similarly, outcome 0 excludes $|\psi_2\rangle$. Although he can't be absolutely certain which state he received, Bob ensures that he won't misidentify them.

> **Exercise 3.24** *Check the other probabilities in Example 3.7 and find out what Bob can do (perhaps nothing) with them.*

Example 3.8 and Exercise 3.25 will enhance your understanding of POVM's and their utilities.

Example 3.8

Suppose Alice wants to send one of three possible messages to Bob using quantum states. She chooses to encode her messages in the nonorthogonal states

$$|\phi_1\rangle = \cos\alpha\,|0\rangle + \sin\alpha\,|1\rangle, \tag{3.122}$$

$$|\phi_2\rangle = \cos\alpha\,|0\rangle + e^{i2\pi/3}\sin\alpha\,|1\rangle, \tag{3.123}$$

$$|\phi_3\rangle = \cos\alpha\,|0\rangle + e^{i4\pi/3}\sin\alpha\,|1\rangle, \tag{3.124}$$

where $0 < \alpha < \frac{\pi}{2}$ and the states are normalized. Clearly, $\langle\phi_i|\phi_j\rangle \neq 0$ when $i \neq j$.

Bob, wishing to decode the messages optimally, opts for a POVM with three elements $E_1, E_2,$ and E_3. For simplicity, let's say these operators are given by

$$E_1 = A\,|\phi_1\rangle\,\langle\phi_1|, \quad E_2 = A\,|\phi_2\rangle\,\langle\phi_2|, \quad E_3 = A\,|\phi_3\rangle\,\langle\phi_3|, \tag{3.125}$$

where A is a normalization constant ensuring $\sum_i E_i = \mathbb{1}$.

Bob calculates the probabilities $p(i \mid j)$ for receiving message j when outcome i is observed:

$$p(i \mid j) = \langle\phi_j|E_i\,|\phi_j\rangle. \tag{3.126}$$

By choosing his POVM elements carefully, Bob can maximize his chances of correctly identifying which state Alice sent, even though he cannot do so with certainty due to the nonorthogonality of the states.

Exercise 3.25 *Based on the above example, design a POVM for the quantum ternary channel such that the POVM elements $E_1, E_2,$ and E_3 are all positive operators satisfying $\sum_i E_i = \mathbb{1}$.*

1. Calculate the normalization constant A in Equation (3.125).
2. Find the probabilities $p(i \mid j)$ for $i, j = 1, 2, 3$.
3. Discuss how Bob can maximize the information he gains from the POVM.

Conclusion

In this chapter, we've journeyed through the enigmatic landscape of quantum mechanics, laying the foundation for deeper explorations. We've dissected the core postulates, probed the subtleties of mixed states, density matrices, and quantum coherence, and familiarized ourselves with key ideas such as separable and entangled states. Our foray into the nuances of generalized measurements has equipped us with a robust understanding of quantum systems. As we've built up this conceptual and mathematical toolkit, we've set the stage for the truly groundbreaking discussions that lie ahead. Now, armed with these foundational insights, we are poised to embark on our expedition into the compelling and rapidly evolving field of quantum computation. The theories and tools elucidated here will serve as our compass and lens, guiding us as we tackle the computational challenges and opportunities that only the quantum realm can offer.

4   Quantum Computation: The Circuit Model

In the burgeoning field of quantum computation, the quantum circuit model stands as a cornerstone, offering a structured approach to harness the power of quantum mechanics for computation. This model represents quantum algorithms as circuits, akin to Boolean circuits in classical computation, but with the transformative capability to process information in superposition and entanglement – two quintessential quantum phenomena. This chapter aims to explore the intricacies of the quantum circuit model, beginning with the fundamental unit known as a qubit. Unlike classical bits, which exist in either a 0 or 1 state, qubits can exist in a superposition of states, a property elegantly visualized using the Bloch sphere and Bloch ball. As we explore how to measure a qubit, we will confront the probabilistic outcomes that make quantum measurement both mysterious and fascinating. Following this foundation, the chapter will provide a comprehensive overview of elementary quantum gates, the building blocks that give the quantum circuit model its expressive power. We will then focus on the pivotal concept of universal quantum computation, sketching a proof for the universality theorem, which asserts that any quantum computation can be realized by a circuit comprising a finite set of specific gates. Lastly, we will take a brief look at the physical realizations of quantum gates, linking the abstract framework to tangible experimental implementations. By weaving together rigorous mathematical formulations and real-world applications, this chapter aims to equip you with a deep understanding of the quantum circuit model and its key role in shaping the future of computation.

4.1   The Qubit

In contrast to a classical bit, which can only assume one of two discrete values 0 or 1, a quantum bit – better known as a **qubit** – is a two-level quantum system characterized by a two-dimensional Hilbert space. Depending on its state, a qubit may be represented in different ways. In a pure state, it is described by a vector in the Hilbert space. In a mixed state, on the other hand, it is represented by a positive operator – the density operator – acting on the Hilbert space. This operator can be expressed as a 2×2 matrix in any basis of the Hilbert space. Both pure and mixed states can be incorporated into a single geometric framework, namely the

Bloch sphere and Bloch ball. To explore this framework in depth, we will consider pure-state and mixed-state qubits separately.

4.1.1 Pure Qubit States

Every two-dimensional Hilbert space has a standard basis, commonly referred to as the **computation basis**:

$$|0\rangle = \begin{pmatrix} 1 \\ 0 \end{pmatrix}, \quad |1\rangle = \begin{pmatrix} 0 \\ 1 \end{pmatrix}. \tag{4.1}$$

These basis states are the $+$ and $-$ eigenstates of the **Pauli matrix**

$$\sigma_z = \begin{pmatrix} 1 & 0 \\ 0 & -1 \end{pmatrix}. \tag{4.2}$$

That is, $\sigma_z |0\rangle = |0\rangle$ and $\sigma_z |1\rangle = -|1\rangle$. Within this computation basis, a pure qubit state $|\psi\rangle$ is expressed as

$$|\psi\rangle = \alpha |0\rangle + \beta |1\rangle, \tag{4.3}$$

where $|\alpha|^2 + |\beta|^2 = 1$ ensures normalization. Because $\alpha, \beta \in \mathbb{C}$ are continuous parameters, a qubit can exist in a superposition of states, in stark contrast to classical bits, which must be either 0 or 1. When measured, for example with respect to the observable σ_z, the qubit collapses to either $|0\rangle$ with probability $|\alpha|^2$ or to $|1\rangle$ with probability $|\beta|^2$.

You may wonder whether any two-level quantum system qualifies to be a qubit. The answer is nuanced; a system must meet the following three criteria, which may sound straightforward but are nontrivial to implement in practice:

- The system can be initialized into a well-defined state, known as the *fiducial state* (e.g., $|0\rangle$).
- Any two states within the system are interconvertible through a unitary transformation. That is, $\forall |\psi\rangle, |\psi'\rangle \in \mathcal{H}$, a unitary operator U exists such that $|\psi'\rangle = U |\psi\rangle$.
- Measurements can be conducted in the computation basis.

Practically, a qubit may also be a two-level subsystem of a multilevel quantum system. This is often the case due to the scarcity of inherently two-level quantum systems. As long as the subsystem satisfies the above criteria, it can function as a qubit. For example, the ground state and the first excited state of an atom could together form a two-level subsystem, thereby serving as a suitable qubit.

True Degrees of Freedom

To pave the way for visualizing the qubit state space, we first examine a generic pure qubit state more closely. The central question is: How many independent degrees of freedom exist in a generic pure qubit state as given by Equation (4.3)? We can express α and β in polar form:

$$|\psi\rangle = ae^{i\varphi_a} |0\rangle + be^{i\varphi_b} |1\rangle, \tag{4.4}$$

where $a, b, \varphi_a, \varphi_b \in \mathbb{R}$. The normalization condition simplifies to $a^2 + b^2 = 1$, enabling us to express a and b as $\cos\left(\frac{\theta}{2}\right)$ and $\sin\left(\frac{\theta}{2}\right)$ respectively.

The appearance of $\frac{\theta}{2}$ rather than θ might seem puzzling at first glance. If you grasp the reason behind it, excellent; if not, fear not, as the mystery will soon be unveiled.

Initially, the qubit state had four real degrees of freedom (two magnitudes and two phases). After applying the normalization condition, this reduces to three real degrees of freedom. An overall phase factor in a quantum state is inconsequential,[1] allowing us to set the relative phase between $|0\rangle$ and $|1\rangle$ as

$$|\psi\rangle = \cos\left(\frac{\theta}{2}\right) |0\rangle + \sin\left(\frac{\theta}{2}\right) e^{i(\varphi_b - \varphi_a)} |1\rangle. \tag{4.5}$$

Redefining $\varphi := \varphi_b - \varphi_a$, the generic pure qubit state becomes

$$|\psi\rangle = \cos\left(\frac{\theta}{2}\right) |0\rangle + \sin\left(\frac{\theta}{2}\right) e^{i\varphi} |1\rangle = \begin{pmatrix} \cos\left(\frac{\theta}{2}\right) \\ \sin\left(\frac{\theta}{2}\right) e^{i\varphi} \end{pmatrix}, \tag{4.6}$$

with $\theta \in [0, \pi]$ and $\varphi \in [0, 2\pi]$ being the only true degrees of freedom for the state, reducing the initial four real degrees of freedom to just two. The ranges of these two parameters will soon be explained along with the presence of $\theta/2$.

Measuring Qubits along Any Direction

If we measure the Pauli matrix σ_z in a generic pure qubit state (4.6), the expectation value would be

$$\langle\psi|\sigma_z|\psi\rangle = \cos^2\left(\frac{\theta}{2}\right) - \sin^2\left(\frac{\theta}{2}\right) = \cos\theta. \tag{4.7}$$

Since $\theta \in [0, \pi]$, this result meets our anticipated requirement, and θ is precisely the angle between the state vector and the z-axis. This understanding partially accounts for the appearance of $\theta/2$ and its range. To unravel the reason completely, let's try to answer the question: What is the expectation value of measuring the qubit along an arbitrary direction, described by a unit three-dimensional vector $\hat{n}$?

To answer this question, we need to learn the geometric role of the three **Pauli matrices**

$$\sigma_x = \begin{pmatrix} 0 & 1 \\ 1 & 0 \end{pmatrix}, \quad \sigma_y = \begin{pmatrix} 0 & -i \\ i & 0 \end{pmatrix}, \quad \sigma_z = \begin{pmatrix} 1 & 0 \\ 0 & -1 \end{pmatrix}. \tag{4.8}$$

---

[1] An overall phase factor like the Berry phase is nontrivial but from a different context and beyond the scope of the book.

They are clearly Hermitian and have the algebraic relations in Equation (4.9):

$$\mathrm{Tr}\sigma_i = 0, \quad \sigma_i^2 = \mathbb{1}_2, \quad i \in \{x, y, z\};$$
$$\sigma_i\sigma_j = i\epsilon_{ijk}\sigma_k \implies [\sigma_i, \sigma_j] = 2\epsilon_{ijk}\sigma_k, \quad i, j, k \in \{x, y, z\}, i \neq j \neq k. \tag{4.9}$$

Here, ϵ_{ijk} is the totally antisymmetric tensor with $\epsilon_{xyz} = 1$. In Eq. (4.9) and subsequent applications of the algebraic relations of the Pauli matrices, the Einstein summation convention – repeated indices are summed over – is adopted. These algebraic relations imply that the three Pauli matrices can serve as the basis vectors of the three-dimensional real space $\mathbb{R}^3$, or in other words, they are a two-dimensional representation of the x-, y-, and z-axes. To wit, any three-dimensional real vector $\vec{V}$ can be represented as a 2×2 matrix:

$$\vec{V} = V_x\sigma_x + V_y\sigma_y + V_z\sigma_z = \begin{pmatrix} V_z & V_x - iV_y \\ V_x + iV_y & -V_z \end{pmatrix}, \tag{4.10}$$

where V_x, V_y, and V_z are the components of $\vec{V}$ along the three axes. The expansion $V_x\sigma_x + V_y\sigma_y + V_z\sigma_z$ can alternatively be written as $\vec{V} \cdot \vec{\sigma}$, where $\vec{\sigma} = (\sigma_x, \sigma_y, \sigma_z)^T$ is regarded as a three-dimensional vector expressed in the usual x-, y-, and z-axes. The superscript T refers to matrix transposition, emphasizing that the vector is a column vector. This formulation in terms of Pauli matrices is critical for understanding measurements along arbitrary directions, as it allows us to express any unit vector $\hat{n}$ in the qubit's Hilbert space in terms of known observables.

In the matrix form, the inner product of two vectors $\vec{V}$ and $\vec{W}$ is computed as

$$\vec{V} \cdot \vec{W} = \frac{1}{2}\mathrm{Tr}\left[\left(\vec{V} \cdot \vec{\sigma}\right)\left(\vec{W} \cdot \vec{\sigma}\right)\right]$$
$$= \frac{1}{2}\mathrm{Tr}\left[\begin{pmatrix} V_z & V_x - iV_y \\ V_x + iV_y & -V_z \end{pmatrix}\begin{pmatrix} W_z & W_x - iW_y \\ W_x + iW_y & -W_z \end{pmatrix}\right] \tag{4.11}$$
$$= V_xW_x + V_yW_y + V_zW_z$$

in view of the relations in (4.9). This inner product enables use to corroborate the orthogonality between two basis vectors:

$$\sigma_i \cdot \sigma_j = \frac{1}{2}\mathrm{Tr}(\sigma_i\sigma_j) = \frac{1}{2}\epsilon_{ijk}\mathrm{Tr}(\sigma_k) = 0. \tag{4.12}$$

Recall that a three-dimensional unit vector $\hat{n}$ can be written as

$$\hat{n} = (\sin\theta\cos\varphi, \sin\theta\sin\varphi, \cos\theta)^T, \tag{4.13}$$

where $\theta \in [0, \pi]$ and $\varphi \in [0, 2\pi]$ are respectively the polar and azimuthal angles. This vector's matrix form reads

$$\sigma_{\hat{n}} := \hat{n} \cdot \vec{\sigma} = \begin{pmatrix} \cos\theta & \sin\theta\cos\varphi - i\sin\theta\sin\varphi \\ \sin\theta\cos\varphi + i\sin\theta\sin\varphi & -\cos\theta \end{pmatrix} = \begin{pmatrix} \cos\theta & \sin\theta e^{-i\varphi} \\ \sin\theta e^{i\varphi} & -\cos\theta \end{pmatrix}. \tag{4.14}$$

What is this matrix as an operator? It is precisely the Pauli operator $\sigma_{\hat{n}}$ that measures the qubit along the $\hat{n}$ direction. What are the eigenstates of this operator?

Rather than directly computing its eigenstates, let's have $\sigma_{\hat{n}}$ act on the generic pure qubit state (4.6) by assuming that the θ and φ in that state coincide with those in $\sigma_{\hat{n}}$:

$$\sigma_{\hat{n}} |\psi\rangle = \begin{pmatrix} \cos\theta & \sin\theta e^{-i\varphi} \\ \sin\theta e^{i\varphi} & -\cos\theta \end{pmatrix} \begin{pmatrix} \cos\left(\frac{\theta}{2}\right) \\ \sin\left(\frac{\theta}{2}\right) e^{i\varphi} \end{pmatrix} = \begin{pmatrix} \cos\left(\frac{\theta}{2}\right) \\ \sin\left(\frac{\theta}{2}\right) e^{i\varphi} \end{pmatrix} = |\psi\rangle. \tag{4.15}$$

That is, the generic state (4.6) is exactly the $+1$ eigenstate of $\sigma_{\hat{n}}$ if $\theta \in [0, \pi]$ and $\varphi \in [0, 2\pi]$ are the polar and azimuthal angles. The answer to our question at the beginning of this subsection is now evident:

$$\langle \psi | \sigma_{\hat{n}} | \psi \rangle = 1. \tag{4.16}$$

This result unravels the appearance of $\theta/2$ in $|\psi\rangle$ and the ranges of the two angular parameters.

Exercise 4.1 *Find the -1 eigenstate of $\sigma_{\hat{n}}$.*

Exercise 4.2 *Now that you have learnt (recalled) the Pauli matrices, can you rewrite the Hamiltonian in Equation (3.26) and its symmetry operator S in Example 3.1 in terms of the Pauli matrices? Hint: You may need the tensor products of two Pauli matrices. What system does the Hamiltonian describe?*

The Bloch Sphere

Equation (4.15) establishes a one-to-one correspondence between a pure qubit state and a three-dimensional unit vector $\hat{n}$:

$$\hat{n}(\theta, \varphi) \xleftrightarrow{1-\text{to}-1} |\psi(\theta, \varphi)\rangle. \tag{4.17}$$

The tips of all such unit vectors form a two-dimensional unit sphere S^2 termed the Bloch sphere [176], on which each point corresponds to a pure qubit state uniquely. Figure 4.1 depicts the Bloch sphere, marked with a few specific pure qubit states. The Bloch sphere offers a visualization of the pure qubit states.

Figure 4.1 Bloch sphere marked with a few specific pure qubit states.

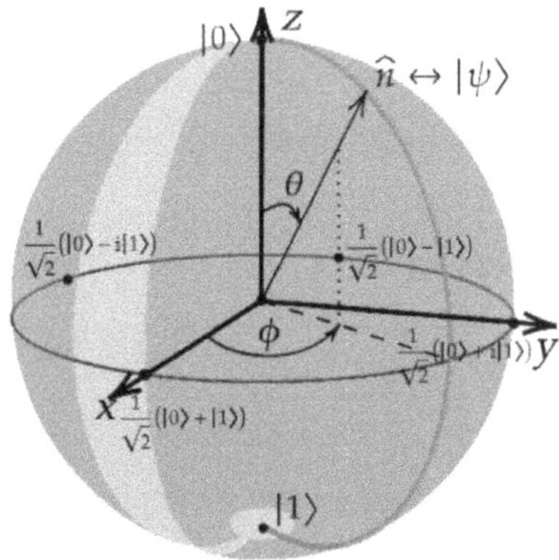

> **Exercise 4.3** *Verify that the marked states in Figure 4.1 are in the right positions.*

Rotating the Bloch Sphere

The Bloch sphere is not merely an intuitive geometrical representation of qubit states; it also serves as a powerful framework for understanding quantum operations on a single qubit. As previously established in Equation (4.15), there exists a one-to-one correspondence between a pure qubit state and a Pauli operator $\sigma_{\hat{n}}$. This Pauli operator is an element of the Lie algebra $\mathfrak{su}(2)$ associated with the Lie group $SU(2)$. These concepts will be explained shortly.

Unitary operations must preserve quantum probabilities. For a single qubit, these operations are described by 2×2 unitary matrices, forming the Lie group $U(2)$, which has a subgroup consisting of 2×2 unitary matrices with unit determinant, namely $SU(2)$. Any $U \in U(2)$ can factorize as

$$U = e^{i\phi}U', \quad \det U' = 1 \tag{4.18}$$

(where $\phi = 0$ if $\det U = 1$), so that $U' \in SU(2)$. This reveals that any unitary operation on a single qubit can be decomposed into an $SU(2)$ operation followed by an overall phase shift. Since an overall phase factor is immaterial to the physics of a quantum state, $SU(2)$ captures all physically relevant operations on a single qubit.

> **Exercise 4.4** *Show that a generic $SU(2)$ matrix is specified by three real continuous parameters. Hence, $SU(2)$ is a three-dimensional Lie group. Describe its topological shape.*

Since $SU(2)$ is a Lie group, so that it is continuous and has its own shape, and is simply connected,[2] any $SU(2)$ matrix is continuously connected to the identity element $\mathbb{1}_2$, the 2×2 identity matrix. One can expand a generic $SU(2)$ matrix U lying near its identity element as $U = \mathbb{1}_2 - i A$, where A is an infinitesimal complex matrix, and $-i$ is a convenient choice for a physical reason to by clear soon. Then, the unitarity condition $UU^\dagger = \mathbb{1}_2$ leads to $A = A^\dagger$, so A must be a Hermitian matrix. Exercise 4.5 shows that the condition $\det U = 1$ results in $\text{Tr} A = 0$. Therefore, A is a traceless Hermitian matrix.

> **Exercise 4.5** *Show that $\det U = 1$ demands that $\text{Tr} A = 0$. Note that A is a first-order approximation of U, so you need to drop any second and higher order terms in your derivation.*

---

[2] The same concept as what you learned in calculus.

It turns out that the set of all (not necessarily infinitesimal) 2×2 traceless Hermitian matrices forms an algebra – the Lie algebra $\mathfrak{su}(2)$ of $SU(2)$. The algebra multiplication is a **Lie bracket**[3] $[A, B] = AB - BA \in \mathfrak{su}(2)$, for all $A, B \in \mathfrak{su}(2)$, that satisfies the **Jacobi identity**:

$$[[A, B], C] + [[B, C], A] + [[C, A], B] = 0. \tag{4.19}$$

Exercise 4.6 *Can you show that the elements of $\mathfrak{su}(2)$ do satisfy the Jacobi identity? You may need Equations (4.9) and (4.20).*

Since a Lie algebra is in the first place a vector space, an element of the algebra can be expanded in a certain basis of the algebra. For $\mathfrak{su}(2)$, it is evident that all 2×2 traceless Hermitian matrices can be expanded in terms of the three Pauli matrices (4.8):

$$A = a\sigma_x + b\sigma_y + c\sigma_z, \tag{4.20}$$

where $a, b, c \in \mathbb{R}$. Interestingly, since the convergence radius of the Taylor series of the exponential function is infinite, and because $SU(2)$ is a *compact*[4] Lie group, any $SU(2)$ matrix U can be written as

$$U = e^{-\hat{\imath}A}. \tag{4.21}$$

The presence of $\hat{\imath}$ in Equation (4.18) ensures that A is Hermitian, and the **exponential map** explicitly guarantees that U is unitary. We will soon address the negative sign in the exponent.

The exponential map, especially the $\hat{\imath}$ in the exponent of Equation (4.21), effectively transforms the real variables a, b, and c into angular variables, such that $a, b, c \in [0, 2\pi]$. Consider the scenario where $a = b = 0$. We have

$$e^{\hat{\imath}c\sigma_z} = \sum_{k=0}^{\infty} \frac{(\hat{\imath}c\sigma_z)^k}{k!} = \sum_{j=0}^{\infty} \frac{(-1)^j c^{2j}}{(2j)!} \mathbb{1}_2 + \hat{\imath} \sum_{j=0}^{\infty} \frac{(-1)^j c^{2j+1}}{(2j+1)!} \sigma_z = \begin{pmatrix} e^{\hat{\imath}c} & 0 \\ 0 & e^{-\hat{\imath}c} \end{pmatrix}, \tag{4.22}$$

which indicates that $c \in [0, 2\pi]$.

Exercise 4.7 *Show that the variables a and b in (4.21) are also both angular variables.*

Let's further examine the effect of an $SU(2)$ element on a pure qubit state. Consider $U = \exp(-\hat{\imath}\frac{\delta}{2}\sigma_z)$ acting on a generic pure qubit state given by Equation (4.6):

---

[3] In this case, the Lie bracket is identified with the commutator of two operators/matrices. The full Lie bracket concept is beyond our scope.

[4] Simply but loosely put, $SU(2)$ has a finite volume.

$$\begin{aligned}
|\psi'\rangle &= e^{-i\frac{\delta}{2}\sigma_z}\,|\psi\rangle = e^{-i\frac{\delta}{2}\sigma_z}\begin{pmatrix}\cos\left(\frac{\theta}{2}\right)\\[4pt]\sin\left(\frac{\theta}{2}\right)e^{i\varphi}\end{pmatrix} = \begin{pmatrix}e^{-i\frac{\delta}{2}} & 0\\[4pt]0 & e^{i\frac{\delta}{2}}\end{pmatrix}\begin{pmatrix}\cos\left(\frac{\theta}{2}\right)\\[4pt]\sin\left(\frac{\theta}{2}\right)e^{i\varphi}\end{pmatrix}\\[8pt]
&= e^{-i\frac{\delta}{2}}\begin{pmatrix}\cos\left(\frac{\theta}{2}\right)\\[4pt]\sin\left(\frac{\theta}{2}\right)e^{i(\varphi+\delta)}\end{pmatrix}.
\end{aligned} \tag{4.23}$$

When we disregard the overall phase factor above, which is physically irrelevant, $|\psi'\rangle$ corresponds to the unit vector $(\sin\theta\cos(\varphi+\delta),\ \sin\theta\sin(\varphi+\delta),\ \cos\theta)^T$ à la Equation (4.15). This vector arises from a counterclockwise rotation of the original unit vector (4.15) by an angle δ about the z-axis, as can be visualized in Figure 4.2. The negative sign in the exponent is chosen to ensure a counterclockwise rotation. This transformation is not just limited to a single point on the Bloch sphere. It affects the entire sphere, leading to what we term as a rotation of the Bloch sphere.

The quantum operation $U = \exp(-i\frac{\delta}{2}\sigma_z)$ introduces a *relative phase* between the computational basis states $|0\rangle$ and $|1\rangle$ and is thus termed a **phase-shift** operation, referred to as a **phase-shift gate** in subsequent discussions. This phase shift manifests as a rotation of the Bloch sphere about the z-axis. If you're curious about executing a rotation around a generic axis, the Pauli operator $\sigma_{\hat{n}}$ comes into play. Specifically, this Pauli operator, which belongs to $\mathfrak{su}(2)$, leads to the $SU(2)$ element $\exp(-i\frac{\delta}{2}\sigma_{\hat{n}})$, which performs a counterclockwise rotation of the Bloch sphere by an angle δ about the unit vector $\hat{n} = (n_x, n_y, n_z)^T$, as visualized by the $\hat{n}$ vector in Figure 4.2.

> **Exercise 4.8** *Have* $U = \exp(-i\frac{\delta}{2}\sigma_{\hat{n}})$ *act on* $|\psi\rangle$ *to see what it does to the state. You may want to consider a few special cases, such as one or two of the components of* $\hat{n}$ *being zero.*

Astute readers may raise a query. Given Equations (4.20) and (4.21), a typical $SU(2)$ element is expressed as $\exp[-i(a\sigma_x + b\sigma_y + c\sigma_z)]$, encompassing the angular parameters a, b, and c. Yet, this doesn't directly translate to an evident

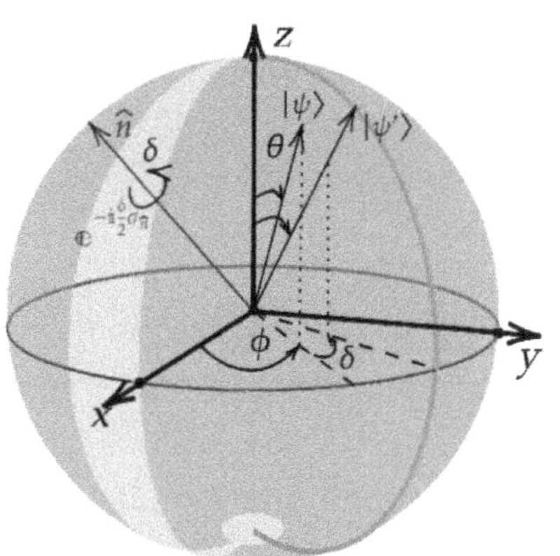

Figure 4.2 Rotation of the Bloch sphere.

rotation of the Bloch sphere about a specific axis. The root of this apparent discrepancy lies in the noncommutative nature of the Pauli matrices: $\exp[-i(a\sigma_x + b\sigma_y + c\sigma_z)] \neq \exp[-ia\sigma_x]\exp[-ib\sigma_y]\exp[-ic\sigma_z]$. Nevertheless, every $SU(2)$ element can be factorized as a sequence of three successive rotations, as will be discussed in Section 4.3.1.

We have walked through the action of an $SU(2)$ element on a pure qubit state and, leveraging the correlation between pure qubit states and three-dimensional unit vectors, have deduced that it indeed enacts a rotation of the Bloch sphere. For readers seeking a deeper affirmation of this $SU(2)$ transformation as a Bloch sphere rotation, or for those unfamiliar with the linkage between $SU(2)$ and $SO(3)$, you may want to read Remark 4.1.

Note to the Reader: Remark 4.1 dives deeper into the mathematical intricacies between $SU(2)$ and $SO(3)$. While it provides additional depth and understanding, it's not essential for the primary topics covered in this book. If you're not interested in these mathematical details, feel free to skip this remark and continue with the main content.

Remark 4.1 Two-dimensional Representation of Rotations

A single-qubit quantum operation $U \in SU(2)$ transforms a pure qubit state $|\psi\rangle$ as $U|\psi\rangle$ and consequently modifies the corresponding bra state as $\langle\psi|U^\dagger$. A natural question arises: How does U transform an observable? The answer lies in the fact that the expectation value $\langle\psi|O|\psi\rangle$ of an observable O remains invariant under a unitary transformation. Explicitly, we have

$$\langle\psi|O|\psi\rangle = \underbrace{\langle\psi|U^\dagger}_{\langle\psi'|}\, \underbrace{UOU^\dagger}_{O'}\, \underbrace{U|\psi\rangle}_{|\psi'\rangle} = \langle\psi'|O'|\psi'\rangle, \tag{4.24}$$

which suggests that

$$O \longrightarrow O' = UOU^\dagger. \tag{4.25}$$

This is also consistent with the observation that an observable, in this context, is generally represented as a matrix, which, having two indices, requires two unitary matrices for its transformation.

Consider the specific case where $O = \sigma_{\hat{n}}$ and $U = \exp(-i\frac{\delta}{2}\sigma_z)$. Computing, we obtain

$$e^{-i\frac{\delta}{2}\sigma_z}\sigma_{\hat{n}}e^{i\frac{\delta}{2}\sigma_z}$$

$$= \begin{pmatrix} e^{-i\frac{\delta}{2}} & 0 \\ 0 & e^{i\frac{\delta}{2}} \end{pmatrix} \begin{pmatrix} n_z & n_x - in_y \\ n_x + in_y & -n_z \end{pmatrix} \begin{pmatrix} e^{i\frac{\delta}{2}} & 0 \\ 0 & e^{-i\frac{\delta}{2}} \end{pmatrix}$$

> **Remark 4.1, continued**
>
> $$= \begin{pmatrix} n_z & \underbrace{(n_x \cos\delta - n_y \sin\delta)}_{n'_x} - \mathrm{i}\underbrace{(n_x \sin\delta + n_y \cos\delta)}_{n'_y} \\ \underbrace{(n_x \cos\delta - n_y \sin\delta)}_{n'_x} + \mathrm{i}\underbrace{(n_x \sin\delta + n_y \cos\delta)}_{n'_y} & -n_z \end{pmatrix}$$
>
> $$= \begin{pmatrix} n_z & n'_x - \mathrm{i}n'_y \\ n'_x + \mathrm{i}n'_y & n_z \end{pmatrix} = \sigma_{\hat{n}'}, \tag{4.26}$$
>
> where $\hat{n}' = (n'_x, n'_y, n_z)^T$. Given the relation (4.10) between a three-dimensional vector and a Pauli operator, this reveals that the $SU(2)$ transformation $\exp(-\mathrm{i}\frac{\delta}{2}\sigma_z)$ induces a counterclockwise rotation of any three-dimensional vector about the z-axis by an angle δ. If you've worked through Exercise 4.7, it should be evident that $\exp(-\mathrm{i}\frac{\delta}{2}\sigma_x)$ and $\exp(-\mathrm{i}\frac{\delta}{2}\sigma_y)$ correspond to counterclockwise rotations about the x- and y-axes, respectively. For subsequent discussions, we define
>
> $$R_x(\delta) := \mathrm{e}^{-\mathrm{i}\frac{\delta}{2}\sigma_x}, \quad R_y(\delta) := \mathrm{e}^{-\mathrm{i}\frac{\delta}{2}\sigma_y}, \quad R_z(\delta) := \mathrm{e}^{-\mathrm{i}\frac{\delta}{2}\sigma_z}, \quad R_{\hat{n}}(\delta) := \mathrm{e}^{-\mathrm{i}\frac{\delta}{2}\sigma_{\hat{n}}}. \tag{4.27}$$
>
> **Exercise 4.9** *Show that a counterclockwise rotation about an arbitrary axis $\hat{n}$ can be expressed as*
>
> $$R_{\hat{n}}(\delta) = \cos\tfrac{\delta}{2}\mathbb{1}_2 - \mathrm{i}\sin\tfrac{\delta}{2}\sigma_{\hat{n}}. \tag{4.28}$$
>
> Equation (4.28) implies that for $\delta \ll 1$, $R_{\hat{n}} \approx R_x(\delta)R_y(\delta)R_z(\delta)$.
>
> **Exercise 4.10** *In this exercise, we derive several common relations. Let $\hat{n}' = R\hat{n}$, where R rotates $\hat{n}$ to $\hat{n}'$. Demonstrate that*
>
> $$R_{\hat{n}'} = RR_{\hat{n}}R^{-1}. \tag{4.29}$$
>
> *With this result in hand, validate the conjugation relations in Equation (4.30):*
>
> $$\begin{aligned} \sigma_x R_y(\delta)\sigma_x &= R_y(-\delta), \\ \sigma_x R_z(\delta)\sigma_x &= R_z(-\delta). \end{aligned} \tag{4.30}$$
>
> Returning to the motivating question behind this remark: How can $SU(2)$ elements, which are 2×2 unitary matrices with unit determinant, rotate three-dimensional vectors? The key lies in the relationship between $SU(2)$ and the Lie group $SO(3)$, as well as the connection between $\mathfrak{su}(2)$ and $\mathfrak{so}(3)$ – the Lie algebra of $SO(3)$, the genuine group of three-dimensional rotations. This relationship is displayed in Equation (4.31):
>
> $$\begin{array}{ccc} \mathfrak{su}(2) & \overset{\cong}{\longleftrightarrow} & \mathfrak{so}(3) \\ {\scriptstyle\exp}\big\downarrow & & \big\downarrow{\scriptstyle\exp} \\ SU(2) & \underset{\text{2-to-1}}{\longrightarrow} & SO(3) \end{array} \tag{4.31}$$

> **Remark 4.1, continued**
>
> The Lie group $SO(3)$ encompasses all 3×3 orthogonal matrices with unit determinant. Such matrices can rotate and maintain the magnitudes of three-dimensional vectors. The correspondence (4.10) between three-dimensional vectors and 2×2 Hermitian traceless matrices means that operations on these vectors mirror operations on the matrices, thereby establishing a mapping from $SO(3)$ to $SU(2)$. An interesting nuance arises when considering the $SU(2)$ parameter $\delta/2$ used to define a counterclockwise rotation by angle δ. Altering δ by 2π results in the $SU(2)$ parameter being $\delta/2 + \pi$. Through the exponential map, this turns to $\exp\left[-i(\frac{\delta}{2} + \pi)\sigma_{\hat{n}}\right]$. Despite this being distinct from $\exp(-i\frac{\delta}{2}\sigma_{\hat{n}})$, both $SU(2)$ elements result in the same rotation of three-dimensional vectors. Therefore, they map to an identical $SO(3)$ element. This realization gives rise to the 2-to-1 mapping from $SU(2)$ to $SO(3)$, labeling $SU(2)$ as the **double cover** group of $SO(3)$. Notwithstanding this 2-to-1 relationship, the Lie algebras $\mathfrak{su}(2)$ and $\mathfrak{so}(3)$ remain isomorphic. This is attributed to the Lie algebra structure, specifically the Lie bracket, which, being a local construct proximate to the identity element, remains oblivious to the global interrelation between the two Lie groups.

4.1.2 Mixed Qubit States

To gain a holistic understanding of qubit states, we must also explore the realm of mixed qubit states. As discussed earlier, these states represent a statistical ensemble of different possible quantum states, adding richness and complexity to our understanding of qubits.

Density Matrices as Pauli Matrices

Earlier, we observed that rotations of the Bloch sphere are conceptualized through Pauli operators and that three-dimensional vectors can be represented by Pauli operators. Expanding upon this idea, any Hermitian matrix (not just traceless ones) associated with a qubit can also be expressed using Pauli matrices, once we allow a modest extension. This includes the density matrix, a positive Hermitian entity.

Let's begin by determining the number of independent degrees of freedom possessed by a typical Hermitian matrix. Astute readers might swiftly recognize this number to be 4: An $\mathfrak{su}(2)$ element, being a traceless Hermitian, has three degrees of freedom, and removing the traceless requirement introduces one additional degree. Even though this insight is valuable, it's instructive to approach the question from first principles.

Since a density matrix is Hermitian, consider a generic Hermitian matrix – a 2×2 complex matrix defined as

$$\begin{pmatrix} \alpha & \beta \\ \gamma & \delta \end{pmatrix} = \begin{pmatrix} \alpha^* & \gamma^* \\ \beta^* & \delta^* \end{pmatrix}, \tag{4.32}$$

where $\alpha, \beta, \gamma, \delta \in \mathbb{C}$. The hermiticity condition requires $\alpha, \delta \in \mathbb{R}$ and $\beta = \gamma^*$. This leads us to express the matrix as

$$
\begin{pmatrix} a & b - \mathring{\imath}c \\ b + \mathring{\imath}c & d \end{pmatrix} = \frac{a+d}{2}\mathbb{1}_2 + b\sigma_x + c\sigma_y + \frac{a-d}{2}\sigma_z, \tag{4.33}
$$

where $a, b, c, d \in \mathbb{R}$. This reveals that the set $\{\mathbb{1}_2, \sigma_x, \sigma_y, \sigma_z\}$ spans all 2×2 Hermitian matrices, with the inclusion of the identity matrix $\mathbb{1}_2$ representing the aforementioned extension.

A 2×2 density matrix can be captured by the right-hand side of Equation (4.33). By recalling the property $\mathrm{Tr}\rho = 1$ and the fact that $\mathrm{Tr}\sigma_i = 0$ for $i = x, y, z$, imposing the condition $a + d = 1$ in Equation (4.33), and defining $x = 2b, y = 2c, z = a - d$, we refine the matrix representation:

$$
\rho = \frac{1}{2}(\mathbb{1}_2 + x\sigma_x + y\sigma_y + z\sigma_z) = \frac{1}{2}\begin{pmatrix} 1+z & x - \mathring{\imath}y \\ x + \mathring{\imath}y & 1 - z \end{pmatrix}, \tag{4.34}
$$

showcasing the most general form of a 2×2 density matrix. That is, a generic density matrix ρ and hence a mixed state is specified by three real parameters x, y, and z.

Familiarity might strike upon viewing the matrix in Equation (4.34), as it draws parallels with a foundational relationship between Pauli matrices and four-dimensional spacetime. To elucidate this relationship, engage with the subsequent exercise.

> **Exercise 4.11** *Utilizing the correspondence between three-dimensional vectors and 2×2 traceless Hermitian matrices, demonstrate that the set $\{\mathbb{1}_2, \sigma_x, \sigma_y, \sigma_z\}$ can be perceived as an orthonormal basis of the four-dimensional Minkowski spacetime. Elaborate on the inner product using the Pauli matrices.*

The Bloch Ball

To achieve a comprehensive understanding of qubit states, it's necessary not only to map the pure qubit states but also to relate the mixed states to the Bloch sphere. Can you foresee where the mixed states might lie? Instead of hazarding a guess, let's derive their locations logically.

A fundamental property of a density matrix is its nonnegativity, or in other words, $\det \rho \geq 0$. By referring to Equation (4.34), the determinant can be expressed as:

$$
\det \rho = \frac{1}{4}\left[1 - (x^2 + y^2 + z^2)\right] \geq 0 \implies x^2 + y^2 + z^2 \leq 1. \tag{4.35}
$$

Since the equality is reserved only for pure states (an exercise: why?), for mixed qubit states inequality (4.36) holds true:

$$
x^2 + y^2 + z^2 < 1. \tag{4.36}
$$

This suggests that mixed qubit states correlate 1-to-1 with points inside a three-dimensional unit ball, which we'll refer to as the **Bloch ball**. Note that this ball is technically an open ball, as it lacks a boundary. When both mixed and pure states are taken into account, the Bloch ball and Bloch sphere coalesce to form a closed ball, with the Bloch sphere being the surface.

> **Exercise 4.12** *Determine the position within the Bloch ball of the mixed qubit state, ρ, with the maximum von Neumann entropy. For this state, compute $\mathrm{Tr}(\rho\sigma_i)$ for $i = x, y, z$, and interpret your results.*

Regardless of whether you've tackled the above exercise, let's compute the expectation value of a general rotation operator $\sigma_{\hat{n}}$ in a generic mixed state ρ. Define $\vec{r} = (x, y, z)^T$. We have:

$$\langle \sigma_{\hat{n}} \rangle = \mathrm{Tr}(\rho\sigma_{\hat{n}})$$

$$= \mathrm{Tr}\left[\frac{1}{2}(\mathbb{1}_2 + \vec{r} \cdot \vec{\sigma})\hat{n} \cdot \vec{\sigma}\right]$$

$$= \mathrm{Tr}\left[\frac{1}{2}(\vec{r} \cdot \vec{\sigma})(\hat{n} \cdot \vec{\sigma})\right]$$

$$= \vec{r} \cdot \hat{n}\,\mathrm{Tr}\left[\frac{1}{2}\mathbb{1}_2\right]$$

$$= \vec{r} \cdot \hat{n}. \tag{4.37}$$

The above derivation employed the algebraic relations (4.9) of the Pauli matrices. Given its applicability in both pure and mixed state scenarios, $\vec{r}$ is aptly referred to as the qubit's polarization.

Incomplete Information of ρ

Throughout Chapter 3, we emphasized that a mixed state ρ offers only partial information about a quantum system. Here, using the Pauli operator $\sigma_{\hat{n}}$, we provide an alternative perspective on this concept. Consider, specifically, the mixed qubit state $\rho = \frac{1}{2}\mathbb{1}_2$. Renaming the + eigenstate $|\psi(\theta, \varphi)\rangle$ of $\sigma_{\hat{n}}$ from Equation (4.15) as $|0_{\hat{n}}\rangle := |\psi(\theta, \varphi)\rangle$, and similarly defining $|1_{\hat{n}}\rangle$ as the − eigenstate of $\sigma_{\hat{n}}$ (assuming you've determined it in Exercise 4.1), we can express

$$\rho = \frac{1}{2}\mathbb{1}_2 = \frac{1}{2}\left(|0_{\hat{n}}\rangle \langle 0_{\hat{n}}| + |1_{\hat{n}}\rangle \langle 1_{\hat{n}}|\right). \tag{4.38}$$

Interestingly, the unit vector $\hat{n}$ offers infinite possibilities, leading to infinite decompositions for ρ. Yet, the expectation value of any observable O in this mixed state is solely $\mathrm{Tr}(\rho O)$, independent of ρ's decomposition. This contrasts starkly with pure states, where, despite infinite potential expansions, the probability amplitude of each state remains deterministic, allowing for the differentiation of expansions through suitable observable measurements.

Rotating the Bloch Ball

The rotation of the Bloch sphere using $SU(2)$ matrices can naturally be extended to the Bloch ball, given that their respective density matrices are both 2×2 Hermitian matrices. Thus, mirroring the rotation of three-dimensional vectors in matrix form, the Bloch ball can be rotated counterclockwise by angle δ about an arbitrary axis $\hat{n}$ using

$$R_{\hat{n}}(\delta)\rho R_{\hat{n}}(-\delta), \tag{4.39}$$

effecting the transformation of the corresponding mixed state ρ.

4.2 Measuring Qubit States

Having comprehended and visualized the single-qubit state space, our next endeavor is to elucidate how to ascertain a given single-qubit state through measurements. As previously underscored, the physical quantities amenable to reliable measurement are the expectation values of observables. When dealing with an observable O, one must conduct repeated measurements on the same state, ideally an infinite number of times, to deduce $\langle O \rangle$. We will explore the cases of pure qubit states and mixed qubit states in due course.

4.2.1 Pure Qubit States

With the Bloch sphere as our guide, the task at hand is to determine the characteristics of a given pure qubit state $|\psi(\theta, \varphi)\rangle$, specifically pinpointing its location on the Bloch sphere, which means finding its coordinates (x, y, z) or (θ, φ). As the Pauli matrices gauge qubit polarizations along the x-, y-, and z-axes, they serve as the appropriate set of observables whose expectation values within the state can unveil the coordinates. Let's perform a straightforward calculation:

$$\langle\psi|\sigma_x|\psi\rangle = \begin{pmatrix} \cos\frac{\theta}{2} & \sin\frac{\theta}{2}e^{i\varphi} \end{pmatrix} \begin{pmatrix} 0 & 1 \\ 1 & 0 \end{pmatrix} \begin{pmatrix} \cos\frac{\theta}{2} \\ \sin\frac{\theta}{2}\exp^{i\varphi} \end{pmatrix} = \sin\theta\cos\varphi = x, \tag{4.40}$$

$$\langle\psi|\sigma_y|\psi\rangle = \begin{pmatrix} \cos\frac{\theta}{2} & \sin\frac{\theta}{2}e^{i\varphi} \end{pmatrix} \begin{pmatrix} 0 & -i \\ i & 0 \end{pmatrix} \begin{pmatrix} \cos\frac{\theta}{2} \\ \sin\frac{\theta}{2}\exp^{i\varphi} \end{pmatrix} = \sin\theta\sin\varphi = y, \tag{4.41}$$

$$\langle\psi|\sigma_z|\psi\rangle = \begin{pmatrix} \cos\frac{\theta}{2} & \sin\frac{\theta}{2}e^{i\varphi} \end{pmatrix} \begin{pmatrix} 1 & 0 \\ 0 & -1 \end{pmatrix} \begin{pmatrix} \cos\frac{\theta}{2} \\ \sin\frac{\theta}{2}\exp^{i\varphi} \end{pmatrix} = \cos\theta = z. \tag{4.42}$$

The tricky question to address now is how to experimentally obtain these expectation values and thereby determine the coordinates. It transpires that acquiring the z coordinate is a simple process, while ascertaining the x and y coordinates necessitates a slight modification. Consequently, we will tackle these two cases separately.

Determining z

Suppose we measure σ_z on $|\psi(\theta, \varphi)\rangle$ many times, let's say $N \gg 1$ times. The outcome $+1$ (indicating state $|0\rangle$) may occur N_0 times, and the outcome -1 (indicating state $|1\rangle$) may occur N_1 times, where $N_0 + N_1 = N$. Let p_0 and p_1 be the probabilities of finding $|0\rangle$ in $|\psi\rangle$ and finding $|1\rangle$ in $|\psi\rangle$. We extract the z coordinate as

$$\frac{N_0}{N} - \frac{N_1}{N} = p_0 - p_1 = \cos^2 \tfrac{\theta}{2} - \sin^2 \tfrac{\theta}{2} = \cos\theta = z. \tag{4.43}$$

The choice of N depends on our desired accuracy of z.

Determining x and y

Why does obtaining the x and y coordinates become somewhat tricky? The reason is that in experiments, it's inconvenient to change the computational basis, which is set to be the eigenbasis of the Pauli matrix σ_z. The measurement apparatus is designed and aligned for measuring qubit polarizations along the z-direction. Therefore, we prefer to perform our computations and measurements consistently within this computational basis, where we always measure σ_z. Nonetheless, our preference poses a challenge: Measuring σ_z on $|\psi(\theta, \varphi)\rangle$ results in obtaining z rather than x. How can we overcome this issue?

The idea is to rotate the Bloch sphere such that the z-axis becomes the x-axis, by means of the rotation operators defined in Equation (4.27). It's essential to note a subtlety: A rotation can be interpreted in a passive or active way. A *passive rotation* rotates the axes of a coordinate system in a certain direction, while an equivalent *active rotation* rotates the vectors in the opposite direction. In experiments, a rotation is an actual quantum operation performed on a quantum state, so we adopt the active perspective on rotations. To rotate the z-axis to the x-axis, we need to rotate the coordinate system about the y-axis counterclockwise by $\pi/2$. Therefore, in an active rotation, we use the rotation operator $R_y(-\pi/2)$, where the minus sign accounts for the clockwise rotation. Consequently, we rotate $|\psi(\theta, \varphi)\rangle$ to be

$$R_y\left(-\tfrac{\pi}{2}\right) |\psi(\theta, \varphi)\rangle. \tag{4.44}$$

Now, by measuring σ_z on this rotated state and calculating as in Equation (4.47), we can determine what we need:

$$\begin{aligned}
p_0' &= \left|\langle 0|R_y\left(-\tfrac{\pi}{2}\right)|\psi(\theta, \varphi)\rangle\right|^2 \\[2mm]
&= \left| \begin{pmatrix} 1 & 0 \end{pmatrix} \frac{1}{\sqrt{2}} \begin{pmatrix} 1 & 1 \\ -1 & 1 \end{pmatrix} \begin{pmatrix} \cos\tfrac{\theta}{2} & \sin\tfrac{\theta}{2}e^{i\varphi} \end{pmatrix} \right|^2 \\[2mm]
&= \frac{1}{2} \left| \cos\tfrac{\theta}{2} + \sin\tfrac{\theta}{2}e^{i\varphi} \right|^2,
\end{aligned} \tag{4.45}$$

$$p_1' = \left| \langle 1 | R_y \left(-\tfrac{\pi}{2} \right) | \psi(\theta, \varphi) \rangle \right|^2$$

$$= \left| \begin{pmatrix} 0 & 1 \end{pmatrix} \frac{1}{\sqrt{2}} \begin{pmatrix} 1 & 1 \\ -1 & 1 \end{pmatrix} \begin{pmatrix} \cos \tfrac{\theta}{2} & \sin \tfrac{\theta}{2} e^{i\varphi} \end{pmatrix} \right|^2$$

$$= \frac{1}{2} \left| - \cos \tfrac{\theta}{2} + \sin \tfrac{\theta}{2} e^{i\varphi} \right|^2, \tag{4.46}$$

so that

$$p_0' - p_1' = \sin \theta \cos \varphi = x. \tag{4.47}$$

We emphasize that p_0' and p_1' are still the probabilities of finding the computational basis states $|0\rangle$ and $|1\rangle$, which aligns with our preference.

As a side note, the procedure above is equivalent to rotating the computational basis from the eigenbasis of σ_z to that of σ_x because

$$\langle 0_x | := \frac{1}{\sqrt{2}} (\langle 0| + \langle 1|) = \langle 0 | R_y \left(-\tfrac{\pi}{2} \right), \tag{4.48}$$

$$\langle 1_x | := \frac{1}{\sqrt{2}} (-\langle 0| + \langle 1|) = \langle 1 | R_y \left(-\tfrac{\pi}{2} \right). \tag{4.49}$$

Then, $p_{0_x} = | \langle 0_x | \psi(\theta, \varphi) \rangle |^2 = p_0'$ and $p_{1_x} = | \langle 1_x | \psi(\theta, \varphi) \rangle |^2 = p_1'$.

> **Exercise 4.13** *Determining y follows a similar procedure. Write it out.*

4.2.2 Mixed Qubit States

Measuring mixed qubit states is analogous to measuring pure qubit states. Let P_0 and P_1 be the projectors onto the computational basis states $|0\rangle$ and $|1\rangle$, respectively. We find that

$$p_0 - p_1 = \mathrm{Tr}(\rho P_0) - \mathrm{Tr}(\rho P_1)$$

$$= \frac{1}{2} \mathrm{Tr} \left[\begin{pmatrix} 1+z & x-iy \\ x+iy & 1-z \end{pmatrix} \begin{pmatrix} 1 & 0 \\ 0 & 0 \end{pmatrix} \right] - \frac{1}{2} \mathrm{Tr} \left[\begin{pmatrix} 1+z & x-iy \\ x+iy & 1-z \end{pmatrix} \begin{pmatrix} 0 & 0 \\ 0 & 1 \end{pmatrix} \right]$$

$$= \frac{1}{2} [(1+z) - (1-z)] = z. \tag{4.50}$$

So, we simply repeatedly measure σ_z to obtain p_0 and p_1, whose difference is z.

As in the case of pure qubit states, finding the x and y coordinates would require rotations of the Bloch ball. We first rotate the Bloch ball to find x. This amounts to rotating the mixed state ρ as

$$R_y \left(-\tfrac{\pi}{2} \right) \rho R_y \left(\tfrac{\pi}{2} \right). \tag{4.51}$$

The cyclic symmetry in tracing a product of matrices motivates us to define the rotated projectors as

$$P_0' = R_y\left(\tfrac{\pi}{2}\right) P_0 R_y\left(-\tfrac{\pi}{2}\right) = \frac{1}{2}\begin{pmatrix} 1 & 1 \\ 1 & 1 \end{pmatrix}, \tag{4.52}$$

$$P_1' = R_y\left(\tfrac{\pi}{2}\right) P_1 R_y\left(-\tfrac{\pi}{2}\right) = \frac{1}{2}\begin{pmatrix} 1 & -1 \\ -1 & 1 \end{pmatrix}. \tag{4.53}$$

Hence,

$$\begin{aligned} p_0' &= \mathrm{Tr}\left[R_y\left(-\tfrac{\pi}{2}\right)\rho R_y\left(\tfrac{\pi}{2}\right) P_0\right] = \mathrm{Tr}\left[\rho R_y\left(\tfrac{\pi}{2}\right) P_0 R_y\left(-\tfrac{\pi}{2}\right)\right] \\ &= \mathrm{Tr}\left(\rho P_0'\right), \end{aligned} \tag{4.54}$$

$$p_1' = \mathrm{Tr}(\rho P_1'), \tag{4.55}$$

such that $p_0' - p_1' = x$.

Exercise 4.14 *Likewise, find the procedure to obtain the y coordinate.*

What we have found so far for single-qubit states would also apply to multi-qubit states owing to the tensor product structure of the multi-qubit Hilbert space.

4.3   The Circuit Model

Armed with our understanding of qubits and qubit measurements, we are poised to expound on the central theme of this chapter: the quantum circuit model. While it may be tempting to view the quantum circuit model as merely a quantum counterpart to classical circuitry, it offers much more. Beyond its classical foundations, the quantum circuit model harnesses the power of quantum superposition and entanglement, unlocking a realm of unprecedented possibilities.

A quantum computer processes n qubits of information. The n-qubit Hilbert space is the tensor product of the individual Hilbert spaces of the n qubits. Hence, a generic n-qubit pure state reads as follows:

$$|\psi\rangle = \sum_{i_1,i_2,\ldots,i_{n-1}=0}^{1} \psi_{i_{n-1}i_{n-2}\ldots i_1 i_0} |i_{n-1}i_{n-2}\cdots i_1 i_0\rangle, \tag{4.56}$$

where $|i_{n-1}i_{n-2}\cdots i_1 i_0\rangle := |i_{n-1}\rangle\otimes|i_{n-2}\rangle\otimes\cdots\otimes|i_1\rangle\otimes|i_0\rangle$. Note that $|\psi\rangle$ is normalized. The qubits are so ordered, as in Equation (4.56), that the n qubits can be interpreted as both an n-qubit string and an n-qubit integer defined in Equation (2.24).

An n-qubit Hilbert space is 2^n-dimensional, encompassing exponentially more information than the fixed single piece of information stored in n classical

Table 4.1 Comparison between quantum parallelism and classical parallelism. Classical parallelism requires an exponential amount of resources compared to quantum parallelism, primarily due to the absence of superposition and entanglement.

	Quantum Parallelism		Classical Parallelism
Levels	2^n	**vs**	2^n
Resource	Linear: n qubits due		Exponential: $n \times 2^n$
required	to superposition and entanglement		bits, no superposition nor entanglement

bits. Thanks to quantum superposition, any quantum operation on an n-qubit state acts on all 2^n basis states simultaneously, rendering quantum computation inherently parallel. But, you might be wondering, how does quantum parallelism differ from its classical counterpart? Well, let's clear that up with Table 4.1.

It's a good time to address a question you might have been pondering over: Why does quantum computation prefer n qubits over a 2^n-level quantum system? One primary reason is the distinctive properties of quantum mechanics employed: Using n qubits leverages not only the principle of superposition but also the phenomenon of quantum entanglement. A single system with 2^n levels doesn't exhibit such entanglement. Additionally, with n qubits, each qubit can be controlled independently without disturbing the remaining qubits, whereas exerting precise control over a particular subspace of a 2^n-level system without influencing the entire system is inherently challenging. Further complicating matters for the 2^n-level system, precisely distinguishing between its energy levels poses significant challenges. In contrast, an n-qubit system, especially if each qubit corresponds to something like an electron with two distinguishable spin states, has fewer energy levels since the different basis states can be distinguished based on qubit properties like spin, even if these states have the same energy.

Quantum computation comprises three fundamental steps:

1. Initialization: The quantum computer begins by preparing the n qubits in the fiducial state $|\psi_i\rangle$, which is typically the state $|00\ldots0\rangle$ with n zeros.
2. Computation: Next, the system undergoes an evolution from the initial state $|\psi_i\rangle$ to the final state $|\psi_f\rangle$ – the desired computational result – using a unitary transformation U: $U|\psi_i\rangle = |\psi_f\rangle$. Note that U is a $2^n \times 2^n$ unitary matrix. This evolution implements the desired computation on the initial state.
3. Measurement: Finally, a standard measurement is performed, which involves measuring the observable σ_z on each qubit to extract the result.

The circuit model allows us to break a $2^n \times 2^n$ unitary matrix U into a set of 2×2 and 4×4 unitary matrices referred to as the elementary quantum gates acting on single qubit and two qubits. These elementary gates are assembled into a quantum circuit that converts n input qubits to n output qubits as the result.

4.3.1 Single-Qubit Gates

As shown in Section 4.1.1, every single-qubit gate belongs to the Lie group $SU(2)$ and can be constructed from rotations on the Bloch sphere or ball, represented by the operators $R_{\hat{n}}(\delta)$ defined in Equation (4.27). In what follows, you will explore a few elementary single-qubit gates and learn how to decompose a generic single-qubit gate into rotations.

The Phase-Shift Gate

In Section 4.1.1, you encountered the rotation operator $R_z(\delta)$, which shifts the relative phase in a generic single-qubit pure state $|\psi(\theta, \varphi)\rangle$ by an angle δ:

$$R_z(\delta)\,|\psi(\theta, \varphi)\rangle = \cos\tfrac{\theta}{2}\,|0\rangle + \sin\tfrac{\theta}{2}e^{i(\varphi+\delta)}\,|1\rangle. \tag{4.57}$$

The quantum gate that accomplishes this phase shift is known as the **phase-shift gate**. In the equation above, an overall phase factor, which is immaterial, is disregarded, so the phase-shift gate is represented as the matrix

$$T(\delta) = \begin{pmatrix} 1 & 0 \\ 0 & e^{i\delta} \end{pmatrix}. \tag{4.58}$$

In quantum circuits, a phase-shift gate is typically depicted as in Figure 4.3.

Two phase-shift gates with specific shift angles are commonly used: The $\pi/4$ and $\pi/2$ phase-shift gates, denoted as the T and S gates, respectively, and drawn as in Figure 4.4.

It's worth noting that $T^2 = S$. As you will discover in Section 4.3.3, the T gate is a fundamental element of the universal set of quantum gates, capable of composing any quantum gate.

The Hadamard Gate

The initial state, often referred to as the fiducial state, for each qubit in a quantum circuit is typically set to $|0\rangle$, a computational basis-state. The **Hadamard gate**, denoted by H, is designed to transform the fiducial state into a superposition:

$$H\,|0\rangle = \frac{1}{\sqrt{2}}(|0\rangle + |1\rangle), \tag{4.59}$$

$$H\,|1\rangle = \frac{1}{\sqrt{2}}(|0\rangle - |1\rangle). \tag{4.60}$$

Figure 4.3 Circuit diagram of the phase-shift gate.

Figure 4.4 Circuit diagrams of the T and S gates.

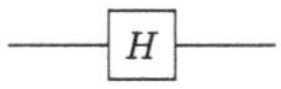

$$-\boxed{H}-$$

Figure 4.5 Circuit diagram of the Hadamard gate.

For future convenience, the two lines of equations can be collectively written as

$$H\,|i\rangle = \frac{1}{2}\left(|0\rangle + (-1)^i\,|1\rangle\right) = \frac{1}{\sqrt{2}}\sum_{i'=0}^{1}(-1)^{ii'}\,|i'\rangle, \tag{4.61}$$

where i is the original qubit state, and i' the resultant qubit state of H's action. The Hadamard gate is represented by the matrix

$$H = \frac{1}{\sqrt{2}}\begin{pmatrix} 1 & 1 \\ 1 & -1 \end{pmatrix}. \tag{4.62}$$

Notably, H is Hermitian: $H^\dagger = H^{-1} = H$. The Hadamard gate is commonly depicted as in Figure 4.5.

The Hadamard gate is indeed a rotation, but which one? To determine this, we can express H in terms of the Pauli matrices:

$$H = \frac{1}{\sqrt{2}}(\sigma_x + \sigma_z). \tag{4.63}$$

Using Equation (4.28) for rotations about an $\hat{n}$ axis, which is provided here for your convenience:

$$R_{\hat{n}}(\delta) = \cos\tfrac{\delta}{2}\mathbb{1}_2 - i\sin\tfrac{\delta}{2}\sigma_{\hat{n}}, \tag{4.64}$$

we find that

$$H = R_{\hat{n}}(\pi), \quad \hat{n} = \left(\tfrac{1}{\sqrt{2}}, 0, \tfrac{1}{\sqrt{2}}\right)^T, \tag{4.65}$$

up to an irrelevant overall phase factor.

To transform the fiducial state of all n qubits into a superposition, we simply apply an H gate to each qubit in parallel:

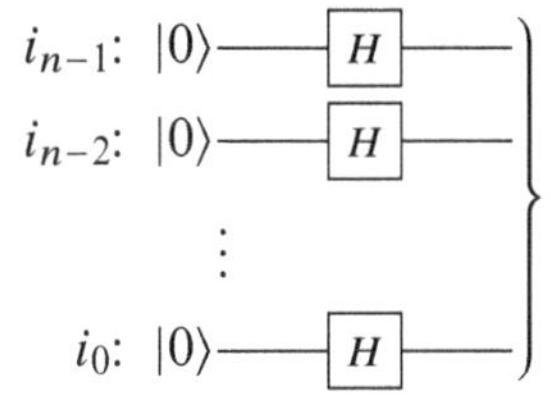

$$\left.\begin{array}{l} i_{n-1}\!:\ |0\rangle-\boxed{H}- \\ i_{n-2}\!:\ |0\rangle-\boxed{H}- \\ \qquad\vdots \\ i_0\!:\ |0\rangle-\boxed{H}- \end{array}\right\}\ H \otimes H \otimes \cdots \otimes H\,|\underset{\underset{i_{n-1}}{\uparrow}}{0}0\ldots\underset{\underset{i_0}{\uparrow}}{0}\rangle. \tag{4.66}$$

Note that in a quantum circuit, the mathematical relation between two parallel circuit lines is represented by the tensor product $\otimes$. The resulting state can be explicitly derived:

$$\underbrace{H \otimes H \otimes \cdots \otimes H}_{n}\,\underbrace{|00\ldots0\rangle}_{n}$$

$$= \frac{1}{\sqrt{2}}(|0\rangle + |1\rangle) \otimes \frac{1}{\sqrt{2}}(|0\rangle + |1\rangle) \otimes \cdots \otimes \frac{1}{\sqrt{2}}(|0\rangle + |1\rangle)$$

$$= \frac{1}{\sqrt{2^n}}\sum_{x=0}^{2^n-1}|x\rangle, \tag{4.67}$$

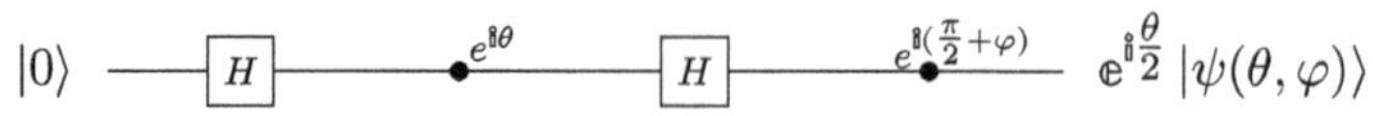

$$|0\rangle \;\longrightarrow\; \boxed{H} \;\xrightarrow{\;\bullet\, e^{i\theta}\;}\; \boxed{H} \;\xrightarrow{\;\bullet\, e^{i(\frac{\pi}{2}+\varphi)}\;}\; e^{i\frac{\theta}{2}}\,|\psi(\theta,\varphi)\rangle$$

Figure 4.6 The circuit creating any state on the Bloch sphere from the state $|0\rangle$.

where the sum runs over all possible n-qubit strings x, ranging from $00\ldots0$ to $11\ldots1$. As expected, this state is a superposition of the 2^n computational basis states of the n-qubit Hilbert space. The unitary transformation described by Equation (4.67) is termed the **Walsh–Hadamard transformation** [21], denoted by

$$W_n := \underbrace{H \otimes H \otimes \cdots \otimes H,}_{n} \qquad (4.68)$$

which we shall encounter again later.

Generating All Single-Qubit Gates

You might be wondering whether it's necessary to design a dedicated physical gate for each single-qubit operation. Fortunately, this is not the case. We have emphasized the phase-shift and Hadamard gates because they can be composed to generate all single-qubit gates, which are essentially rotations of the Bloch sphere. Do you know why? Take a moment to ponder before continuing.

Here's the reason: The phase-shift gate represents a rotation about the z-axis, which is an exponentiation of the σ_z matrix. The Hadamard gate is an exponentiation of a specific combination of σ_z and σ_x. When you multiply or take the commutator of σ_x and σ_z, you obtain σ_y. Composing the Hadamard and phase-shift gates with different shift angles involves a sufficient number of products and commutators of the Pauli matrices to generate the entire $SU(2)$ group, and consequently, all single-qubit gates. To gain a deeper understanding of this, observe how the composition of phase-shift and Hadamard gates in Equation (4.69) acts on the fiducial state $|0\rangle$ to create any state on the Bloch sphere, with the exception of an overall phase factor:

$$T\left(\tfrac{\pi}{2} + \varphi\right) HT(\theta)H\,|0\rangle = e^{i\frac{\theta}{2}}\,|\psi(\theta,\varphi)\rangle. \qquad (4.69)$$

In this expression, the overall phase factor is retained, as it may be relevant when considering the qubit as one of the n qubits in a quantum circuit. An overall phase of the qubit does not equate to an overall phase of an n-qubit state. The corresponding quantum circuit representation is depicted in Figure 4.6:

This is an appropriate place to mention how to interpret a circuit line in a quantum circuit. The input is always positioned at the left end of the line, while the output is at the right end.[5] Therefore, the circuit line should be read from left to right. In computational terms, the quantum gates on the same circuit line should be multiplied, following the principles of matrix multiplication, in a manner that ensures the

---

[5] One can also choose to place the input (output) at the right (left) end of a circuit; however, this would be opposite to the modern reading habit, which is from left to right.

$$X := \sigma_x$$
$$Y := -i\sigma_y$$
$$Z := \sigma_z$$

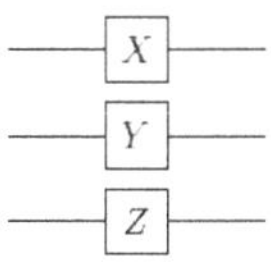

Figure 4.7 Circuit diagrams of the X, Y, and Z gates.

leftmost gate acts on the qubit first, and the rightmost gate acts last. This is why the gates in Equation (4.69) appear in the opposite order as they appear in the circuit in Figure 4.6.

While the Hadamard and phase-shift gates are universal for single-qubit operations, it is often practical in applications to design specific gates, such as in Figure 4.7.

Exercise 4.15 derives some useful relations involving the X, Y, Z, and H gates.

Exercise 4.15 *Verify the conjugation relations*

$$XYX = -Y, \qquad YXY = -X, \tag{4.70}$$
$$ZXZ = -X, \qquad XZX = -Z, \tag{4.71}$$
$$YZY = -Z, \qquad ZYZ = -Y, \tag{4.72}$$

$$HXH = Z, \quad HZH = X, \quad HYH = -Y. \tag{4.73}$$

Decomposing Single-Qubit Gates

As previously promised, we now show that any $SU(2)$ element, or in other words, any single-qubit gate, can be decomposed into a sequence of three rotations about the standard axes. This decomposition, known as the Euler-angle decomposition, is named after Euler, who introduced it.

Euler angles specify the orientation of a *rotated coordinate system* (the X-Y-Z system in Figure 4.8) with respect to the *fixed x-y-z coordinate system*. A rotation can be viewed from both an active and a passive perspective. In the passive view, it is the coordinate system that is rotated, not the vectors themselves. Therefore, a three-dimensional rotation can be seen as rotating the fixed coordinate system to align with the rotated one. The Euler angles are those by which the fixed coordinate system must be rotated to match the rotated one. This process involves three sequential rotations: first, a rotation of the fixed system about the z-axis by an angle α; second, a rotation about the normal vector N by an angle β; and finally, a rotation about the Z-axis by an angle γ. This sequence of rotations is visualized in Figure 4.8 and expressed mathematically as

$$R(\alpha, \beta, \gamma) = R_Z(\gamma)R_N(\beta)R_z(\alpha), \tag{4.74}$$

parameterizing and decomposing the corresponding $SO(3)$ element $R(\alpha, \beta, \gamma)$. Notably, the ranges of the Euler angles are $0 \leq \alpha, \gamma \leq 2\pi$ and $0 \leq \beta \leq \pi$ because β is a polar angle.

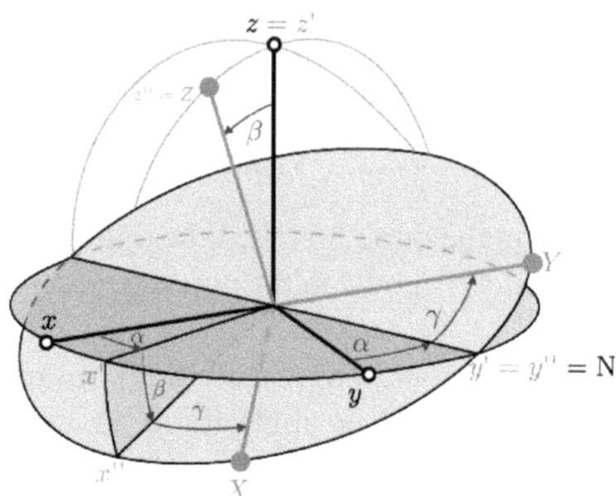

Figure 4.8 Euler angles $\alpha, \beta,$ and γ, which specify the orientation of the X-Y-Z (rotated) frame with respect to the x-y-z (fixed) frame. The vector N is the normal vector of the z-Z plane and lies along the intersection of the x-y and X-Y planes. Intermediate axes $x', x'', y', y'', z', z''$ are also drawn.

We are not finished yet, however, as $R(\alpha, \beta, \gamma)$ is not decomposed into a sequence of rotations solely about the fixed axes. To address this, we can utilize Equation (4.29) to express R_Z and R_N as rotations with respect to the fixed axes. Given that the normal vector N is obtained by rotating the y-axis about the z-axis by the Euler angle α, and the Z-axis is rotated from the z-axis about the vector N by the Euler angle β, we can write:

$$R_N(\beta) = R_z(\alpha)R_y(\beta)R_z(-\alpha), \tag{4.75}$$

$$R_Z(\gamma) = R_N(\beta)R_z(\gamma)R_N(-\beta). \tag{4.76}$$

Substituting these expressions into Equation (4.74) results in

$$R(\alpha, \beta, \gamma) = R_z(\alpha)R_y(\beta)R_z(\gamma), \tag{4.77}$$

which genuinely decomposes $R(\alpha, \beta, \gamma)$ into a sequence of rotations about the fixed axes, specifically, the z and y axes. Because of the correspondence between $SU(2)$ and $SO(3)$, this leads us to the **Euler-angle decomposition** of any $U \in SU(2)$,

$$U = R_z(\alpha)R_y(\beta)R_z(\gamma), \tag{4.78}$$

where $\alpha, \gamma \in [0, 2\pi]$ and $\beta \in [0, \pi]$ are the Euler angles.

It's important to note that the definition of Euler angles in Figure 4.8 is not unique, and alternative definitions exist. For example, α may be defined as the angle between the x-axis and the normal vector N, resulting in a z-x-z decomposition of R instead of the z-y-z decomposition in Equation (4.77).

The Euler-angle decomposition (4.78) of single-qubit gates gives rise to another useful decomposition of single-qubit gates: For any $U \in SU(2)$, there exist $A, B, C \in SU(2)$ satisfying $ABC = \mathbb{1}_2$, the 2×2 identity matrix, such that

$$U = AXBXC, \tag{4.79}$$

where the X gate is defined in Figure 4.7. This decomposition can be proven straightforwardly:

$$U = R_z(\alpha)R_y(\beta)R_z(\gamma)$$
$$= R_z(\alpha)R_y(\tfrac{\beta}{2})R_y(\tfrac{\beta}{2})R_z(\tfrac{\alpha+\gamma}{2})R_z(-\tfrac{\alpha-\gamma}{2})$$

$$= \underbrace{R_z(\alpha)R_y(\tfrac{\beta}{2})}_{A} \; X \; \underbrace{XR_y(\tfrac{\beta}{2})X}_{R_y(-\tfrac{\beta}{2})} \; \underbrace{XR_z(\tfrac{\alpha+\gamma}{2})X}_{R_z(-\tfrac{\alpha+\gamma}{2})} \; X \; \underbrace{X \, R_z(-\tfrac{\alpha-\gamma}{2})}_{C}$$

$$\underbrace{\hspace{4cm}}_{B}$$

$$= AXBXC, \tag{4.80}$$

where $X^2 = \mathbb{1}_2$ and Equation (4.30) were used. Clearly, by definition,

$$ABC = R_z(\alpha)R_y(\tfrac{\beta}{2})R_y(-\tfrac{\beta}{2})R_z(-\tfrac{\alpha+\gamma}{2})R_z(-\tfrac{\alpha-\gamma}{2}) = \mathbb{1}_2. \tag{4.81}$$

We shall make good use of the seemingly clumsy decomposition (4.79) soon.

4.3.2 Two-Qubit Gates

As pointed out and emphasized earlier, entanglement sets a fundamental difference between quantum and classical computation. To create entanglement and utilize it in quantum computation, we need two-qubit gates, especially the entangling gates – including the controlled gates – to be defined. Here, we shall explore the elementary gates and some important two-qubit gates.

The CNOT Gate

The classical CNOT gate may alter only the value of a classical bit, as seen in Section 2.4.4. In contrast, the quantum CNOT gate can also generate entanglement, and is therefore also known as an *entangling gate*. To see this, let's present the CNOT gate as a 4×4 unitary matrix in the two-qubit computational basis. The basis reads

$$|0\rangle := |00\rangle = \begin{pmatrix} 1 \\ 0 \\ 0 \\ 0 \end{pmatrix}, \quad |1\rangle := |01\rangle = \begin{pmatrix} 0 \\ 1 \\ 0 \\ 0 \end{pmatrix}, \quad |2\rangle := |10\rangle = \begin{pmatrix} 0 \\ 0 \\ 1 \\ 0 \end{pmatrix}, \quad |3\rangle := |11\rangle = \begin{pmatrix} 0 \\ 0 \\ 0 \\ 1 \end{pmatrix}. \tag{4.82}$$

The last two vector components in the two-qubit basis states $|0\rangle$ and $|1\rangle$ are both zero, so they coincide with the single-qubit basis states and should not cause any confusion. We may use the single digit two-qubit basis $\{|0\rangle, |1\rangle, |2\rangle, |3\rangle\}$ and $\{|00\rangle, |01\rangle, |10\rangle, |11\rangle\}$ interchangeably. In this basis the CNOT gate has the matrix form

$$\text{CNOT} = \begin{pmatrix} 1 & 0 & 0 & 0 \\ 0 & 1 & 0 & 0 \\ 0 & 0 & 0 & 1 \\ 0 & 0 & 1 & 0 \end{pmatrix}. \tag{4.83}$$

Figure 4.9 Circuit diagram of the CNOT gate.

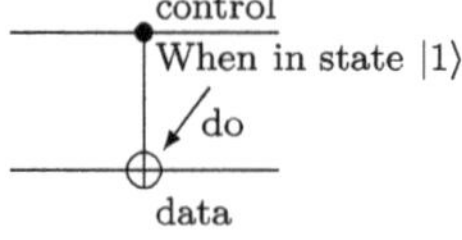

Figure 4.10 Circuit diagram of the E gate.

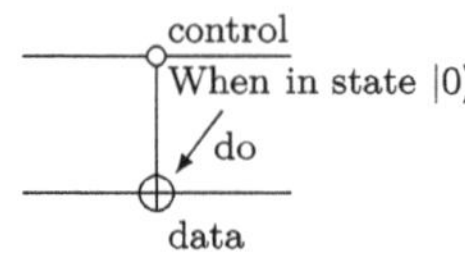

Figure 4.11 Circuit diagram of the F gate.

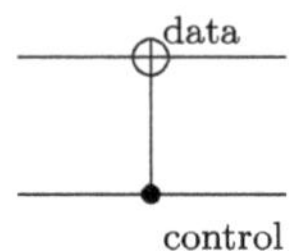

It is easy to verify that $\mathrm{CNOT}^2 = \mathbb{1}_4$. The CNOT gate is block diagonalized: The upper left block is $\mathbb{1}_2$ and acts trivially on the right qubit of the two (or the top one in the circuit). The lower right block is σ_x and acts on the left qubit (or the bottom one in the circuit). Keep in mind that σ_x flips state $|0\rangle$ to $|1\rangle$ and vice versa. Figure 4.9 depicts the CNOT gate.

Here, the *control qubit* is represented by a thick black dot, whereas the *data qubit* is represented by the $\oplus$. Analogous to the classical CNOT gate, the quantum CNOT gate decides what to do with the bottom qubit – the data qubit – according to the state of the top qubit – the control qubit. By Equations (4.82) and (4.83), the quantum CNOT gate flips (retains) the bottom qubit if the control bit is in state $|1\rangle$ ($|0\rangle$). The simple calculation below shows that the CNOT gate generates entanglement; it turns a separable state to an entangled state:

$$\mathrm{CNOT}\underbrace{(\alpha\,|0\rangle + \beta\,|1\rangle)\,|0\rangle}_{\text{separable}} = \underbrace{\alpha\,|00\rangle + \beta\,|11\rangle}_{\text{entangled}}. \tag{4.84}$$

The CNOT gate has three variations:

$$E = \begin{pmatrix} 0 & 1 & 0 & 0 \\ 1 & 0 & 0 & 0 \\ 0 & 0 & 1 & 0 \\ 0 & 0 & 0 & 1 \end{pmatrix}, \quad F = \begin{pmatrix} 1 & 0 & 0 & 0 \\ 0 & 0 & 0 & 1 \\ 0 & 0 & 1 & 0 \\ 0 & 1 & 0 & 0 \end{pmatrix}, \quad G = \begin{pmatrix} 0 & 0 & 1 & 0 \\ 0 & 1 & 0 & 0 \\ 1 & 0 & 0 & 0 \\ 0 & 0 & 0 & 1 \end{pmatrix}. \tag{4.85}$$

Take a moment to think about what they do on two-qubit states before reading on. Keeping the basis (4.82) in mind, it's easy to see:

- The E gate flips the data (bottom) qubit if the control (top) qubit is in state $|0\rangle$. Figure 4.10 depicts the circuit with the control bit being a hollow dot.
- The F gate flips the data (top) qubit if the control (bottom) qubit is in state $|1\rangle$. In the circuit, the top (bottom) qubit is the data (control) qubit; see Figure 4.11.

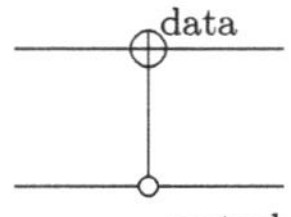

Figure 4.12 Circuit diagram of the G gate.

- The G gate flips the data (top) qubit if the control (bottom) qubit is in state $|0\rangle$. In the circuit, the top (bottom) qubit is the data (control) qubit; see Figure 4.12.

> **Exercise 4.16** *Show that the E, F, and G gates also generate entanglement. You can demonstrate by examples.*

In fact, the E, F, and G gates can be constructed from the CNOT and Hadamard gates. For example,

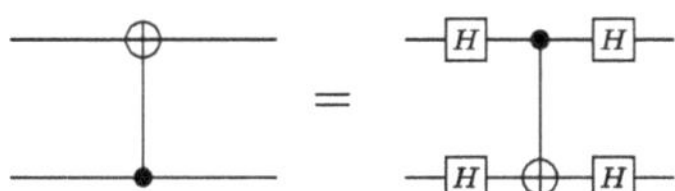

Figure 4.13 F gate is equal to the original CNOT gate conjugated by the Hadamard gate.

where the RHS reads $(H \otimes H)\,\mathrm{CNOT}(H \otimes H)$.

> **Exercise 4.17** *Verify Figure 4.13. Construct the E and G gates from the CNOT and Hadamard gates. Hint: You may need the conjugation relation in Figure 4.14.*

Since the CNOT gate generates entanglement, it is possible to create all four two-qubit Bell states, which are crucial states in quantum computation, from the computational basis states by means of the CNOT and Hadamard gate. The latter creates superposition. Let's first record the four Bell states:

$$
\begin{aligned}
|\phi^+\rangle &= \frac{1}{\sqrt{2}}(|00\rangle + |11\rangle), & |\phi^-\rangle &= \frac{1}{\sqrt{2}}(|00\rangle - |11\rangle); \\
|\psi^+\rangle &= \frac{1}{\sqrt{2}}(|01\rangle + |10\rangle), & |\psi^-\rangle &= \frac{1}{\sqrt{2}}(|01\rangle - |10\rangle).
\end{aligned}
\tag{4.86}
$$

We claim that the circuit in Figure 4.15 can generate the four Bell states from the input state $|i_1 i_0\rangle$, depending on the values of i_1 and i_0.

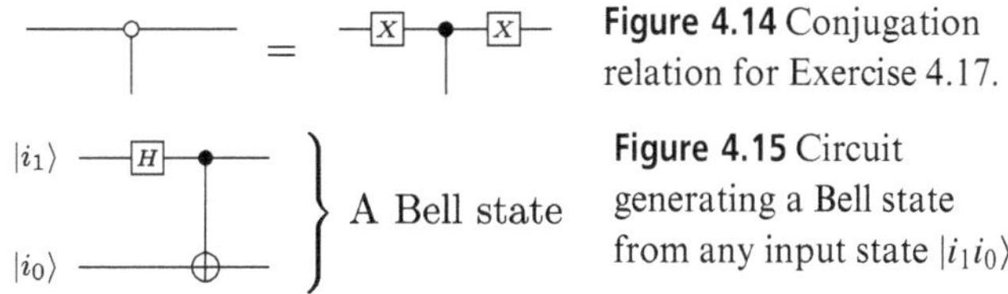

Figure 4.14 Conjugation relation for Exercise 4.17.

Figure 4.15 Circuit generating a Bell state from any input state $|i_1 i_0\rangle$.

To prove this claim, you are encouraged to first complete Exercise 4.18.

Exercise 4.18 *Show that*

$$\mathrm{CNOT} = |0\rangle\langle 0| \otimes \mathbb{1}_2 + |1\rangle\langle 1| \otimes X. \tag{4.87}$$

Keeping Equation (4.87) in mind and recalling that $H = (X + Z)/\sqrt{2}$, we can express the circuit in Figure 4.15 mathematically and derive as follows:

$$
\begin{aligned}
&\overbrace{(|0\rangle\langle 0| \otimes \mathbb{1}_2 + |1\rangle\langle 1| \otimes X)}^{\text{The circuit}} \overbrace{H \otimes \mathbb{1}_2 |i_1\rangle \otimes |i_0\rangle}^{\text{The input}}\\
&=(|0\rangle\langle 0| \otimes \mathbb{1}_2 + |1\rangle\langle 1| \otimes X)H |i_1\rangle \otimes |i_0\rangle\\
&= |0\rangle\langle 0|H |i_1\rangle \otimes |i_0\rangle + |1\rangle\langle 1|H |i_1\rangle \otimes X |i_0\rangle\\
&\overset{X|i_0\rangle=|\overline{i_0}\rangle}{=\!=\!=\!=\!=} \langle 0|H |i_1\rangle |0i_0\rangle + \langle 1|H |i_1\rangle |1\overline{i_0}\rangle\\
&= \frac{1}{\sqrt{2}}\Big[(\underbrace{\langle 0|X |i_1\rangle}_{\equiv i_1} + \underbrace{\langle 0|Z |i_1\rangle}_{\equiv 1-i_1}) |0i_0\rangle + (\underbrace{\langle 1|X |i_1\rangle}_{\equiv 1-i_1} + \underbrace{\langle 1|Z |i_1\rangle}_{\equiv -i_1}) |1\overline{i_0}\rangle \Big]\\
&= \frac{1}{\sqrt{2}}\big[|0i_0\rangle + (1 - 2i_1) |1\overline{i_0}\rangle \big]\\
&= \begin{cases} |\phi^+\rangle, & i_1 i_0 = 00;\\ |\psi^+\rangle, & i_1 i_0 = 01;\\ |\phi^-\rangle, & i_1 i_0 = 10;\\ |\psi^-\rangle, & i_1 i_0 = 11. \end{cases}
\end{aligned}
\tag{4.88}
$$

The SWAP Gate

Similar to the classical CROSSOVER (SWAP) gate, the quantum SWAP gate exchanges the states of the two qubits it acts on; however, its effect on a generic two-qubit state is tremendously different from its classical counterpart because of quantum superposition. The SWAP gate can be expressed as

$$\mathrm{SWAP} = |00\rangle\langle 00| + |01\rangle\langle 10| + |10\rangle\langle 01| + |11\rangle\langle 11| \tag{4.89}$$

$$
= \begin{pmatrix} 1 & 0 & 0 & 0\\ 0 & 0 & 1 & 0\\ 0 & 1 & 0 & 0\\ 0 & 0 & 0 & 1 \end{pmatrix}.
\tag{4.90}
$$

and depicted as in Figure 4.16.

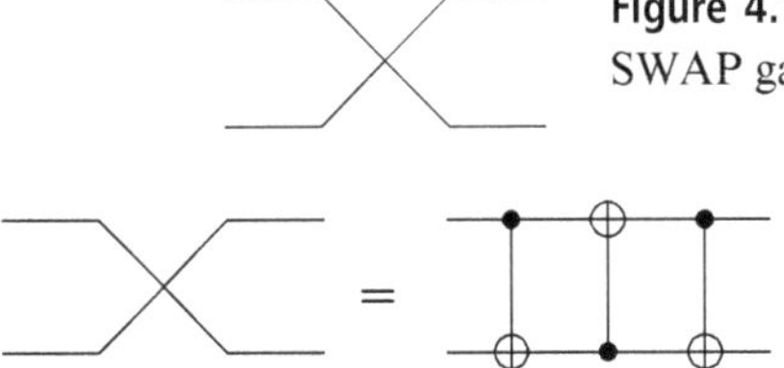

Figure 4.16 Circuit diagram of the SWAP gate.

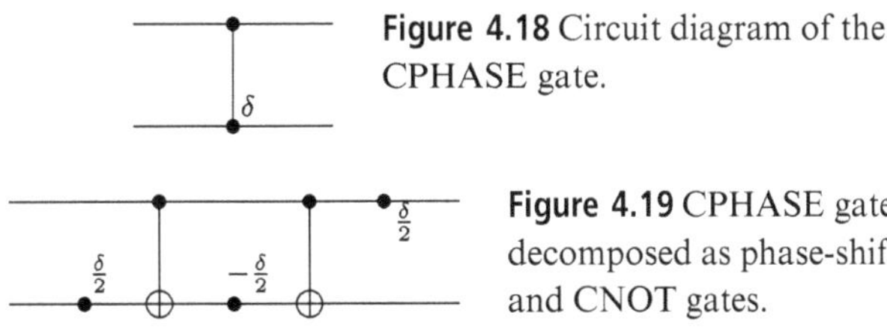

Figure 4.17 SWAP gate can be factorized as a combination of the CNOT and F gates.

Figure 4.18 Circuit diagram of the CPHASE gate.

Figure 4.19 CPHASE gate decomposed as phase-shift and CNOT gates.

> **Exercise 4.19** *Show the equivalence of the two circuit diagrams in Figure 4.17: The SWAP gate can be factorized into a combination of the CNOT and F gates.*

The CPHASE Gate

The idea of altering a data qubit according to the state of a control qubit can be extended beyond simply flipping the state of the data qubit. An important though simple such gate is the controlled phase (CPHASE) gate, denoted as $C\Phi(\delta)$ and defined by

$$
C\,\Phi(\delta) = \begin{pmatrix} 1 & 0 & 0 & 0 \\ 0 & 1 & 0 & 0 \\ 0 & 0 & 1 & 0 \\ 0 & 0 & 0 & e^{i\delta} \end{pmatrix},
\tag{4.91}
$$

which is block diagonalized in the same way as the CNOT gate and easy to comprehend. Its circuit diagram is depicted in Figure 4.18.

> **Exercise 4.20** *Show that a CPHASE gate can be composed by phase-shift and CNOT gates as in Figure 4.19.*
>
> **Exercise 4.21** *Show that the CPHASE gate is also an entangling gate.*

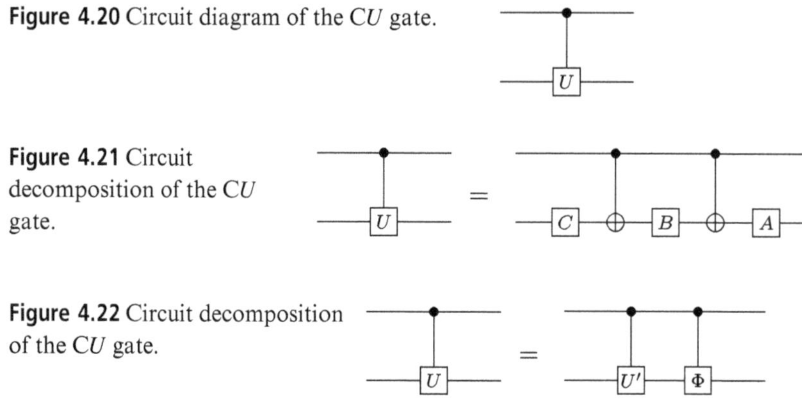

Figure 4.20 Circuit diagram of the CU gate.

Figure 4.21 Circuit decomposition of the CU gate.

Figure 4.22 Circuit decomposition of the CU gate.

The CU Gate

More generally, the σ_x can be replaced as the lower right block of the CNOT gate by any unitary single-qubit gate U, which acts on the data qubit. This yields a controlled U gate, or CU gate. Namely,

$$CU = |0\rangle\langle 0| \otimes \mathbb{1}_2 + |1\rangle\langle 1| \otimes U = \begin{pmatrix} 1 & 0 & 0 \\ 0 & 1 & \\ 0 & & U \end{pmatrix}, \tag{4.92}$$

whose circuit diagram in shown in Figure 4.20.

You may think that there should be a dedicated circuit for each U; however, this is untrue. The decomposition (4.79) of any $SU(2)$ element enables factorizing a CU gate with $U \in SU(2)$ as in Figure 4.21.

Here, the single-qubit gates A, B, and C are defined in Equation (4.79). This fact reveals to a certain extent the key role of the CNOT gate in quantum computation.

Astute readers may notice that the unitary gate U in a CU gate is not necessarily an $SU(2)$ element. Indeed, the U can be an element of $U(2)$; that is, an $SU(2)$ element multiplied with a phase factor. This phase factor is relevant because it is attached to the data qubit, not an overall phase factor of the entire two-qubit state. The question now is whether a CU gate with $U \in U(2)$ can be factorized into the CNOT and single-qubit gates. It can be, as we now show.

Any $U \in U(2)$ can be written as $U = e^{i\phi}U'$, where $U' \in SU(2)$ and $\phi \in [0, 2\pi]$. The multiplication $e^{i\phi}U'$ acts on the data qubit and can thus be factorized into the product of two gates: U' and a rather trivial gate $\Phi := e^{i\phi}\mathbb{1}_2$. Since U' and Φ commute, we can place the Φ gate on either side of U'. Hence, we can choose to have the circuit decomposition in Figure 4.22.

That is, the CU gate factorizes into a CU' gate and a $C\Phi$ gate. Note that this $C\Phi$ gate is not the CPHASE gate $C\Phi(\delta)$ defined in Equation (4.91). While the $C\Phi$ may assign the data qubit an overall phase factor, the $C\Phi(\delta)$ may shift the data qubit's

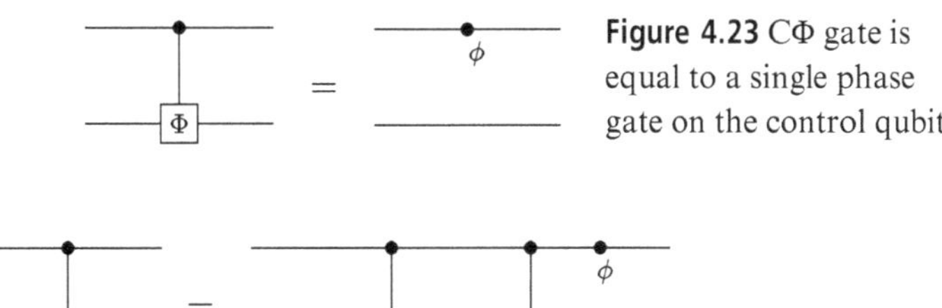

Figure 4.23 CΦ gate is equal to a single phase gate on the control qubit.

Figure 4.24 Decomposition of the CU gate for $U \in U(2)$.

relative phase. Having already factorized the CU' gate in Figure 4.21, we need only to factorize the CΦ gate. The calculation in Equation (4.93) leads to the answer.

$$
\begin{aligned}
C\Phi &= |0\rangle\langle 0| \otimes \mathbb{1}_2 + |1\rangle\langle 1| \otimes \Phi \\
&= |0\rangle\langle 0| \otimes \mathbb{1}_2 + e^{i\phi}|1\rangle\langle 1| \otimes \mathbb{1}_2 \\
&= \underbrace{\left(|0\rangle\langle 0| + e^{i\phi}|1\rangle\langle 1|\right)}_{=T(\phi)} \otimes \mathbb{1}_2 \\
&= T(\phi) \otimes \mathbb{1}_2,
\end{aligned}
\tag{4.93}
$$

which is merely a single-qubit phase-shift gate $T(\phi)$ on the original control qubit, as shown in Figure 4.23.

Therefore, combining factorizations in Figures 4.22, 4.22, and 4.23 gives for any $U \in U(2)$ the decomposition in Figure 4.24.

Here, the RHS consists of at most four single-qubit gates and two CNOT gates. We say "at most" because in certain cases, some of the single-qubit gates may be trivial and/or the two CNOT gates multiply to the identity when B is trivial.

4.3.3 Universal Quantum Gates

We have gained a sufficient sense of things to tackle the quest for universal quantum gates. A quantum computation on n qubits is in general a $2^n \times 2^n$ unitary matrix $U \in U(2^n)$ acting on an n-qubit state $|\psi\rangle \in \mathcal{H}^n$. We need not design a dedicated n-qubit gate for each U, thanks to Theorem 4.1.

Theorem 4.1. *The Universality Theorem [177, 178]: The set of single-qubit gates and the CNOT gate form a universal set of quantum gates, in the sense that any n-qubit unitary gate can be constructed by composing a selection of these gates.*

The proof of the theorem involves a few lemmas and theorems. We shall not go through all of them in detail but will outline the proof and elaborate on certain key points.

Note that each single-qubit gate acts nontrivially only on one qubit, and so too does the CNOT gate, which alters only the data qubit. The idea is to first decompose a generic $U \in U(2^n)$ gate into the product of a certain number of unitary matrices, to be termed two-level unitaries, each of which appears to act on a certain two-dimensional subspace of $\mathcal{H}^n$. Then we try to decompose each two-level unitary into the CNOT and single-qubit gates acting on the corresponding two-dimensional subspaces.

2-Level Unitaries

On a 2^n-dimensional Hilbert space, a two-level unitary is a $2^n \times 2^n$ unitary matrix that acts nontrivially only on two particular vector components of any state vector in the Hilbert space. To understand this abstract definition, let's see an example. Consider a two-qubit, so four-dimensional, Hilbert space; the unitary matrix

$$U = \begin{pmatrix} \alpha^* & 0 & 0 & \beta \\ 0 & 1 & 0 & 0 \\ 0 & 0 & 1 & 0 \\ -\beta^* & 0 & 0 & \alpha \end{pmatrix}, \tag{4.94}$$

where $|\alpha|^2 + |\beta|^2 = 1$, is a **two-level unitary** because the central block of the matrix is $\mathbb{1}_2$ and

$$\tilde{U} = \begin{pmatrix} \alpha^* & \beta \\ -\beta^* & \alpha \end{pmatrix} \tag{4.95}$$

is a 2×2 unitary matrix, called the *reduced matrix* of U. An important lemma follows:

Lemma 4.2. *Let $U \in U(N)$. Then, there exist $M \leq \frac{N(N-1)}{2}$ two-level unitaries $U_1, U_2, \ldots, U_M$, such that*

$$U = U_1 U_2 \ldots U_M \tag{4.96}$$

We will not prove this lemma here because it's straightforward and recursive. You can try to prove it yourself or look it up in the literature [21].

Now that we can factorize a generic n-qubit unitary gate into the product of certain two-level unitaries, it suffices to show that any two-level unitary can be constructed from the CNOT and single-qubit gates. We shall achieve this in a few steps, utilizing several new concepts to be introduced along the way.

Qubit Strings

As mentioned at the beginning of this section, a basis state of an n-qubit Hilbert space $\mathcal{H}^n$ is a binary string or a binary number $i_{n-1}i_{n-2}\cdots i_2 i_1 i_0$, where $i_k \in \{0, 1\}$.

We can order these binary numbers from the least to the greatest, viz from $00\cdots000$ to $11\cdots111$, such that a generic unitary gate is expressed as an $2^n \times 2^n$ unitary matrix in this ordered basis.

Then, a two-level unitary U has only at most four nontrivial elements indexed by two binary strings, and the rest of the matrix is effectively an identity matrix. For example, the two-level unitary (4.94) has the nontrivial components $U_{00,00}, U_{00,11}, U_{11,00},$ and $U_{11,11}$.

Let s and t, $s < t$ as numbers, be the two binary strings indexing the nontrivial elements of a two-level unitary U. We can assemble the nontrivial elements into a 2×2 matrix – the reduced matrix

$$\tilde{U} := \begin{pmatrix} U_{ss} & U_{st} \\ U_{ts} & U_{tt} \end{pmatrix}, \tag{4.97}$$

such as the $\tilde{U}$ in Equation (4.95). It's tempting to think that $\tilde{U}$ acts on the two-dimensional subspace $\mathrm{span}\{|s\rangle, |t\rangle\}$ only, and the entire two-level unitary seems to be a $C\tilde{U}$ gate. But not yet, because the binary strings s and t may not be next to each other in the code space – the space of the ordered n-qubit binary strings. That is, s and t may differ by more than one qubit, so the corresponding basis states $|s\rangle$ and $|t\rangle$ cannot be regarded as virtually the basis states of a qubit. We thus need the concept and technique of a gray code.

Gray Code

For two binary strings s and t, $s < t$, the **Gray code** [179] connecting s and t is a sequence of binary strings $\{g_1 = s, g_2, g_3, \ldots, g_{m-1}, g_m = t\}$, where g_k and g_{k+1} differ by exactly one bit! Example 4.1 will aid our understanding.

Example 4.1

A Gray code: Consider $s = 1000010$ and $t = 1101000$. The Gray code in this case reads

$$\left. \begin{array}{l} s = g_1 = 1\,0\,0\,0\,0\,1\,0 \\ g_2 = 1\,1\,0\,0\,0\,1\,0 \\ g_3 = 1\,1\,0\,1\,0\,1\,0 \\ t = g_4 = 1\,1\,0\,1\,0\,0\,0 \end{array} \right\} \text{The Gray code.} \tag{4.98}$$

As inferred by this example, if s and t differ by m bits, $m \leq n$, the corresponding Gray code contains $m + 1$ binary strings and is said to be a distance-m code.

Gray codes seem to be a neat binary game, but what are they good for in our task? Here is the idea. Given a two-level unitary U and its reduced matrix $\tilde{U}$, indexed by two binary strings s and t, $s < t$, we can use a sequence of gates, one at a time, to turn the string $s = g_1$ to g_2, g_2 to g_3 ... and finally to g_{m-1}. Since now g_{m-1} and $g_m = t$ differ by only a single bit (code distance 1), $\tilde{U}$ can act on $\text{span}\{|g_{m-1}\rangle, |g_m = t\rangle\}$. Afterwards, we reverse what's been done by a sequence of gates taking g_{m-1} to g_{m-2}, then to g_{m-3}, and all the way to $g_1 = s$, one at a time. In the end, the overall effect on the n qubits is the same as that of U. The moral here is that the sequence of gates needed are simple and can be constructed from the CNOT and single-qubit gates. Let's comprehend this procedure through Example 4.2.

Example 4.2

Consider the two-level unitary in Equation (4.99) acting on three qubits:

$$U = \begin{pmatrix} a & 0 & 0 & 0 & 0 & 0 & 0 & b \\ 0 & 1 & 0 & 0 & 0 & 0 & 0 & 0 \\ 0 & 0 & 1 & 0 & 0 & 0 & 0 & 0 \\ 0 & 0 & 0 & 1 & 0 & 0 & 0 & 0 \\ 0 & 0 & 0 & 0 & 1 & 0 & 0 & 0 \\ 0 & 0 & 0 & 0 & 0 & 1 & 0 & 0 \\ 0 & 0 & 0 & 0 & 0 & 0 & 1 & 0 \\ c & 0 & 0 & 0 & 0 & 0 & 0 & d \end{pmatrix}, \tag{4.99}$$

where $a, b, c, d \in \mathbb{C}$. The reduced matrix

$$\tilde{U} = \begin{pmatrix} a & b \\ c & d \end{pmatrix} \tag{4.100}$$

is indexed by $s = 000$ and $t = 111$, differing by three bits. The Gray code connecting s and t reads

$$\begin{aligned} s = g_1 &= 0\,0\,0 \\ g_2 &= 0\,0\,1 \\ g_3 &= 0\,1\,1 \\ t = g_4 &= 1\,1\,1. \end{aligned} \tag{4.101}$$

Now, we need the gates to bring $|g_1\rangle$ to $|g_3\rangle$. Since each time we just flip 1 bit of the string, we can achieve this by a kind of generalized control gate, specifically the controlled-controlled NOT (C^2NOT) gates. A C^2NOT gate acting on two control and one data qubit flips the data qubit if the two control qubits are both in state $|0\rangle$. Note that given three qubits, each of them can be the data qubit, and the two control qubits have four different states in total that determine whether the data qubit should be flipped. Hence, there are 12 possible C^2NOT gates, but we won't need all of them in this example. By means of the C^2NOT gates, Figure 4.25 shows the circuit that achieves our goal.

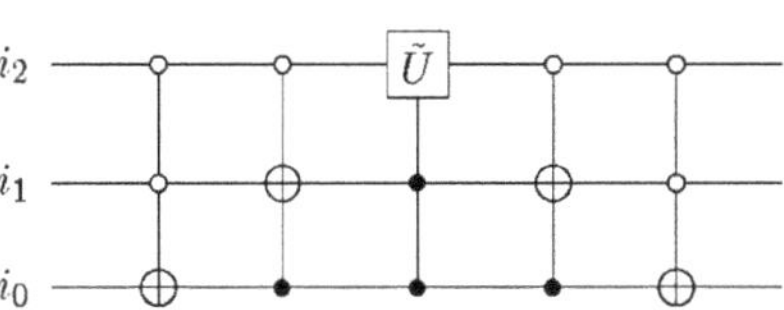

Figure 4.25 Circuit that realizes the unitary U.

This circuit indeed acts trivially on any basis state outside the Gray code. For instance, Figure 4.26 shows that $|110\rangle$ is preserved by the circuit.

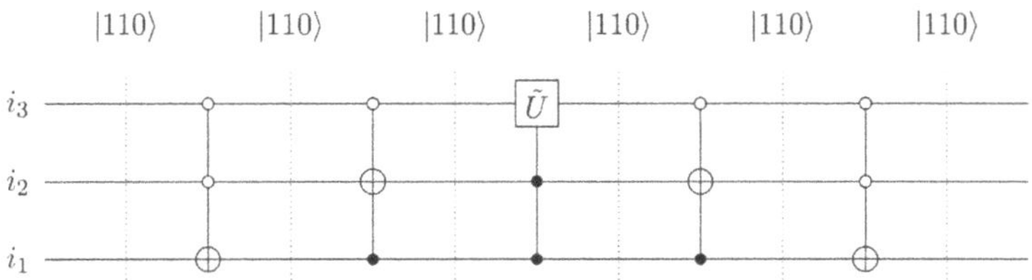

Figure 4.26 State $|110\rangle$ is not in the Gray code and is thus preserved by the circuit.

Although you can try other basis states, we can directly prove that the circuit functions as the U. Let's first check how the U acts on a generic state spanned by $|s\rangle = |000\rangle$ and $|t\rangle = |111\rangle$:

$$U(\alpha\,|000\rangle + \beta\,|111\rangle) = (\alpha a + \beta b)\,|000\rangle + (\alpha c + \beta d)\,|111\rangle. \qquad (4.102)$$

Figure 4.27 illustrates how the circuit in Figure 4.25 acts on $\alpha\,|000\rangle + \beta\,|111\rangle$, which indeed outputs $(\alpha a + \beta b)\,|000\rangle + (\alpha c + \beta d)\,|111\rangle$, in agreement with the action of U.

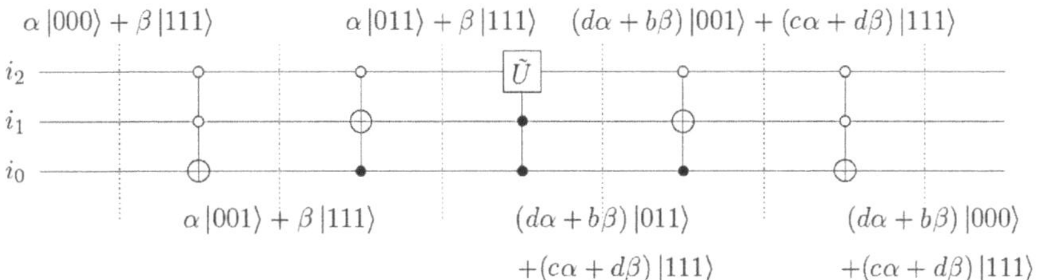

Figure 4.27 Circuit in Figure 4.25 transforms input state $\alpha\,|000\rangle + \beta\,|111\rangle$ to output $(\alpha a + \beta b)\,|000\rangle + (\alpha c + \beta d)\,|111\rangle$.

The steps in the example above generalize to n qubits. But in this case, we would need the $C^p\text{NOT}$ gates and $C^p\tilde{U}$ gates, where $p < n$, to factorize a $U \in U(2^n)$. These gates are not the CNOT and single-qubit gates. Hence, the question boils down to constructing the $C^p\text{NOT}$ gates and $C^p\tilde{U}$ gates from the CNOT and single-qubit gates. Since the decompositions will be general, we shall not restrict to any reduced matrix $\tilde{U}$ in the sequel but write U instead.

Exercise 4.22 *Consider the case of a two-level unitary U on three qubits:*

$$U = \begin{pmatrix} 1 & 0 & 0 & 0 & 0 & 0 & 0 & 0 \\ 0 & a & 0 & 0 & 0 & 0 & b & 0 \\ 0 & 0 & 1 & 0 & 0 & 0 & 0 & 0 \\ 0 & 0 & 0 & 1 & 0 & 0 & 0 & 0 \\ 0 & 0 & 0 & 0 & 1 & 0 & 0 & 0 \\ 0 & 0 & 0 & 0 & 0 & 1 & 0 & 0 \\ 0 & c & 0 & 0 & 0 & 0 & d & 0 \\ 0 & 0 & 0 & 0 & 0 & 0 & 0 & 1 \end{pmatrix}. \tag{4.103}$$

*Decompose this two-level unitary by following the procedure in Example 4.2.
Draw the circuit diagram.*

Decomposing the C^2NOT and C^2U Gates

We first tackle the case of $p = 2$ to build up our intuition. The C^2NOT gate is decomposed as in Figure 4.28.

Exercise 4.23 *Verify the decomposition in Figure 4.28.*

A C^2U gate is decomposed in the circuit diagram shown in Figure 4.29. A specific CV gate with $V^2 = U$ is needed to accomplish this. The intuition behind this decomposition is straightforward: We require two CV gates, each with a different control qubit but sharing the same data qubit as the C^2U gate. Additionally, we introduce a $CV^\dagger$ gate, which can potentially cancel out one of the two CV gates. We construct the decomposition in such a way that one of the two CV gates always remains

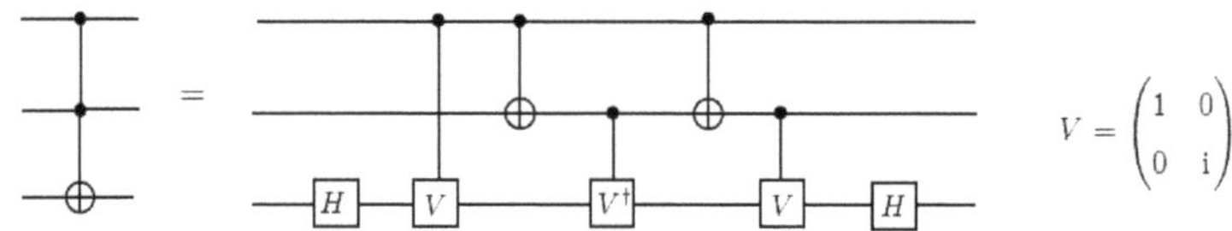

Figure 4.28 Decomposition of the C^2NOT gate by Hadamard, CNOT, and CV gates.

Figure 4.29
Decomposition of a
C^2U gate, where
$V^2 = U$ is required.

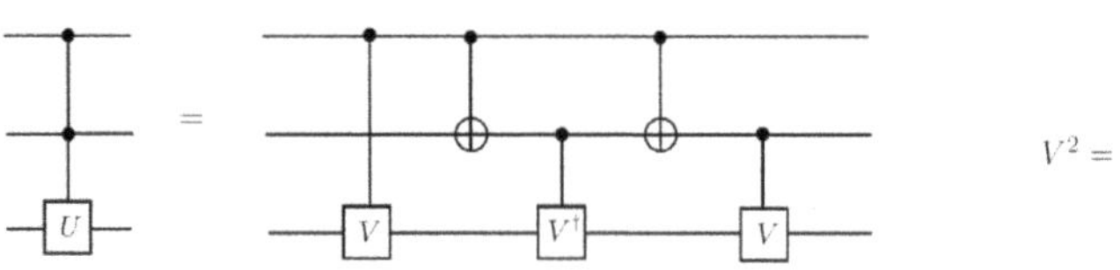

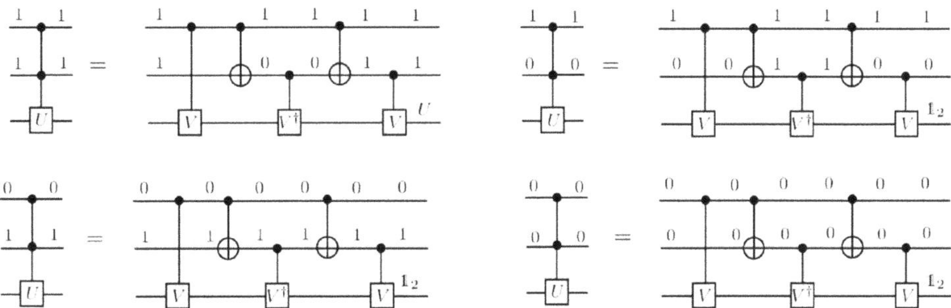

Figure 4.30 Work flow of the decomposition of a C^2U gate .

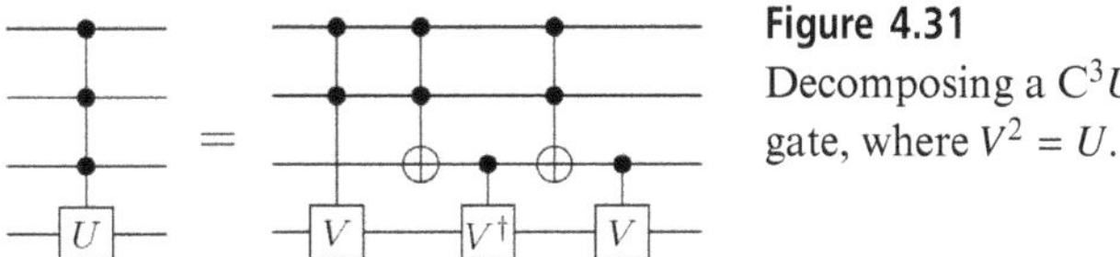

Figure 4.31
Decomposing a C^3U
gate, where $V^2 = U$.

active, while either the $CV^\dagger$ gate or the remaining CV gate may take effect. In the former case, the actions of V and $V^\dagger$ negate each other, resulting in no action of U. In the latter case, U is applied. Importantly, the determination of whether U acts or not depends on the control qubits in the same manner as it does in the C^2U gate, ensuring the correctness of the decomposition. Figure 4.30 illustrates how the circuit works depending on the four possible states of the two control qubits, verifying the intuition.

Exercise 4.24 *Decompose a $C^2\Phi$ gate and draw the decomposed circuit diagram.*

Decomposing the C^pU Gates

The intuition in decomposing the C^2U gates easily extends to decomposing the C^pU gates. Consider $p = 3$ first. We can find the single-qubit unitary V, such that $V^2 = U$. Then, we take a CV gate, a $CV^\dagger$ gate, and a C^2V gate. We then simply add one more circuit line on top of the circuit in Figure 4.29, and replace the rightmost CV gate there with our new C^2V gate, and the CNOT gates there with C^2NOT gates. Finally, we end up with the decomposition in Figure 4.31.

Should you worry about the C^2V gates and C^2NOT gates in the decomposition? No, because they have already been decomposed in Figures 4.28 and 4.29. Therefore, one can decompose any C^pU gate recursively, as in Figure 4.32.

It remains to decompose the C^pNOT gate. We shall not do it here because it's rather lengthy and tedious. The motivated reader is encouraged to explore the original literature [180]. It is a theorem [180] that to implement a generic unitary $U \in U^{2^n}$, the number of single-qubit and CNOT gates required is $\mathcal{O}(n^2 4^n)$. This number is exponential, so the decomposition is inefficient; however, it suffices for our proof.

Figure 4.32 Decomposing a $C^p U$ gate, where $V^2 = U$.

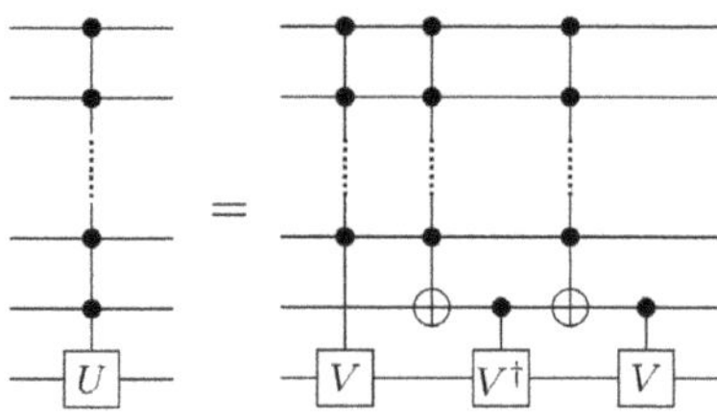

Approximating the Universal Set

The previously established universal set of elementary gates is indeed exact. Nevertheless, this exactness comes at a cost. As mentioned earlier, it leads to inefficient decompositions of large unitary gates. Another limitation is that the universal set comprises uncountably many single-qubit gates. This raises the question: Can we achieve the capability to construct arbitrary unitaries using only a finite, discrete set of single-qubit gates along with the CNOT gate? The answer is affirmative, but only in an approximate sense.

Theorem 4.3. *Solovay–Kitaev theorem: Let $\mathcal{G}$ be a finite set of elements in $SU(2)$ that includes their inverses and generates a dense subgroup of $SU(2)$. For any $\varepsilon > 0$, there exists[6] a constant c such that for any $U \in SU(2)$, there exists a sequence S of gates from $\mathcal{G}$ with length $O\left(\log^c\left(1/\varepsilon\right)\right)$ such that $\|S - U\| \leq \varepsilon$. [181, 182]*

Using this theorem, one can show that a quantum circuit consisting of m constant-qubit gates can be approximated to any accuracy ϵ by

$$\mathcal{O}\left(m \log^c \frac{m}{\epsilon}\right) \tag{4.104}$$

CNOT, Hadamard, and T gates. Here $c \approx 2$. Therefore, an arbitrary n-qubit unitary U can be approximated to any accuracy ϵ by

$$\mathcal{O}\left(n^2 4^n \log^c \frac{n^2 4^n}{\epsilon}\right) \tag{4.105}$$

CNOT, Hadamard, and T gates.

While the exponential scaling $\mathcal{O}(n^2 4^n)$ for decomposing an arbitrary n-qubit unitary seems daunting, this is a worst-case scenario that does not invalidate the efficiency of quantum computation: Practical quantum algorithms utilize structured unitaries that can be compiled into polynomial-sized circuits using advanced techniques that go far beyond the basic Solovay-Kitaev construction. Furthermore, modern compilation strategies, including machine learning and algorithm-specific optimizations, exploit the inherent symmetries and connectivity of problems like those in Shor's or Grover's algorithms (see Chapter 5) to achieve efficient decompositions with a gate count that scales polynomially with the number of qubits, thus preserving the quantum advantage.

---

[6] The value of c does not depend on ε, and in fact can be taken to be $\log_{(1+\sqrt{5})/2} \delta$ for any fixed $\delta > 0$.

4.4 Physical Realization of Quantum Gates

Having understood the quantum circuit model, you might misconceive that a quantum gate, a unitary transformation, can be performed instantly, like a miracle. In reality, a quantum gate is always realized through a unitary time evolution driven by the Hamiltonian of the underlying quantum system that supplies the qubits. As a result, a quantum gate always requires a finite amount of time to complete its operation. In this section, we will briefly explain how a quantum gate is realized through such a Hamiltonian, focusing on a typical scenario where the qubits are represented by spin-half particles, such as electrons and nuclear spins.

4.4.1 Single-Qubit Hamiltonians

A typical single-qubit Hamiltonian reads

$$H = -\mu \left\{ h_0 \sigma_z + h_1 \left[\cos(\omega t + \phi) \sigma_x \right] \right\}, \tag{4.106}$$

where h_0 and h_1 are the strengths of the static and oscillating magnetic fields.

Exercise 4.25 *Solve the Schrödinger equation with the Hamiltonian (4.106), and show that at the resonance frequency $\omega = \omega_0 = -2\mu h_0/\hbar$, the solution becomes*

$$|\psi(t)\rangle = U(t) |\psi(0)\rangle = e^{-i\frac{\Omega t}{2}(\cos\phi\,\sigma_x + \sin\phi\,\sigma_y)} |\psi(0)\rangle$$

$$= \begin{pmatrix} \cos\frac{\Omega t}{2} & -ie^{-i\phi}\sin\frac{\Omega t}{2} \\ -ie^{i\phi}\sin\frac{\Omega t}{2} & \cos\frac{\Omega t}{2} \end{pmatrix} |\psi(0)\rangle, \tag{4.107}$$

where $\Omega = \frac{\mu h_1}{\hbar}$, and the computational basis is assumed. Hint: You may need the rotating-wave approximation.

A **Rabi pulse** is a period τ of the oscillating magnetic field at the resonance frequency $\omega_0 = -2\mu h_0/\hbar$. It's easy to see that Rabi pulses of specific periods serve as specific single-qubit quantum gates. See Examples 4.3 and 4.4.

Example 4.3

To realize a NOT gate, namely the X gate, we can set $\Omega\tau = \pi$ and $\phi = 0$; then

$$U(\tau) = - \begin{pmatrix} 0 & 1 \\ 1 & 0 \end{pmatrix} = -\sigma_x = X.$$

Example 4.4

We need two pulses to realize a Hadamard gate. We set $\Omega\tau = \frac{\pi}{2}$ and $\phi = \frac{\pi}{2}$ for the first pulse, and set $\Omega\tau = \pi$ and $\phi = 0$ for the second pulse. Then,

$$U_{\text{Had}} = \begin{pmatrix} 0 & -i \\ -i & 0 \end{pmatrix} \begin{pmatrix} \frac{1}{\sqrt{2}} & -\frac{1}{\sqrt{2}} \\ \frac{1}{\sqrt{2}} & \frac{1}{\sqrt{2}} \end{pmatrix} = \frac{-i}{\sqrt{2}} \begin{pmatrix} 1 & 1 \\ 1 & -1 \end{pmatrix} \tag{4.108}$$

Exercise 4.26 *How can a Z gate be implemented?*

Exercise 4.27 *Can any rotation gate be realized by Hamiltonian (4.106)?*

4.4.2 Two-Qubit Gates

If a two-qubit gate doesn't generate entanglement, it essentially reduces to two independent single-qubit gates acting on each qubit separately. Therefore, our focus here is on entangling gates. To implement two-qubit entangling gates, such as the *CU* gates, we need a Hamiltonian that couples the two qubits. Typically, this Hamiltonian consists of two parts: a static part that induces coupling between the two qubits and a time-dependent part that enables control in a *CU* gate. In this discussion, we'll primarily address the static coupling part.

The coupling between two qubits is often represented by the term $\sum_i \sigma_i \otimes \sigma_i$, where $i = x, y, z$. For simplicity, we will consider only the Ising-type coupling, namely $\sigma_z \otimes \sigma_z$. The corresponding Hamiltonian, including the single-qubit terms, takes the form

$$H_I = -h_0 \left(\mu_1 \sigma_z^{(1)} + \mu_2 \sigma_z^{(2)} \right) + J \sigma_z^{(1)} \otimes \sigma_z^{(2)}. \tag{4.109}$$

Here, h_0 represents the strength of the static magnetic field, J is the coupling strength, and the superscripts on the Pauli matrices indicate the qubit to which they belong. The energy levels of this Hamiltonian correspond to the computational basis states $|00\rangle$, $|01\rangle$, $|10\rangle$, and $|11\rangle$:

$$\begin{aligned}
E_{00} &= -(\mu_1 + \mu_2)h_0 + J, \quad E_{10} = (\mu_1 - \mu_2)h_0 - J, \\
E_{01} &= -(\mu_1 - \mu_2)h_0 - J, \quad E_{11} = (\mu_1 + \mu_2)h_0 + J.
\end{aligned} \tag{4.110}$$

The energy differences between these levels are crucial for understanding the behavior of two-qubit gates:

$$E_{00} - E_{01} = \begin{cases} -2\mu_2 h_0/\hbar, & J = 0, \\ -2(\mu_2 h_0 + J)/\hbar, & J \neq 0. \end{cases} \tag{4.111}$$

$$E_{10} - E_{11} = \begin{cases} -2\mu_2 h_0/\hbar, & J = 0, \\ -2(\mu_2 h_0 - J)/\hbar, & J \neq 0. \end{cases} \tag{4.112}$$

As observed, when $J \neq 0$, the two energy differences $E_{00} - E_{01}$ and $E_{10} - E_{11}$ can be distinguished, enabling gates like CNOT, as in Example 4.5.

Example 4.5

Suppose we want to realize the CNOT gate that takes the first spin as the control qubit and the second as the data qubit. We can inject a pulse at the frequency $\omega(J) = -2(\mu_2 h_0 - J)$. This pulse will trigger the transition between states $|10\rangle$ and $|11\rangle$, effectively flipping the second qubit while keeping the first qubit in the state $|1\rangle$. This precisely accomplishes the desired CNOT operation on the two spins.

Clearly, when $J = 0$, however, the energy differences become identical, making such gates impossible.

Physical platforms that implement quantum gates, such as NMR [23], ion traps [20], neutral atoms, and optical lattices [22], are subjects of extensive research and practical applications. For a more in-depth exploration of these quantum computing technologies, we recommend consulting the relevant literature.

Exercise 4.28 *Try to find a Hamiltonian of two interacting qubits that can realize a CPHASE gate.*

Conclusion

In this chapter, we explored the quantum circuit model. Beginning with the concept of a qubit, we examined the unique properties of quantum states, including superposition and entanglement, which form the foundation of quantum information processing. The chapter discussed single-qubit states represented on the Bloch sphere and introduced measurement as a probabilistic operation central to quantum mechanics.

Building on this foundation, we studied quantum gates, emphasizing their role as the building blocks of quantum circuits. We focused on essential single-qubit gates, such as the Hadamard and phase-shift gates, and decomposed more complex operations using the universal set of gates. Two-qubit gates, particularly the CNOT gate, were highlighted for their capacity to create entangled states, which distinguish quantum computation from classical approaches.

Finally, we addressed the practical challenges and theoretical considerations in realizing quantum circuits physically, including the decomposition of circuits into basic gates. As we transition to the next chapter, we will expound on the mathematical structure and algorithms that leverage these principles, such as Grover's and Shor's algorithms, exemplifying the power of quantum computation in solving complex problems.

5   Quantum Computation: Algorithms

If the models of (quantum) computation or the physical realization of the models constitute the body of (quantum) computation, then algorithms can be thought of as its soul. Figure 5.1 is adapted from the influential textbook on quantum computation and information by Nielsen and Chuang [13]. This figure portrays the fundamental quantum algorithms and their interrelations with various applications. Even though over two decades have passed since the book's publication, and quantum algorithms have seen considerable evolution, branching out into domains such as quantum simulation, quantum error correction, and quantum machine learning, the foundational aspects illustrated in the figure still hold true. In this chapter, we will expound on two key algorithms in quantum computation: Grover's search algorithm [36] and Shor's algorithm [35] (primarily focusing on its quantum component – the period-finding algorithm). As a primer, we will begin with two fundamental and pioneering quantum algorithms: the Deutsch [34] and Deutsch–Josza algorithms [183]. After this initial exposition, we will explore quantum integral transforms, which form the backbone of both Grover's and Shor's algorithms.

5.1   Deutsch Algorithm

Imagine being given a binary function of a single bit:

$$f : \{0, 1\} \rightarrow \{0, 1\}, \tag{5.1}$$

and tasked with determining whether this function is constant or balanced. For clarity, there are only four possible functions of this kind:

$$\text{Constant:} \quad f_1(0) = f_1(1) = 0; \qquad f_2(0) = f_2(1) = 1; \tag{5.2}$$

$$\text{Balanced:} \quad f_3(0) = 0, f_3(1) = 1; \qquad f_4(0) = 1, f_4(1) = 0. \tag{5.3}$$

At the outset, it's unclear which of these functions you're dealing with. The challenge is to devise an algorithm to discern this for you.

From a classical standpoint, any algorithm you design would necessitate testing the provided function f on both inputs, 0 and 1, to draw a conclusion. In contrast, Deutsch introduced an ingenious quantum algorithm that capitalizes on quantum superposition, enabling the problem to be resolved with just a single evaluation.

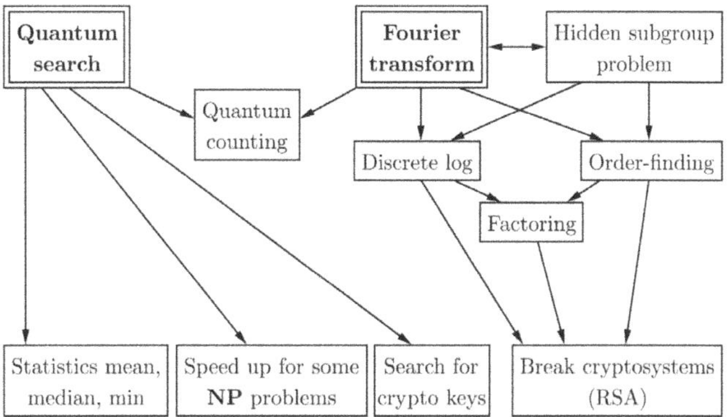

Figure 5.1 Fundamental quantum algorithms and their applications. (Adapted from Figure 4.1 in "Quantum Computation and Quantum Information" by Nielsen and Chuang.)

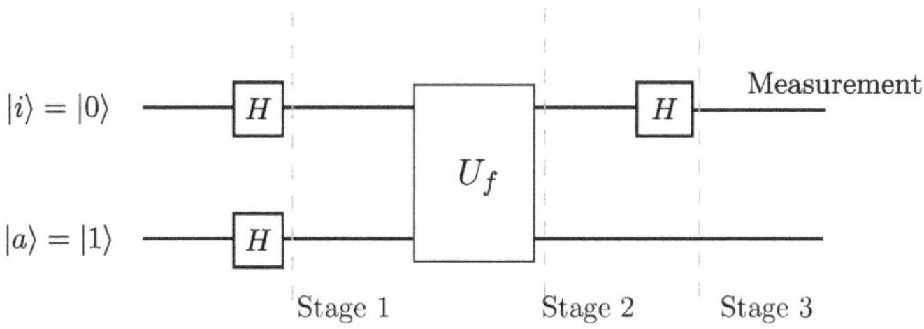

Figure 5.2 Circuit of the Deutsch algorithm.

5.1.1 Deutsch Oracle

In the realm of quantum computation, how do we represent a function? A function f is realized through a unitary transformation (or matrix) U_f, defined as

$$U_f : |\mathbf{i}, \mathbf{a}\rangle \mapsto |\mathbf{i}, \mathbf{a} \oplus f(\mathbf{i})\rangle, \tag{5.4}$$

where $\mathbf{i} = i_{n-1} i_{n-2} \cdots i_0$ represents the input qubit string and $\mathbf{a}$ is an ancillary qubit string.[1] Does U_f evoke thoughts of reversible classical computation? Since f is unknown to you, U_f acts as a black box, which is conventionally termed an **oracle**. Specifically, the U_f used in the Deutsch problem is known as the **Deutsch oracle**.

5.1.2 Circuit of the Algorithm

The Deutsch algorithm that tackles the problem is implemented by the quantum circuit in Figure 5.2:

Let's evaluate this circuit to see whether the Deutsch algorithm truly solves the problem as promised. We shall follow the flow of the qubits along the circuit.

- Initialization: In this case, we have only one input qubit i and an *ancillary qubit* a. The initial state is set to be $|ia\rangle = |01\rangle$.

---

[1] The boldface letters are used to emphasize that they label qubit strings, binary numbers greater than 1, and/or n-dimensional binary vectors, while regular letters label single qubits.

- Stage 1: Walsh–Hadamard transformation:

$$|\psi_1\rangle = H \otimes H \,|01\rangle = \frac{1}{2}(|00\rangle - |01\rangle + |10\rangle - |11\rangle). \tag{5.5}$$

- Stage 2: Oracle action:

$$U_f\,|\psi_1\rangle = \frac{1}{2}\big(|0\rangle\,|0 \oplus f(0)\rangle - |0\rangle\,|1 \oplus f(0)\rangle + |1\rangle\,|0 \oplus f(1)\rangle - |1\rangle\,|1 \oplus f(1)\rangle\big)$$

$$= \frac{1}{2}\left(|0\rangle\,|f(0)\rangle - |0\rangle\,\big|\overline{f(0)}\big\rangle + |1\rangle\,|f(1)\rangle - |1\rangle\,\big|\overline{f(1)}\big\rangle\right)$$

$$= \frac{1}{2}\begin{cases} |0\rangle\,|0\rangle - |0\rangle\,|1\rangle & f(0) = 0 \\ |0\rangle\,|1\rangle - |0\rangle\,|0\rangle & f(0) = 1 \end{cases} + \frac{1}{2}\begin{cases} |1\rangle\,|0\rangle - |1\rangle\,|1\rangle & f(1) = 0 \\ |1\rangle\,|1\rangle - |1\rangle\,|0\rangle & f(1) = 1 \end{cases}$$

$$= \frac{1}{\sqrt{2}}(-1)^{f(0)}\,|0\rangle \otimes \frac{1}{\sqrt{2}}(|0\rangle - |1\rangle) + \frac{1}{\sqrt{2}}(-1)^{f(1)}\,|1\rangle \otimes \frac{1}{\sqrt{2}}(|0\rangle - |1\rangle)$$

$$= \frac{1}{\sqrt{2}}\Big[(-1)^{f(0)}\,|0\rangle + (-1)^{f(1)}\,|1\rangle\Big] \otimes \underbrace{\frac{1}{\sqrt{2}}(|0\rangle - |1\rangle)}_{\text{Now useless}}. \tag{5.6}$$

On the RHS of the fourth equality, the phase factors $(-1)^{f(0)}$ and $(-1)^{f(1)}$ are due to the ancillary qubit a but are in front of the state of qubit i. This phenomenon is referred to as the *phase kick-back* [13, 21, 184]. On the RHS of the last equality above, the state of qubit a is completely uncorrelated with that of qubit i; it is thus useless and can be forgotten in any subsequent computation if necessary, justifying its name – ancillary qubit.

- Stage 3: We can forget about ancillary qubit a from now on. Focusing on qubit i we have

$$H\frac{1}{\sqrt{2}}\Big[(-1)^{f(0)}\,|0\rangle + (-1)^{f(1)}\,|1\rangle\Big]$$

$$= \frac{1}{2}\left\{\Big[(-1)^{f(0)} + (-1)^{f(1)}\Big]\,|0\rangle + \Big[(-1)^{f(0)} - (-1)^{f(1)}\Big]\,|1\rangle\right\}$$

$$= \begin{cases} |0\rangle & f \text{ is constant,} \\ |1\rangle & f \text{ is balanced.} \end{cases} \tag{5.7}$$

- Readout: if the outcome of measuring qubit i is 0 (1) with certainty, f is constant (balanced).

The gist of the Deutsch algorithm is that it calls upon f only once! This is possible only because of superposition.

The Deutsch oracle may appear too dark a black box to you. In fact, the oracle can be further dissected. To find how, let's scrutinize Stage 2 above. We can contrast the state $U_f\,|\psi_1\rangle$ before separating the ancillary qubit a with state $|\psi_1\rangle$ in Stage 1:

$$U_f|\psi_1\rangle \; \propto \; |0\rangle\,|f(0)\rangle - |0\rangle\,|\overline{f(0)}\rangle + |1\rangle\,|f(1)\rangle - |1\rangle\,|\overline{f(1)}\rangle. \tag{5.8}$$

$$|\psi_1\rangle \; \propto \; |0\rangle|0\rangle \quad - \quad |0\rangle|1\rangle \quad + \quad |1\rangle|0\rangle \quad - \quad |1\rangle|1\rangle$$

This comparison indicates that the Deutsch oracle acts like a controlled gate in the following sense: The ancillary qubit state $|a\rangle \xrightarrow{U_f} |\bar{a}\rangle$ if $f(i) = 1$; otherwise, $|a\rangle$

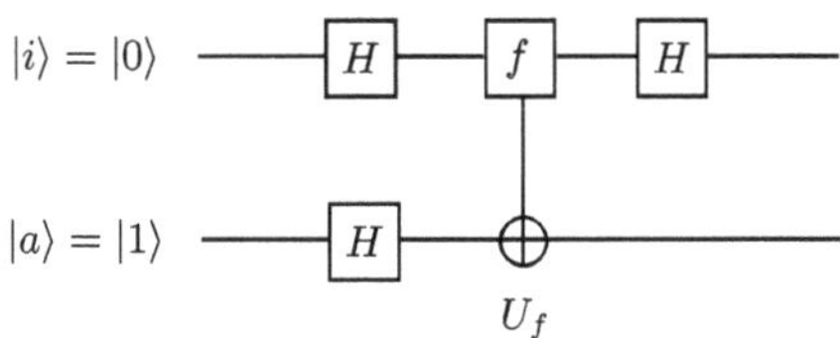

Figure 5.3 Circuit of the Deutsch algorithm with the refined oracle.

remains intact. That is, the ancillary qubit is the data qubit, whereas the control qubit is $f(i)$. We can denote such a controlled gate by C_fNOT gate, depicted as in Figure 5.3, which replaces the Deutsch oracle in Figure 5.2 with the C_fNOT gate.

Well, you might not be impressed so much because after all the Deutsch algorithm handles only functions whose input is a single qubit. But lo and behold, as we now generalize the Deutsch algorithm to deal with functions whose input is an arbitrary qubit string.

5.2 Deutsch–Josza Algorithm

The Deutsch–Josza algorithm generalizes the Deutsch algorithm to solve the Deutsch–Josza problem: whether an n-bit binary function

$$f : s_n = \{0, 1, 2, \ldots, 2^n - 1\} \rightarrow \{0, 1\} \tag{5.9}$$

is constant or balanced; is, whether $f(\mathbf{i})$ is constant for all $\mathbf{i} \in s_n$, or $f(\mathbf{i}) = 0 \, (1)$ for half (the other half) of the input strings in s_n. Note that n can be any nonnegative integer because 2^n is always even.

5.2.1 Circuit of the Algorithm

The circuit of the Deutsch–Josza algorithm should involve $n + 1$ qubits, of which n are the input qubits, denoted by an input string $\mathbf{i}$, and one ancillary qubit a. The refined Deutsch oracle motivates us to conjecture that the **Deutsch–Josza oracle** should be represented by a C_f^nNOT gate, in which the ancillary qubit a is the data qubit, whose final value is controlled by $f(\mathbf{i})$. Figure 5.4 shows the circuit diagram of the algorithm.

Let's verify that with this circuit, the Deutsch–Josza algorithm truly does its job by evaluating the circuit as we did in the case of Deutsch algorithm.

- Initialization: $|\mathbf{i}\rangle |a\rangle = |i_{n-1} i_{n-2} \cdots i_0\rangle |a\rangle = |00 \cdots 0\rangle |1\rangle$.
- Stage 1: Walsh–Hadamard transformation: By Equation (4.67), we have

$$|\psi_1\rangle = \underbrace{H \otimes H \otimes \cdots H \otimes H}_{W_n} |00 \cdots 0\rangle |1\rangle$$

$$= \frac{1}{\sqrt{2^n}} \sum_{i=0}^{2^n - 1} |\mathbf{i}\rangle \otimes \frac{1}{\sqrt{2}} (|0\rangle - |1\rangle). \tag{5.10}$$

- Stage 2: Oracle action:

$$|\psi_2\rangle = U_f |\psi_1\rangle$$

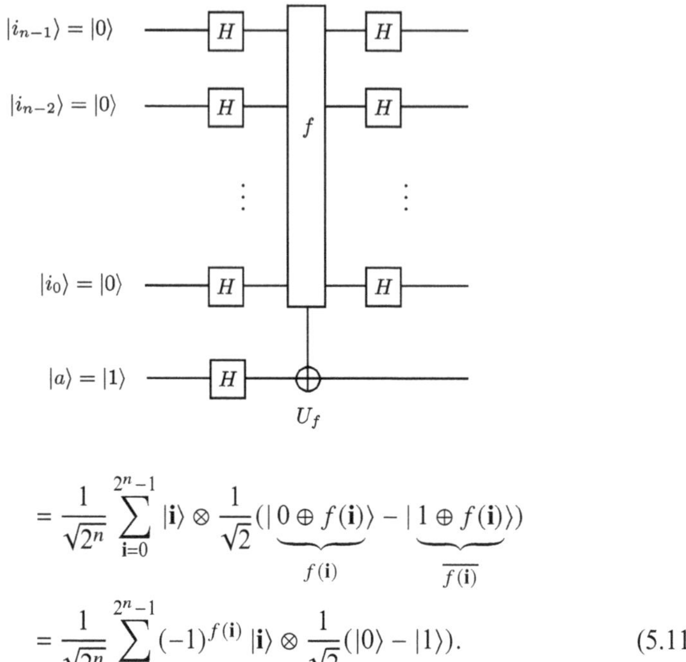

Figure 5.4 Circuit of the Deutsch–Josza algorithm.

$$= \frac{1}{\sqrt{2^n}} \sum_{i=0}^{2^n-1} |\mathbf{i}\rangle \otimes \frac{1}{\sqrt{2}} (|\underbrace{0 \oplus f(\mathbf{i})}_{f(\mathbf{i})}\rangle - |\underbrace{1 \oplus f(\mathbf{i})}_{\overline{f(\mathbf{i})}}\rangle)$$

$$= \frac{1}{\sqrt{2^n}} \sum_{i=0}^{2^n-1} (-1)^{f(\mathbf{i})} |\mathbf{i}\rangle \otimes \frac{1}{\sqrt{2}} (|0\rangle - |1\rangle). \tag{5.11}$$

The phase kick-back is evident.

- Stage 3: By Equation (4.61),

$$|\psi_3\rangle = W_n \otimes \mathbb{1}_2 |\psi_2\rangle$$

$$= \frac{1}{\sqrt{2^n}} \sum_{i=0}^{2^n-1} (-1)^{f(\mathbf{i})} W_n |\mathbf{i}\rangle \otimes \frac{1}{\sqrt{2}} (|0\rangle - |1\rangle)$$

$$= \frac{1}{\sqrt{2^n}} \sum_{i=0}^{2^n-1} (-1)^{f(\mathbf{i})} H |i_{n-1}\rangle \otimes H |i_{n-2}\rangle \otimes \cdots \otimes H |i_0\rangle \otimes \frac{1}{\sqrt{2}} (|0\rangle - |1\rangle)$$

$$= \frac{1}{\sqrt{2^n}} \sum_{i=0}^{2^n-1} (-1)^{f(\mathbf{i})} \frac{1}{\sqrt{2^n}} \left[\bigotimes_{k=n-1}^{0} \sum_{i'_k=0}^{1} (-1)^{i_k i'_k} |i'_k\rangle \right] \otimes \frac{1}{\sqrt{2}} (|0\rangle - |1\rangle)$$

$$= \frac{1}{2^n} \sum_{i=0}^{2^n-1} \sum_{i'=0}^{2^n-1} (-1)^{f(\mathbf{i})} (-1)^{\mathbf{i} \cdot \mathbf{i}'} |\mathbf{i}'\rangle \otimes \underbrace{\frac{1}{\sqrt{2}} (|0\rangle - |1\rangle)}_{\text{Useless}}. \tag{5.12}$$

In the last equality above, we regard the bit strings $\mathbf{i}$ and $\mathbf{i}'$ as n-dimensional binary vectors, so

$$\mathbf{i} \cdot \mathbf{i}' = i_{n-1} i'_{n-1} + i_{n-2} i'_{n-2} + \cdots + i_0 i'_0. \tag{5.13}$$

Again, the ancillary qubit is completely separated out.

- Readout: The result can be read out by measuring the qubit string $\mathbf{i}$, namely the top n qubits.

– If f is constant, $f(\mathbf{i}) \equiv f(\mathbf{0})\ \forall\mathbf{i}$, the post-measurement state would be

$$\frac{(-1)^{f(\mathbf{0})}}{2^n} \sum_{\mathbf{i}=0}^{2^n-1} \sum_{\mathbf{i}'=0}^{2^n-1} (-1)^{\mathbf{i}\cdot\mathbf{i}'} |\mathbf{i}'\rangle$$

$$= \frac{(-1)^{f(\mathbf{0})}}{2^n} \sum_{\mathbf{i}'=0}^{2^n-1} \underbrace{\sum_{\mathbf{i}=0}^{2^n-1} e^{\mathring{\mathbf{i}}\frac{2\pi\mathbf{i}\cdot\mathbf{i}'}{2}}}_{\delta_{\mathbf{i}'0}} |\mathbf{i}'\rangle$$

$$\propto |00\cdots0\rangle, \tag{5.14}$$

where the last summand vanishes if $\mathbf{i}' \neq \mathbf{0}$.

– If f is balanced, the probability amplitude of the outcome $\mathbf{i}' = \mathbf{0}$ of the top n qubits is

$$\frac{1}{2^n} \sum_{\mathbf{i}=0}^{2^n-1} (-1)^{f(\mathbf{i})}(-1)^{\mathbf{i}\cdot\mathbf{0}} \equiv 0 \tag{5.15}$$

because the signs $(-1)^{f(\mathbf{i})}$ cancel out exactly.

Therefore, upon measuring the top n qubits sufficiently many times, if they are found in state $|00\cdots0\rangle$ with certainty, f is constant; otherwise, f is balanced.

So neat, this algorithm! The moral is again that the algorithm calls upon the oracle U_f only once (once and for all) owing to quantum superposition. In contrast, a classical algorithm targeting the Deutsch–Josza problem would have to call upon the function f at least $n/2$ times to find the answer.

Exercise 5.1 *Simon's algorithm [185]: Consider a function $f : \{0,1\}^n \rightarrow \{0,1\}^n$ with the following property: There exists a sequence $s \in \{0,1\}^n$, such that $s \neq 0^n$, and for any $x, y \in \{0,1\}^n$, $f(x) = f(y)$ if and only if $x = y$ or $x \oplus y = s$. Here, $\oplus$ represents the bitwise XOR operation.*

Try to devise a quantum algorithm to identify the sequence s. We have a black box operator U_f designed with $2n$ input qubits and $2n$ output qubits. Given an input state

$$|k_1 k_2 \cdots k_n\rangle \otimes |0\rangle^{\otimes n}, \tag{5.16}$$

the black box transforms it as follows:

$$U_f \left(|k_1 k_2 \cdots k_n\rangle \otimes |0\rangle^{\otimes n}\right) = |k_1 k_2 \cdots k_n\rangle \otimes |f(k_1, k_2, \cdots, k_n)\rangle. \tag{5.17}$$

5.3 Quantum Integral Transforms

The Deutsch and Deutsch–Josza algorithms serve as appetizers, setting the stage for the entrées of quantum algorithms that exemplify the computational prowess of quantum mechanics. Central to these advanced algorithms is an elementary form of unitary transformation known as a quantum integral transform, the focus of this section.

5.3.1 Discrete Integral Transforms

Let $n \in \mathbb{N}$, $N = 2^n$, and $s_n = \{0, 1, 2, \ldots, N - 1\}$. Given any function $f : s_n \to \mathbb{C}$, its **discrete integral transform (DIT)** $\tilde{f} : s_n \to \mathbb{C}$ is defined by

$$\tilde{f}(\mathbf{j}) = \sum_{\mathbf{i}=0}^{N-1} K(\mathbf{j}, \mathbf{i}) f(\mathbf{i}). \tag{5.18}$$

Here, $K : s_n \times s_n \to \mathbb{C}$ is the **integral kernel**, whose precisely form depends on the specific function f being transformed. Since s_n is a discrete set, both f and $\tilde{f}$ can be regarded as N-dimensional vectors:

$$f = \begin{pmatrix} f(0) \\ f(1) \\ \vdots \\ f(N-1) \end{pmatrix}, \quad \tilde{f} = \begin{pmatrix} \tilde{f}(0) \\ \tilde{f}(1) \\ \vdots \\ \tilde{f}(N-1) \end{pmatrix}. \tag{5.19}$$

So, the kernel K is an $N \times N$ complex matrix with entries $K_{\mathbf{ji}} = K(\mathbf{j}, \mathbf{i})$. The DIT (5.18) then takes the form of a matrix product:

$$\tilde{f} = K f. \tag{5.20}$$

If K is unitary,

$$f = K^{-1} \tilde{f} = K^{\dagger} \tilde{f} \tag{5.21}$$

is the corresponding inverse DIT.

5.3.2 Quantum Computation of the DIT

The DIT described above is purely classical. We can render it quantum in the following sense. We first represent each element in s_n by an n-qubit string $\mathbf{i} = i_{n-1} i_{n-2} \cdots i_0$, such that the states $\{|\mathbf{i}\rangle \, | \, \mathbf{i} \in s_n\}$ form a basis of this N-dimensional Hilbert space. Any unitary operator U is represented in this basis as

$$U |\mathbf{i}\rangle = \sum_{j=0}^{N-1} |\mathbf{j}\rangle \underbrace{\langle \mathbf{j}|U|\mathbf{i}\rangle}_{U_{\mathbf{ji}}} = \sum_{j=0}^{N-1} U_{\mathbf{ji}} |\mathbf{j}\rangle . \tag{5.22}$$

Given any DIT with kernel K, there exists a unitary operator U, with $U_{\mathbf{ji}} = K_{\mathbf{ji}}$, such that by Equations (5.18) and (5.22),

$$U \sum_{\mathbf{i}=0}^{N-1} f(\mathbf{i}) |\mathbf{i}\rangle = \sum_{\mathbf{i}=0}^{N-1} f(\mathbf{i}) U |\mathbf{i}\rangle$$

$$= \sum_{i=0}^{N-1} f(\mathbf{i}) \sum_{j=0}^{N-1} K_{\mathbf{ji}} |\mathbf{j}\rangle$$

$$= \sum_{j=0}^{N-1} \sum_{i=0}^{N-1} K_{\mathbf{ji}} f(\mathbf{i}) |\mathbf{j}\rangle$$

$$= \sum_{j=0}^{N-1} \tilde{f}(\mathbf{j}) |\mathbf{j}\rangle . \tag{5.23}$$

This unitary operator U is said to compute the corresponding DIT. Such a unitary matrix U is also termed a **quantum integral transform (QIT)**.

> **Exercise 5.2** Let $\tilde{f} = Kf$ be a DIT with unitary kernel K. Prove Parseval's theorem:
>
> $$\sum_{i=0}^{N-1} |f(\mathbf{i})|^2 = \sum_{j=0}^{N-1} |\tilde{f}(\mathbf{j})|^2. \tag{5.24}$$

Let's take a look at a simple case of a QIT, Example 5.1.

Example 5.1

The n-qubit Walsh–Hadamard transformation (4.67) is in fact a QIT with the kernel

$$W_n(\mathbf{j}, \mathbf{i}) = \frac{1}{\sqrt{N}}(-1)^{\mathbf{j}\cdot\mathbf{i}}, \tag{5.25}$$

where $\mathbf{i}, \mathbf{j} \in s_n$ are regarded as two n-dimensional binary vectors, such that

$$U_{W_n} |\mathbf{i}\rangle = \frac{1}{\sqrt{N}} \sum_{j=0}^{N-1} (-1)^{\mathbf{j}\cdot\mathbf{i}} |\mathbf{j}\rangle , \tag{5.26}$$

where U_{W_n} is the unitary matrix identified with W_n.

5.3.3 Quantum Fourier Transforms

A **quantum Fourier transform** [186] **(QFT)** is a particular QIT that has a variety of applications in quantum algorithms, such as Shor's factorization algorithm. A QFT computes a **discrete Fourier transform (DFT)**. A DFT is a DIT:

$$\tilde{f}(\mathbf{j}) = \frac{1}{\sqrt{N}} \sum_{i=0}^{N-1} e^{-\hat{\imath}\frac{2\pi\mathbf{i}\mathbf{j}}{N}} f(\mathbf{i}) \tag{5.27}$$

with the kernel

$$K(\mathbf{j}, \mathbf{i}) = \frac{1}{\sqrt{N}} e^{-\hat{\imath}\frac{2\pi\mathbf{i}\mathbf{j}}{N}}, \tag{5.28}$$

where $\mathbf{i}, \mathbf{j} \in s_n$ are regarded as two nonnegative integers. This kernel is unitary, $KK^\dagger = \mathbb{1}_N$, because

$$\sum_{i=0}^{N-1} K(\mathbf{j}, \mathbf{i}) K(\mathbf{i}, \mathbf{j}')^* = \frac{1}{N} \sum_{i=0}^{N-1} e^{-\hat{\mathbb{i}}\frac{2\pi\mathbf{i}(\mathbf{j}-\mathbf{j}')}{N}} = \delta_{\mathbf{j}\mathbf{j}'}. \tag{5.29}$$

The QFT of this DFT is an n-qubit unitary operator U_n, such that

$$U_n |\mathbf{i}\rangle = \sum_{j=0}^{N-1} K(\mathbf{j}, \mathbf{i}) |\mathbf{j}\rangle = \frac{1}{\sqrt{N}} \sum_{j=0}^{N-1} e^{-\hat{\mathbb{i}}\frac{2\pi\mathbf{i}\mathbf{j}}{N}} |\mathbf{j}\rangle. \tag{5.30}$$

Example 5.2

Consider $n = 1$, so $N = 2$. The corresponding QFT U_1 reads

$$U_1 |\mathbf{i}\rangle = \frac{1}{\sqrt{2}} \sum_{j=0}^{1} e^{-\hat{\mathbb{i}}\pi\mathbf{i}\mathbf{j}} |\mathbf{j}\rangle. \tag{5.31}$$

See what this U_1 is? It is the Hadamard gate! Recall Equation (4.61) if you forgot.
Consider arbitrary n but let $\mathbf{i} = \mathbf{0}$, then

$$U_n |\mathbf{0}\rangle = \frac{1}{\sqrt{N}} \sum_{j=0}^{N-1} |\mathbf{j}\rangle, \tag{5.32}$$

which is precisely what a Walsh–Hadamard transformation does on $|\mathbf{0}\rangle$. See Equation (5.26). Note however that U_n is not the Walsh–Hadamard transformation.

To emphasize again, a generic n-qubit state $|\psi\rangle$ is a superposition of the basis states $\{|\mathbf{i}\rangle\}$, so a QFT U_n acts simultaneously on all the $|\mathbf{i}\rangle$'s. This is a great advantage of quantum computation over classical computation. Nevertheless, this advantage may sometimes engender a challenge: In solving certain problems, the solution may lie in only one of the terms in the output superposed state; however, we don't have a choice of which terms are to be observed in measurements, as they appear per their probabilities. This challenge can be overcome by a pertinent QIT that can amplify the probability of the term that encodes the problem's solution to be much larger than those of all the other terms in the superposition. We shall encounter such a situation later, in Section 5.6, which discusses Grover's search algorithm.

5.3.4 QFT Circuits

Having understood the mathematics of QFTs, it's time to implement QFTs on quantum circuits. Let's begin with the simplest case, $n = 1$, and then increase the number of qubits to 2 and 3 to build up our intuition for the case of arbitrary n.

When $n = 1$, as we saw in Example 5.2, the QFT U_n is simply the Hadamard gate, so there is no need for a dedicated circuit.

Two-Qubit QFT

To design the circuit for a two-qubit QFT, so that $n = 2$, we shall examine in detail how U_2 acts on two-qubit state basis states $|i_1 i_0\rangle$ to decompose U_2 into a composition of elementary gates. Setting $n = 2$ in Equation (5.30) leads to

$$U_2\,|i_1 i_0\rangle = \frac{1}{2} \sum_{j_1, j_0 = 0}^{1} e^{-\mathrm{i}\frac{2\pi \mathrm{i}(2j_1 + j_0)}{4}}\,|j_1 j_0\rangle, \tag{5.33}$$

where $\mathbf{i} = i_1 i_0$ and $\mathbf{j} = j_1 j_0$ are bit strings, while $\mathbf{i} = 2i_1 + i_0$ and $\mathbf{j} = 2j_1 + j_0$ represent integers. Let's rewrite the above action step by step:

$$
\begin{aligned}
U_2\,|i_1 i_0\rangle &= \frac{1}{2} \sum_{j_1=0}^{1} e^{-\mathrm{i}\frac{2\pi \mathrm{i} j_1}{2}}\,|j_1\rangle \otimes \sum_{j_0=0}^{1} e^{-\mathrm{i}\frac{2\pi \mathrm{i} j_0}{2^2}}\,|j_0\rangle \\[2mm]
&= \frac{1}{2}\left(|0\rangle + e^{-\mathrm{i}\frac{2\pi(2i_1 + i_0)}{2}}\,|1\rangle\right) \otimes \left(|0\rangle + e^{-\mathrm{i}\frac{2\pi(2i_1 + i_0)}{2^2}}\,|1\rangle\right) \\[2mm]
&= \frac{1}{2}\left(|0\rangle + e^{-\mathrm{i}\frac{2\pi i_0}{2}}\,|1\rangle\right) \otimes \left(|0\rangle + e^{-\mathrm{i}\frac{2\pi(2i_1 + i_0)}{2^2}}\,|1\rangle\right) \\[2mm]
&= \frac{1}{\sqrt{2}}\,\frac{1}{\sqrt{2}}\,\underbrace{\left(|0\rangle + (-1)^{i_0}\,|1\rangle\right)}_{H|i_0\rangle} \otimes \left(|0\rangle + e^{-\mathrm{i}\frac{2\pi i_0}{2^2}}(-1)^{i_1}\,|1\rangle\right) \\[2mm]
&= \frac{1}{\sqrt{2}}(H \otimes \mathbb{1}_2)\left[\,|i_0\rangle \otimes \left(|0\rangle + \underbrace{e^{-\mathrm{i}\frac{2\pi i_0}{2^2}}}_{\parallel}(-1)^{i_1}\,|1\rangle\right)\right]
\end{aligned} \tag{5.34}
$$

$$1, i_0 = 0;\ \ \phi_{01} := \tfrac{\pi}{2}, i_0 = 1$$

$$
\begin{aligned}
&= \underbrace{(H \otimes \mathbb{1}_2)\,C\Phi^{i_0}(\phi_{01})(\mathbb{1}_2 \otimes H)\,|i_0\rangle\,|i_1\rangle}_{C\Phi^{i_0}(\phi_{01})(\mathbb{1}_2 \otimes H)|i_0\rangle|i_1\rangle} \\[2mm]
&= (H \otimes \mathbb{1}_2)\,C\Phi^{i_0}(\phi_{01})(\mathbb{1}_2 \otimes H)\,\underbrace{|i_0\rangle\,|i_1\rangle}_{\text{SWAP}|i_1\rangle|i_0\rangle} \\[2mm]
&= (H \otimes \mathbb{1}_2)\,C\Phi^{i_0}(\phi_{01})(\mathbb{1}_2 \otimes H)\,\text{SWAP}\,|i_1 i_0\rangle. \tag{5.35}
\end{aligned}
$$

In the derivation above, we defined a CPHASE gate $C\Phi^{i_0}(\phi_{01})$, which generates a relative phase $\exp(-\mathrm{i}\phi_{01})$ on qubit i_1 if the control qubit i_0 is in state $|1\rangle$. The rotation angle is denoted that way to be generalized to cases of $n > 2$:

$$\phi_{pq} = \frac{2\pi}{2^{q-p+1}}, \quad 0 \le p < q \le n - 1, \tag{5.36}$$

which agrees with the ϕ_{01} when $p = 0$ and $q = 1$. The CPHASE gate $C\Phi^{i_p}(\phi_{pq})$ is defined accordingly. The expression (5.35) for U_2 manifests the circuit of the two-qubit QFT, as depicted in Figure 5.5.

Exercise 5.3 *Show that collectively we can write*

$$C\Phi^{i_p}(\phi_{pq})\,|i_p i_q\rangle = e^{-\mathrm{i}\phi_{pq} i_p i_q}\,|i_p i_q\rangle. \tag{5.37}$$

$$
\mathrm{C}\Phi^{i_p}(\phi_{pq})\left|i_p i_q\right\rangle = \mathrm{C}\Phi^{i_q}(\phi_{pq})\left|i_p i_q\right\rangle .
$$

A consequence is the symmetry

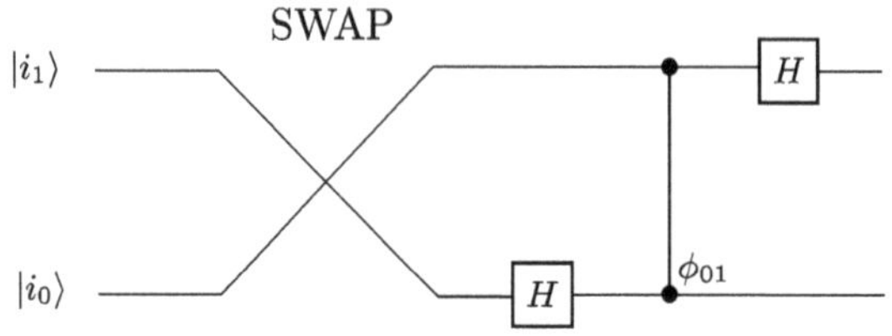

Figure 5.5 The quantum circuit for a two-qubit QFT. $|i_1\rangle$ SWAP H $|i_0\rangle$ H ϕ_{01}

Note that $\mathrm{C}\Phi^{i_p}(\phi_{pq})$ is a two-qubit gate acting on $\left|i_p i_q\right\rangle$; however, for convenience in later derivations, the exercise above motivates us to define a single-qubit gate phase-shift operator $\Phi^{i_p}(\phi_{pq})$ that acts on $\left|i_q\right\rangle$ only as

$$
\Phi^{i_p}(\phi_{pq})\left|i_q\right\rangle = \mathrm{e}^{-\hat{\imath}\phi_{pq}i_p i_q}\left|i_q\right\rangle, \tag{5.38}
$$

such that

$$
\mathrm{C}\Phi^{i_p}(\phi_{pq})\left|i_p i_q\right\rangle = \left|i_p\right\rangle \otimes \Phi^{i_p}(\phi_{pq})\left|i_q\right\rangle. \tag{5.39}
$$

We don't refer to $\Phi^{i_p}(\phi_{pq})$ as a gate but rather as an operator because it depends on the state $\left|i_p\right\rangle$ and is not really a single-qubit gate. It is simply a handy expression for $\mathrm{C}\Phi^{i_p}(\phi_{pq})$ that allows us to extract the effect of $\mathrm{C}\Phi^{i_p}(\phi_{pq})$ on $\left|i_q\right\rangle$ in derivations.

Three-Qubit QFTs

Now we move on to the case of $n = 3$. The second RHS of Equation (5.34) prompts us to directly express $U_3\left|i_2 i_1 i_0\right\rangle$ as follows and continue the derivation:

$$
\begin{aligned}
U_3 &\left|i_2 i_1 i_0\right\rangle \\
&= \frac{1}{\sqrt{2^3}}\left[|0\rangle + \mathrm{e}^{-\hat{\imath}\frac{2\pi i_0}{2}}|1\rangle\right] \otimes \left[|0\rangle + \mathrm{e}^{-\hat{\imath}2\pi\left(\frac{i_1}{2}+\frac{i_0}{2^2}\right)}|1\rangle\right] \otimes \left[|0\rangle + \mathrm{e}^{-\hat{\imath}2\pi\left(\frac{i_2}{2}+\frac{i_1}{2^2}+\frac{i_0}{2^3}\right)}|1\rangle\right] \\
&= \frac{1}{\sqrt{2^3}}\left[|0\rangle + (-1)^{i_0}|1\rangle\right] \otimes \Phi^{i_0}(\phi_{01})\left[|0\rangle + (-1)^{i_1}|1\rangle\right] \\
&\quad \otimes \Phi^{i_0}(\phi_{02})\Phi^{i_1}(\phi_{12})\left[|0\rangle + (-1)^{i_2}|1\rangle\right] \\
&= H\left|i_0\right\rangle \otimes \Phi^{i_0}(\phi_{01})H\left|i_1\right\rangle \otimes \Phi^{i_0}(\phi_{02})\Phi^{i_1}(\phi_{12})H\left|i_2\right\rangle \\
&= (H \otimes \mathbb{1}_2 \otimes \mathbb{1}_2)\left(\left|i_0\right\rangle \otimes \Phi^{i_0}(\phi_{01})H\left|i_1\right\rangle \otimes \Phi^{i_0}(\phi_{02})\Phi^{i_1}(\phi_{12})H\left|i_2\right\rangle\right) \\
&= (H \otimes \mathbb{1}_2 \otimes \mathbb{1}_2)\mathrm{C}\Phi^{i_0}(\phi_{01})(\mathbb{1}_2 \otimes H \otimes \mathbb{1}_2)\mathrm{C}\Phi^{i_0}(\phi_{02})\mathrm{C}\Phi^{i_1}(\phi_{12}) \\
&\quad \times (\mathbb{1}_2 \otimes \mathbb{1}_2 \otimes H)\left|i_0 i_1 i_2\right\rangle \\
&= (H \otimes \mathbb{1}_2 \otimes \mathbb{1}_2)\mathrm{C}\Phi^{i_0}(\phi_{01})(\mathbb{1}_2 \otimes H \otimes \mathbb{1}_2)\mathrm{C}\Phi^{i_0}(\phi_{02})\mathrm{C}\Phi^{i_1}(\phi_{12}) \\
&\quad \times (\mathbb{1}_2 \otimes \mathbb{1}_2 \otimes H)P_3\left|i_2 i_1 i_0\right\rangle, \tag{5.40}
\end{aligned}
$$

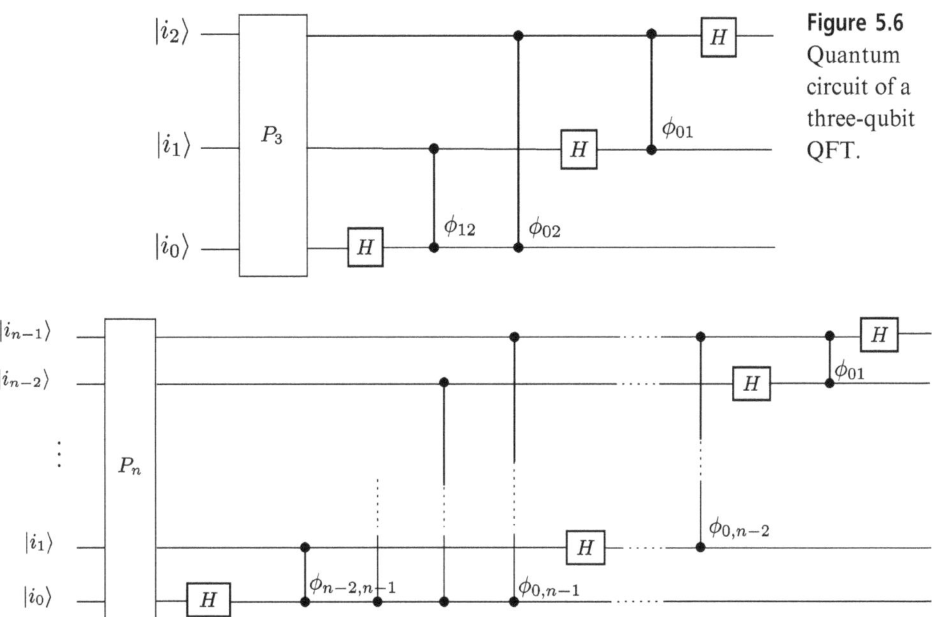

Figure 5.6 Quantum circuit of a three-qubit QFT.

Figure 5.7 The quantum circuit of n-qubit QFT.

where Equations (5.38) and (5.39) are used to get the second equality, and a permutation gate P_3 is defined in the last equality. The gate P_3 reverses the ordering of the three qubits and is realized by a single SWAP gate on qubits i_2 and i_0. This result can immediately be depicted as the circuit in Figure 5.6.

n-Qubit QFT

With our experience of building the circuits of two-qubit and three-qubit QFTs, we can easily write down the decomposition of a generic n-qubit QFT U_n:

$$U_n |i_{n-1}i_{n-2}\cdots i_0\rangle = H_0 \left\{ \overrightarrow{\prod_{r=1}^{n-1}} \left[\overrightarrow{\prod_{m=0}^{r-1}} C\Phi^{i_m}(\phi_{mr}) H_r \right] \right\} P_n |i_{n-1}i_{n-2}\cdots i_0\rangle, \qquad (5.41)$$

where

$$H_r := \underbrace{\mathbb{1}_2 \otimes \cdots \otimes \mathbb{1}_2}_{r-1} \otimes H \otimes \underbrace{\mathbb{1}_2 \otimes \cdots \otimes \mathbb{1}_2}_{n-r}. \qquad (5.42)$$

We also defined the ordered product $\overrightarrow{\prod}$ of gates, in which the gates are ordered from left to right with increasing indices. See the RHS of Equation (5.40) for an example. The permutation gate P_n reverses the order of the n input qubits and can be realized by $\lfloor \frac{n}{2} \rfloor$ SWAP gates. Figure 5.7 depicts the circuit diagram corresponding to Equation (5.41).

Efficiency of QFT

Despite a formidable looking circuit, QFT is an efficient algorithm. Let's analyze. An n-qubit QFT requires a P_n gate, n Hadamard gates, which are already elementary, and $n(n-1)/2$ $C\Phi_{pq}$ gates.

- A P_n gate $= \lfloor \frac{n}{2} \rfloor$ SWAP gates. But by Figure 4.17, a SWAP gate $=$ three CNOT gates. Hence,

$$P_n = \Theta(n) \text{ CNOTs.} \tag{5.43}$$

Here $\Theta(n)$ is a tight bound.[2]
- Recall Figure 4.24 showing that any CU gate can be decomposed into at most six elementary gates? So,

$$\frac{n(n-1)}{2} C\Phi_{pq} = \Theta(n^2) \text{ elementary gates.} \tag{5.44}$$

Therefore, the complexity of an n-qubit QFT is $\Theta(n^2)$, which is quadratic in n. The efficiency of QFT is mainly due to quantum superposition and entanglement. While the role of superposition is obvious and has been repeatedly emphasized, the role of entanglement is more subtle. The next section will discuss a quantum algorithm – period finding, where QFT plays is the crux component. Period finding showcases the role of quantum entanglement in increasing the efficiency.

5.4 Period Finding

To pave the way for understanding Shor's algorithm, it is imperative to explore period finding. Period finding as a quantum algorithm is the quintessential element of Shor's algorithm, representing its sole quantum component. Moreover, this algorithm stands as a testament to the capabilities of quantum Fourier transforms (QFTs) and finds utility in diverse applications beyond Shor's algorithm, such as:

- **Simon's algorithm**: Preceding many quantum algorithms, Simon's algorithm (recall Exercise 5.1) is adept at swiftly resolving a black-box problem by unearthing a concealed string. The crux of its efficiency lies in the adept use of period finding.
- **Hidden subgroup problem (HSP)** [187]: Both Shor's and Simon's algorithms can be categorized as instances of the HSP, which aims to discern a concealed subgroup within a known group via a black-box function. The incorporation of period finding is pivotal in addressing the HSP across diverse groups.

---

[2] The Θ notation represents both an upper and lower bound on the growth rate of a function. Specifically, if $T(n)$ is $\Theta(f(n))$, it means there exist constants $c_1, c_2 > 0$ and a value n_0 such that for all $n \geq n_0$: $c_1 \cdot f(n) \leq T(n) \leq c_2 \cdot f(n)$. In other words, $T(n)$ grows asymptotically at least as fast as $f(n)$ and at most as fast as $f(n)$. This is in contrast to $\mathcal{O}(f(n))$, which only provides an upper bound, meaning $T(n)$ grows at most as fast as $f(n)$.

- **Quantum algorithms for the discrete logarithm problem** [184]: Mirroring Shor's algorithm, these algorithms capitalize on period finding to resolve the discrete logarithm problem within polynomial time.
- **Certain quantum cryptanalysis techniques** [188]: A subset of quantum cryptanalysis methodologies harness the power of period finding to probe cryptographic constructs. While Shor's algorithm poses a threat to RSA encryption due to its ability to factorize efficiently, an array of cryptographic protocols might be vulnerable to quantum onslaughts, particularly those orchestrated using period-finding tactics.

The challenge of period finding revolves around identifying the period of a function $f : \mathbb{N} \to \mathbb{N}$, which translates to discerning a $\mathfrak{p} < N \in \mathbb{N}$ such that $f(x+\mathfrak{p}) = f(x)$ for all x. In a classical computation, roughly $\mathcal{O}(N) = \mathcal{O}(2^n)$ operations are required to pinpoint the period, translating to exponential complexity[3] relative to the input magnitude n. As we now explore, the QFT proves instrumental in accelerating period finding, metamorphosing it into a potent quantum algorithm.

To proceed, it's essential to confine the functions under consideration. Given a finite number of qubits, denoted by n, we focus on functions $f : s_n \to s_n$, where s_n retains its previous definition. The period of f is a natural number $\mathfrak{p}$ such that $\mathfrak{p} < N - 1 = 2^n - 1$. Initially, we'll concentrate on the instance where $n = 3$ and subsequently address broader scenarios. Why $n = 3$? It is the most rudimentary non-trivial example, highlighting the efficacy of QFT. When $n = 1$, a periodic function is essentially constant. As for $n = 2$? Consider this an engaging exercise for you.

Additionally, it's beneficial to introduce a handy terminology tailored to our context. A collection of input qubits is frequently termed an n-qubit **quantum register** or simply an n-qubit register. Given that in quantum computation a function f translates to an oracle U_f (as defined in Equation (5.4)), period finding necessitates two n-qubit registers. The premier register accommodates the n input qubits, and in a quantum circuit, it's conventionally positioned above the secondary register, which houses the n ancillary qubits.

5.4.1 Case of $n = 3$

In this case, $N = 8$ and $s_3 = \{0, 1, 2, \ldots, 7\}$. The oracle encodes a function $f : s_3 \to s_3$. We need two three-qubit registers $|\mathbf{i}\rangle\,|\mathbf{a}\rangle = |i_2 i_1 i_0\rangle\,|a_2 a_1 a_0\rangle$. The circuit in Figure 5.8 delineates the algorithm.

Let's evaluate the circuit to see how and why it works as anticipated.

- Initialization: $|\mathbf{i}\rangle\,|\mathbf{a}\rangle = |\mathbf{0}\rangle\,|\mathbf{0}\rangle$.
- Walsh–Hadamard transformation of register 1:

$$W_3 \otimes \mathbb{1}_8\,|\mathbf{0}\rangle\,|\mathbf{0}\rangle = \frac{1}{\sqrt{8}} \sum_{i=0}^{7} |\mathbf{i}\rangle\,|\mathbf{0}\rangle. \tag{5.45}$$

---

[3] While marginally more proficient classical algorithms are available, their efficiency remains bounded within the exponential realm.

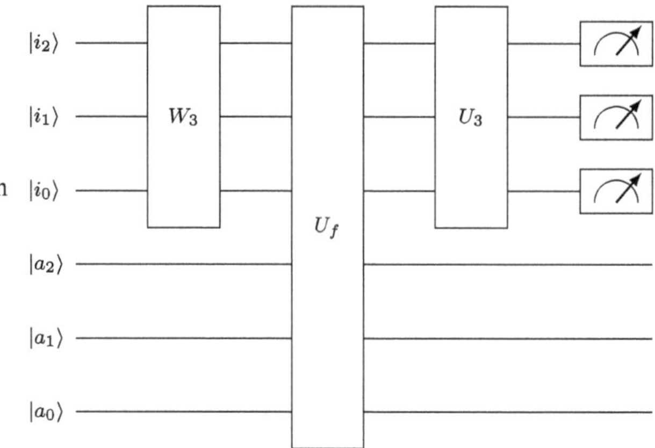

Figure 5.8 Circuit of period finding with $n = 3$. The three-qubit Walsh–Hadamard transformation, oracle, and three-qubit quantum Fourier transform are respectively labeled by W_3, U_f, and U_3.

- Oracle action:

$$|\psi\rangle = U_f \frac{1}{\sqrt{8}} \sum_{\mathbf{i}=0}^{7} |\mathbf{i}\rangle\, |\mathbf{0}\rangle = \frac{1}{\sqrt{8}} \sum_{\mathbf{i}=0}^{7} |\mathbf{i}\rangle\, |f(\mathbf{i})\rangle. \tag{5.46}$$

- Three-qubit QFT on register 1:

$$
\begin{aligned}
|\psi'\rangle = U_3\,|\psi\rangle &= \frac{1}{8} \sum_{\mathbf{i},\mathbf{j}=0}^{7} \mathrm{e}^{-\hat{\imath}\frac{2\pi \mathbf{i}\mathbf{j}}{8}}\, |\mathbf{j}\rangle\, |f(\mathbf{i})\rangle \\
&= \frac{1}{8}\,|\mathbf{0}\rangle\,(|f(\mathbf{0})\rangle + |f(\mathbf{1})\rangle + \cdots + |f(\mathbf{7})\rangle) \\
&\quad + \frac{1}{8}\,|\mathbf{1}\rangle\,(|f(\mathbf{0})\rangle + \mathrm{e}^{-\hat{\imath}\frac{\pi}{4}}\,|f(\mathbf{1})\rangle + \cdots + \mathrm{e}^{-\hat{\imath}\frac{7\pi}{4}}\,|f(\mathbf{7})\rangle) \\
&\quad + \cdots \\
&\quad + \frac{1}{8}\,|\mathbf{7}\rangle\,(|f(\mathbf{0})\rangle + \mathrm{e}^{-\hat{\imath}\frac{7\pi}{4}}\,|f(\mathbf{1})\rangle + \cdots + \mathrm{e}^{-\hat{\imath}\frac{7\cdot7\pi}{4}}\,|f(\mathbf{7})\rangle). \tag{5.47}
\end{aligned}
$$

Note that each ket state on the RHS is the tensor product of three qubit states. For example, $|\mathbf{6}\rangle = |110\rangle$.

- Readout: Suppose $f(x + \mathfrak{p}) = f(x)$, where $p \in \mathbb{N}$ and $\mathfrak{p} < 7$; we can determine $\mathfrak{p}$ by measuring register 1.

The readout stage described above may appear enigmatic to you. To prevent you from going astray, let's examine a concrete example where $\mathfrak{p} = 2$.

Example 5.3

For $\mathfrak{p} = 2$, since

$$f(0) = f(2) = f(4) = f(6), \tag{5.48}$$

$$f(1) = f(3) = f(5) = f(7), \tag{5.49}$$

then the state $|\psi'\rangle$ in Equation (5.47) becomes

$$|\psi'\rangle = \boxed{\frac{1}{2}\,|\mathbf{0}\rangle\,(|f(\mathbf{0})\rangle + |f(\mathbf{1})\rangle)} \tag{5.50}$$

$$+ \frac{1}{8}\,|\mathbf{1}\rangle\left[\left(1 + \mathrm{e}^{-\hat{\imath}\frac{2\pi}{4}} + \mathrm{e}^{-\hat{\imath}\frac{4\pi}{4}} + \mathrm{e}^{-\hat{\imath}\frac{6\pi}{4}}\right)|f(\mathbf{0})\rangle + \left(\mathrm{e}^{-\hat{\imath}\frac{\pi}{4}} + \mathrm{e}^{-\hat{\imath}\frac{3\pi}{4}} + \mathrm{e}^{-\hat{\imath}\frac{5\pi}{4}} + \mathrm{e}^{-\hat{\imath}\frac{7\pi}{4}}\right)|f(\mathbf{1})\rangle\right]$$

$$+ \cdots$$

$$+ \boxed{\frac{1}{8}\,|4\rangle\left[\left(1+e^{-\hat{\imath}\frac{8\pi}{4}}+e^{-\hat{\imath}\frac{16\pi}{4}}+e^{-\hat{\imath}\frac{24\pi}{4}}\right)|f(0)\rangle +\left(e^{-\hat{\imath}\frac{4\pi}{4}}+e^{-\hat{\imath}\frac{12\pi}{4}}+e^{-\hat{\imath}\frac{20\pi}{4}}+e^{-\hat{\imath}\frac{28\pi}{4}}\right)|f(1)\rangle\right]}$$

$$+ \cdots$$

$$+ \frac{1}{8}\,|7\rangle\left[\left(1+e^{-\hat{\imath}\frac{14\pi}{4}}+e^{-\hat{\imath}\frac{28\pi}{4}}+e^{-\hat{\imath}\frac{42\pi}{4}}\right)|f(0)\rangle +\left(e^{-\hat{\imath}\frac{7\pi}{4}}+e^{-\hat{\imath}\frac{21\pi}{4}}+e^{-\hat{\imath}\frac{35\pi}{4}}+e^{-\hat{\imath}\frac{49\pi}{4}}\right)|f(1)\rangle\right],$$

where all terms but the two boxed lines cancel out due to destructive interference. Hence,

$$|\psi'\rangle = \frac{1}{2}(|0\rangle\,|f(0)\rangle + |0\rangle\,|f(1)\rangle + |4\rangle\,|f(0)\rangle - |4\rangle\,|f(1)\rangle). \tag{5.51}$$

Therefore, if measuring register 1 always results in outcome 0 and 4, we know that $\mathfrak{p} = 2$. This pattern cannot be shared by any other case with $\mathfrak{p} \neq 2$. Try Exercise 5.4 to convince yourself.

> **Exercise 5.4** *Follow Example 5.3 to evaluate the circuit of period finding in Figure 5.8 for* $\mathfrak{p} = 3$.

The intrinsic strength of QFT lies in its ability to introduce phase factors that exhibit periodic behavior across the terms in a superposition. In period finding, the primary function of QFT is to exploit these phase factors, facilitating the cancellation of particular terms. Consequently, this leaves behind specific entangled states spanning the two registers. Notably, each period $\mathfrak{p}$ is intrinsically linked with a distinct entangled state of the two registers.

Now you can see that entanglement in QFT plays a pivotal role in amplifying the probability of specific outcomes and canceling out others. When the QFT is applied to a quantum register, it intricately entangles the qubits, transforming computational bases into frequency bases. This shift enables the efficient extraction of periodicity information from quantum states, which is crucial for algorithms like period finding and thus Shor's factorization method. The entanglement generated effectively collapses the computational space, narrowing down to only those outcomes that are coherent with the underlying periodicity or phase of the function in question. This coherence leads to constructive interference for the "right" answers and destructive interference for the "wrong" ones. This interference pattern, a direct consequence of entanglement, is what allows the QFT to give meaningful outputs and contributes to the quantum speed-up observed in algorithms that utilize the QFT. Insights into the nuanced role of entanglement in quantum computational processes can be found in Jozsa & Linden's research (2003) [189] and the exploration by Ekert & Jozsa (1996) [190] of how entanglement is harnessed as a computational resource.

5.4.2 Case of Any n

We are now set to discuss and comprehend the case of an arbitrary n. To elucidate the fundamental concepts, it is unnecessary to dwell on the most intricate general scenarios. Instead, we can focus on cases where the period p of the function f divides $N = 2^n$, or $N/\mathrm{p} = m \in \mathbb{N}$. Although more complex situations exist, they don't diverge from the conclusions drawn here.

Starting with the post-Oracle state, we have

$$|\psi\rangle = \frac{1}{\sqrt{N}} \sum_{i=0}^{N-1} |i\rangle \, |f(i)\rangle, \tag{5.52}$$

where $i, f(i) \in s_n$. The lesson from the case of $n = 3$ and $\mathrm{p} = 2$ suggests that after measuring register 1, the post-measurement state of register 2 must collapse into a specific state:

$$|f(i_0)\rangle = |f(i_0 + \mathrm{p})\rangle = \cdots = |f(i_0 + (m-1)\mathrm{p})\rangle, \tag{5.53}$$

where $i_0 \in s_n$ is arbitrary. Remind yourself of Equation (5.51) to get the intuition.

Hence, for a function f with period p, the state $|\psi\rangle$ in Equation (5.52) can be written as

$$|\psi\rangle = \frac{1}{\sqrt{\mathrm{p}}} \sum_{i_0=0}^{\mathrm{p}-1} \frac{1}{\sqrt{m}} \sum_{l=0}^{m-1} |i_0 + l\mathrm{p}\rangle \, |f(i_0)\rangle. \tag{5.54}$$

At this juncture, it becomes clear that for a fixed i_0, the state of register 2 remains invariant with varying l. Thus, the role of entanglement is pivotal in distinguishing the periodicity p from measurements on register 1 alone. In other words, we need only to consider any m terms with the same i_0 in Equation (5.54). That is, we can examine the state with a generic $i_0 < \mathrm{p}$:

$$|\psi'\rangle = \frac{1}{\sqrt{m}} \sum_{l=0}^{m-1} |i_0 + l\mathrm{p}\rangle \, |f(i_0)\rangle. \tag{5.55}$$

Now that $|f(i_0)\rangle$ is separated, we can disregard it. Executing a QFT on register 1 results in

$$
\begin{aligned}
|\psi''\rangle &= \frac{1}{\sqrt{m}} \frac{1}{\sqrt{N}} \sum_{l=0}^{m-1} \sum_{j=0}^{N-1} e^{-i\frac{2\pi(i_0+l\mathrm{p})j}{N}} |j\rangle \\
&= \frac{1}{\sqrt{mN}} \sum_{j=0}^{N-1} e^{-i\frac{2\pi i_0 j}{N}} \sum_{l=0}^{m-1} e^{-i\frac{2\pi l \mathrm{p} j}{N}} |j\rangle \\
&= \frac{1}{\sqrt{mN}} \sum_{j=0}^{N-1} e^{-i\frac{2\pi i_0 j}{N}} \underbrace{\sum_{l=0}^{m-1} e^{-i\frac{2\pi l j}{m}}}_{= \begin{cases} m, & j = km, k = 0, 1, \dots, (\mathrm{p}-1); \\ 0, & j \neq km \ \forall k. \end{cases}} |j\rangle \\
&= \sqrt{\frac{m}{N}} \sum_{k=0}^{\mathrm{p}-1} e^{-i\frac{2\pi i_0 km}{N}} |km\rangle \\
&= \frac{1}{\sqrt{\mathrm{p}}} \sum_{k=0}^{\mathrm{p}-1} e^{-i\frac{2\pi i_0 k}{\mathrm{p}}} \left|k\frac{N}{\mathrm{p}}\right\rangle,
\end{aligned}
\tag{5.56}
$$

where $m = N/\mathfrak{p}$ is acknowledged. The key takeaway? If we measure register 1, the outcomes will be $kN/\mathfrak{p}$, $k = 0, 1, \ldots, \mathfrak{p} - 1$. All these outcomes are equally likely:

$$p\left(k\frac{N}{\mathfrak{p}}\right) \equiv \frac{1}{\mathfrak{p}}, \quad \forall k = 0, 1, \ldots, \mathfrak{p} - 1. \tag{5.57}$$

> **Exercise 5.5** *Reexamine the case with $N = 8$, or $n = 3$, using the general treatment in Equations (5.52) through (5.56). Consider respectively $p = 2$ and $p = 3$.*

Nevertheless, since we have no prior knowledge of $\mathfrak{p}$ and k, it's inefficient and unreliable to find $\mathfrak{p}$ directly from the probabilities of the measurement outcomes. Hence, we need to extract the period $\mathfrak{p}$ from the outcomes:

- If an outcome is 0, no information about k and $\mathfrak{p}$ can be extracted. So, we need to repeat the algorithm and measure again.
- If an outcome is nonzero, it must be some integer c, which is rounded up from the experimental result – a real number very close to c. In this case, we know that there must be a $k \neq 0$, such that $kN/\mathfrak{p} = c$, or equivalently $k/\mathfrak{p} = c/N$. Because c and N are known, we can extract k and $\mathfrak{p}$ from the ratio c/N.
  To do so, we need to reduce c/N to $k/\mathfrak{p}$; however, this process relies on a classical algorithm – *the continued fraction algorithm*. We shall not dwell on this classical algorithm here but direct the reader to the literature, for example Section 8.5 in Nakahara & Ohmi [21], and Section 5.3.1 in Nielsen & Chuang [13]. There is a catch however: If $\gcd(k, \mathfrak{p}) \neq 1$, the reduced ratio may become $k'/\mathfrak{p}'$ with $\mathfrak{p}' < \mathfrak{p}$, which is wrong, so the period finding failed. We should run the entire algorithm again, obtain a new measurement outcome, and reduce the ratio again. Shor showed that it takes

$$\mathcal{O}(\log\log \mathfrak{p}) \tag{5.58}$$

runs of the algorithm to keep the probability of getting the correct $k/\mathfrak{p}$ arbitrarily close to 1.

5.5 Shor's Algorithm

Factorizing an integer as the product of prime numbers has been an exponentially difficult problem for classical computers. The RSA (Rivest–Shamir–Adleman) encryption [25], commonly adopted by banks and other financial institutions, is based on this very challenge. RSA encryption works by using two large prime numbers to produce a public key, which can be shared openly and used to encrypt messages. The safety of RSA encryption lies in the product of these primes. Decrypting messages encrypted with the public key requires a private key, which is derived from the two original prime numbers. Without knowledge of these prime numbers, one would have to factorize this large product to derive the private key – a task that's immensely difficult and impossible practically with current classical computing methods. This difficulty ensures the security of RSA. Nonetheless, it's this exact challenge of factorization that Shor's algorithm seeks to address efficiently with quantum computation.

Shor's algorithm of factorization stands out as perhaps the most remarkable discovery in the development of quantum computation. The algorithm reduces the complexity from exponential on classical computers to polynomial on quantum computers. Shor's algorithm revolves around two pivotal steps:

- The task of determining the factors of an integer can be equivalently framed as ascertaining the period of a specific integer function.
- This period determination can be efficiently tackled with QFT on quantum computers, an algorithm we have previously explored.

Indeed, period finding remains the sole quantum component within Shor's algorithm. Given our prior examination of this component, we will succinctly outline Shor's algorithm in this section without extensive elaboration.

It is essential to remember that each iteration of Shor's algorithm identifies two factors of an integer, which might not necessarily be prime. Should this situation arise, one must reiterate the algorithm on the previously identified nonprime factors until all the determined factors are prime. If the integer to be factored possesses precisely two prime factors, then Shor's algorithm can determine both simultaneously and at once. We shall now delineate Shor's algorithm using a flow chart, accompanied by requisite clarifications.

5.5.1 The Algorithm

It suffices to consider an odd integer $N \in \mathbb{N}$ because if it were even, we can take its odd factor $N/2$. Naively, it would take $\sqrt{N}$ trials to find all the prime factors of N. Since $N = 2^n$ for some $n \in \mathbb{N}$, the complexity is

$$\mathcal{O}(\sqrt{N}) = \mathcal{O}(\sqrt{2^n}) = \mathcal{O}(e^{\frac{n}{2}\ln 2}), \tag{5.59}$$

which is exponential. Cleverer classical algorithms exist to improve this complexity but it remains exponential. Shor's algorithm reduced this complexity to polynomial. Figure 5.9 depicts the flow chart of Shor's algorithm, where necessary explanations are inserted in dashed text boxes.

5.5.2 The Complexity Class

As elucidated in the flow chart, Shor's algorithm may fail if p is odd or $m^{p/2} + 1 = 0$ mod N, and may need to be reiterated with a different $m < N$. This indicates that Shor's algorithm (period finding too) is probabilistic in nature and belongs to the class of probabilistic algorithms. The efficient computation of the order of m on a quantum computer ensures that Shor's algorithm runs in polynomial

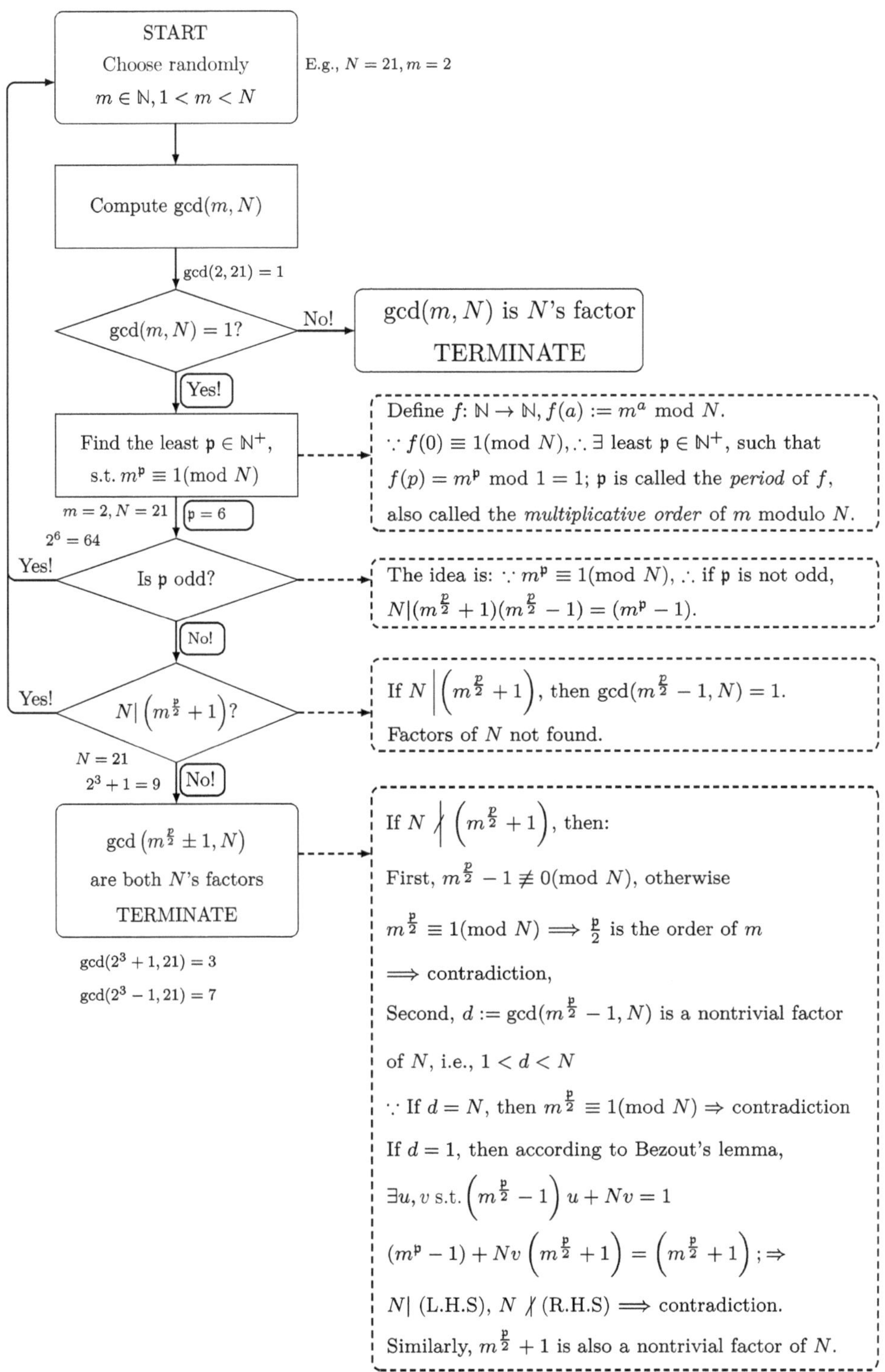

Figure 5.9 The flow chart of Shor's algorithm.

time. Consequently, Shor's algorithm belongs to the class of BQP (bounded-error quantum probabilistic) algorithms. The precise complexity of Shor's algorithm is given by

$$O((\log N)^2 (\log \log N)(\log \log \log N)). \tag{5.60}$$

As a concluding remark, Shor's algorithm dramatically enhances the efficiency of integer factorization due to the inherent mathematical structure that facilitates the reduction of integer factorization to period finding. Nonetheless, there are problems that do not possess such mathematical structures. For these problems, the speedup provided by quantum algorithms may not be as significant as that achieved by Shor's algorithm. In Section 5.6, we will explore a representative structureless problem and the quantum algorithm designed to address it.

5.6   Grover's Search Algorithm

Grover's search algorithm offers a quantum solution to a classic challenge: searching for a specific file within a pile of unordered files. Formally, amongst $N = 2^n$ unordered files, the task is to locate the one that meets a predetermined condition. The disorderliness of the problem arises from the random arrangement of the files. Mathematically, let us consider a binary function $f : s_n \rightarrow \{0, 1\}$ defined such that

$$f(\mathbf{i}) = \begin{cases} 1 & \text{if } \mathbf{i} = \mathbf{z}; \\ 0 & \text{otherwise.} \end{cases} \tag{5.61}$$

The mission is to determine the unique $\mathbf{z} \in s_n$ for which $f(\mathbf{z}) = 1$. Classically, solving this demands a complexity of $\mathcal{O}(N) = \mathcal{O}(2^n)$, which grows exponentially with the input size n. One might wonder: Can a quantum algorithm be more efficient?

To approach this problem with quantum intuition, begin by encoding the N possible values of the function's variable $\mathbf{i}$ onto an n-qubit register. This allows each variable value to correspond with the basis states $|0\rangle, |1\rangle, \ldots, |\mathbf{z}\rangle, \ldots, |N-1\rangle$. The next step involves transitioning the register's state into an equal probability superposition, expressed as

$$|\psi_0\rangle = \frac{1}{\sqrt{N}} \sum_{\mathbf{i}=0}^{N-1} |\mathbf{i}\rangle. \tag{5.62}$$

As hinted right before Section 5.3.4, the crux of the solution lies in identifying a QIT that can significantly amplify the probability of the desired basis state $|\mathbf{z}\rangle$ within $|\psi_0\rangle$. This specific QIT is known as the **selective phase rotation transform (SPRT)**.

5.6.1   Selective Phase Rotation Tranform

A classical SPRT is a DIT with the kernel

$$K_n(\mathbf{j}, \mathbf{i}) = \mathrm{e}^{\mathrm{i}\theta_\mathbf{i}} \delta_{\mathbf{ij}} \tag{5.63}$$

for all $\mathbf{i}, \mathbf{j} \in s_n$, such that

$$\tilde{f}(\mathbf{j}) = \sum_{\mathbf{i}=0}^{N-1} K_n(\mathbf{j}, \mathbf{i}) f(\mathbf{i}) = e^{i\theta_\mathbf{j}} f(\mathbf{j}). \tag{5.64}$$

A quantum SPRT is a unitary matrix R, with matrix elements $R_{\mathbf{j}\mathbf{i}} := K_n(\mathbf{j}, \mathbf{i})$, such that

$$R |\mathbf{i}\rangle = \sum_{\mathbf{j}=0}^{N-1} K_n(\mathbf{j}, \mathbf{i}) |\mathbf{j}\rangle = e^{i\theta_\mathbf{i}} |\mathbf{i}\rangle. \tag{5.65}$$

Note that a quantum SPRT attaches an dedicated overall phase factor to each basis state $|\mathbf{i}\rangle$ of the register but different phase factors to each component of a generic state, which is a superposition of the basis states.

How can we implement quantum SPRTs on quantum circuits? Although we have learnt the general procedure of decomposing an arbitrary n-qubit unitary gate, we can take special care of quantum SPRTs, which are simple n-qubit gates.

Consider $n = 1$. The corresponding SPRT is the 2×2 unitary matrix

$$R_1 = \begin{pmatrix} e^{i\theta_0} & 0 \\ 0 & e^{i\theta_1} \end{pmatrix} = e^{i\theta_0} \begin{pmatrix} 1 & 0 \\ 0 & e^{i(\theta_1 - \theta_0)} \end{pmatrix} \in U(2), \tag{5.66}$$

which is just a phase-shift gate with an extra overall phase factor.

For $n = 2$, the corresponding SPRT is the 4×4 unitary matrix

$$R_2 = \begin{pmatrix} e^{i\theta_3} & 0 & 0 & 0 \\ 0 & e^{i\theta_2} & 0 & 0 \\ 0 & 0 & e^{i\theta_1} & 0 \\ 0 & 0 & 0 & e^{i\theta_0} \end{pmatrix} = \underbrace{\begin{pmatrix} 1 & 0 & 0 & 0 \\ 0 & 1 & 0 & 0 \\ 0 & 0 & e^{i\theta_1} & 0 \\ 0 & 0 & 0 & e^{i\theta_0} \end{pmatrix}}_{\tilde{R}_2} \underbrace{\begin{pmatrix} e^{i\theta_3} & 0 & 0 & 0 \\ 0 & e^{i\theta_2} & 0 & 0 \\ 0 & 0 & 1 & 0 \\ 0 & 0 & 0 & 1 \end{pmatrix}}_{\tilde{R}'_2}. \tag{5.67}$$

The two matrices $\tilde{R}_2$ and $\tilde{R}'_2$ are precisely two CU gates, with the U's being two single-qubit SPRTs. Namely,

$$\tilde{R}_2 = |0\rangle \langle 0| \otimes \mathbb{1}_2 + |1\rangle \langle 1| \otimes R_1, \tag{5.68}$$
$$\tilde{R}'_2 = |0\rangle \langle 0| \otimes R'_1 + |1\rangle \langle 1| \otimes \mathbb{1}_2, \tag{5.69}$$

where R_1 is defined in (5.66), and

$$R'_1 := \begin{pmatrix} e^{i\theta_3} & 0 \\ 0 & e^{i\theta_2} \end{pmatrix}. \tag{5.70}$$

Hence, the quantum circuit implementing the two-qubit SPRT using elementary gates is as shown in Figure 5.10.

Figure 5.10 The circuit of the two-qubit SPRT.

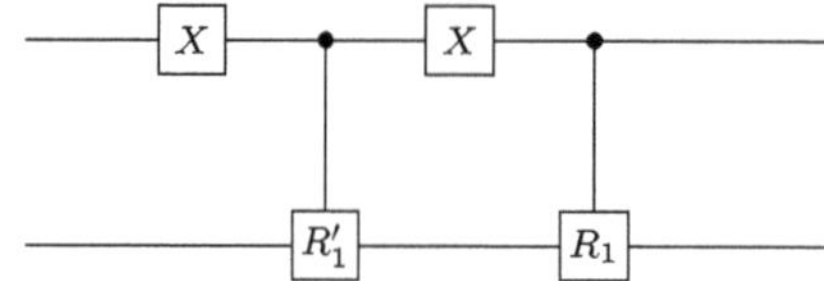

Exercise 5.6 *Why are there two X gates in Figure 5.10?*

5.6.2 Quantum Search Algorithm

Now that we have understood quantum SPRTs, we can redirect our attention to the primary task at hand: amplifying the probability of $|\mathbf{z}\rangle$ in the state $|\psi_0\rangle$, as described in Equation (5.62), in comparison to any other basis state $|\mathbf{i} \neq \mathbf{z}\rangle$.

The Algorithm

The concept is straightforward. To amplify the probability of outcome $\mathbf{z}$, it's vital to distinguish its probability amplitude from those of any other outcome $\mathbf{i} \neq \mathbf{z}$. The most rudimentary way to achieve this is by flipping the sign. Thus, the state $|\psi_0\rangle$ from Equation (5.62) is transformed to

$$\frac{1}{\sqrt{N}} \left(\sum_{i=0, i \neq z}^{N-1} |\mathbf{i}\rangle - |\mathbf{z}\rangle \right). \tag{5.71}$$

The core idea of Grover's algorithm is to iteratively increase the amplitude of the target state $|\mathbf{z}\rangle$ in Equation (5.71) by reflecting all amplitude coefficients about their mean value. Inspecting Equation (5.71), the amplitude of $|\mathbf{z}\rangle$ is $-\frac{1}{\sqrt{N}}$, while those of the other states remain $\frac{1}{\sqrt{N}}$. The mean of the amplitudes of $|\psi\rangle$ is

$$\text{mean} = \frac{1}{N} \left(-\frac{1}{\sqrt{N}} + (N-1) \cdot \frac{1}{\sqrt{N}} \right) = \frac{1}{\sqrt{N}} \cdot \frac{N-2}{N}. \tag{5.72}$$

By reflecting all amplitudes around this mean, we increase the amplitude of $|\mathbf{z}\rangle$ while slightly decreasing the amplitudes of the other states. For the target state $|\mathbf{z}\rangle$, reflection around the mean increases its amplitude to

$$\frac{1}{\sqrt{N}} + 2 \left(\frac{1}{\sqrt{N}} \cdot \frac{N-2}{N} \right) = \frac{1}{\sqrt{N}} \left(\frac{2N-4}{N} + 1 \right). \tag{5.73}$$

For large N, the increment $(2 - 4/N)/\sqrt{N} \ll 1$. So, to amplify the probability of $|\mathbf{z}\rangle$ sufficiently, iterative reflection around the mean is necessary. This iterative process allows Grover's algorithm to converge toward the target state with optimal efficiency.

With this understanding, we can construct the algorithm by defining the necessary operations:

- Initialization: Prepare the n-qubit register to the fiducial state $|\mathbf{0}\rangle$.

- Apply the Walsh–Hadamard transformation:

$$|\psi_0\rangle = W_n\,|\mathbf{0}\rangle = \frac{1}{\sqrt{N}}\sum_{i=0}^{N-1}|\mathbf{i}\rangle\,. \tag{5.74}$$

- Use a quantum SPRT to contrast the signs: We need to define a special SPRT that flips the sign in front of state $|\mathbf{z}\rangle$ only. Recall that our target oracle function f has the property $f(\mathbf{z}) = 1$ and $f(\mathbf{i} \neq \mathbf{z}) = 0$. So, we can design a quantum SPRT R_f realizing the kernel

$$K_f(\mathbf{j},\mathbf{i}) = e^{i\pi f(\mathbf{i})}\delta_{\mathbf{ij}} = (-1)^{f(\mathbf{i})}\delta_{\mathbf{ij}}, \tag{5.75}$$

such that

$$|\psi\rangle = R_f\,|\psi_0\rangle = \frac{1}{\sqrt{N}}\left(\sum_{i=0,i\neq z}^{N-1}|\mathbf{i}\rangle - |\mathbf{z}\rangle\right). \tag{5.76}$$

Clearly, as a unitary matrix, R_f can be written as

$$R_f = \mathbb{1}_N - 2\,|\mathbf{z}\rangle\langle\mathbf{z}|. \tag{5.77}$$

- Let A be the unitary operator that can reflect all amplitudes in $|\psi\rangle$ about the mean value (5.72). This operator A can be constructed as follows. In view of Equation (5.73), we wish to have

$$|\psi_1\rangle = A\,|\psi\rangle = \frac{1}{\sqrt{N}}\left[\sum_{i=0,i\neq z}^{N-1}\left(\frac{2N-4}{N}-1\right)|\mathbf{i}\rangle + \left(\frac{2N-4}{N}+1\right)|\mathbf{z}\rangle\right]. \tag{5.78}$$

Now, we can reverse-engineer to obtain A:

$$
\begin{aligned}
A\,|\psi\rangle &= \frac{1}{\sqrt{N}}\left[\sum_{i=0,i\neq z}^{N-1}\left(\frac{2N-4}{N}-1\right)|\mathbf{i}\rangle + \left(\frac{2N-4}{N}+1\right)|\mathbf{z}\rangle\right]\\[2mm]
&= \frac{1}{\sqrt{N}}\left(-\sum_{i=0,i\neq z}^{N-1}|\mathbf{i}\rangle + |\mathbf{z}\rangle\right) + \frac{1}{\sqrt{N}}\sum_{i=0}^{N-1}|\mathbf{i}\rangle\,\frac{2(N-2)}{N}\\[2mm]
&= \frac{1}{\sqrt{N}}\left(-\sum_{i=0,i\neq z}^{N-1}|\mathbf{i}\rangle + |\mathbf{z}\rangle\right) + \frac{1}{\sqrt{N}}\sum_{i=0}^{N-1}|\mathbf{i}\rangle\,\frac{2}{N}\left[\underbrace{\sum_{j=0}^{N-1}\sum_{i'=0,i'\neq z}^{N-1}\underbrace{\langle\mathbf{j}|\mathbf{i}'\rangle}_{\delta_{ji'}}}_{=N-1} - \underbrace{\sum_{j=0}^{N-1}\underbrace{\langle\mathbf{j}|\mathbf{z}\rangle}_{\delta_{jz}}}_{=1}\right]\\[2mm]
&= \underbrace{\frac{1}{\sqrt{N}}\left(-\sum_{i=0,i\neq z}^{N-1}|\mathbf{i}\rangle + |\mathbf{z}\rangle\right)}_{-|\psi\rangle} + \frac{2}{N}\sum_{i,j=0}^{N-1}|\mathbf{i}\rangle\langle\mathbf{j}|\,\underbrace{\frac{1}{\sqrt{N}}\left(\sum_{i'=0,i'\neq z}^{N-1}|\mathbf{i}'\rangle - |\mathbf{z}\rangle\right)}_{|\psi\rangle}\\[2mm]
&= \left(-\mathbb{1}_N + \frac{2}{N}\sum_{i,j=0}^{N-1}|\mathbf{i}\rangle\langle\mathbf{j}|\right)|\psi\rangle\,,
\end{aligned}
$$

indicating that

$$A := -\mathbb{1}_N + \frac{2}{N} \sum_{i,j=0}^{N-1} |i\rangle\langle j| = -\mathbb{1}_N + 2\,|\psi_0\rangle\langle\psi_0|. \tag{5.79}$$

> **Exercise 5.7** *Recall Equation (5.25) that expresses the n-qubit Walsh–Hadamard transform as a QIT. Now, if we define another quantum SPRT*
>
> $$\mathcal{R}_0(j, i) = (-1)^{1-\delta_{i0}}\delta_{ij}, \tag{5.80}$$
>
> *show that*
>
> $$A = W_n \mathcal{R}_0 W_n. \tag{5.81}$$

But we are not done yet; as previously explained, we need to iterate. To do so, let's rewrite $|\psi_1\rangle$ in Equation (5.78) in a more convenient manner as follows.
- Let $\sin\theta = 1/\sqrt{N}$ and $\cos\theta = \sqrt{N-1}/\sqrt{N}$. Then

$$\sin(2\theta) = 2\frac{\sqrt{N-1}}{N}, \quad \cos(2\theta) = \frac{N-2}{N}. \tag{5.82}$$

- The coefficient in front of $|z\rangle$ in $|\psi_1\rangle$ reads

$$\frac{1}{\sqrt{N}}\left(\frac{2N-4}{N}-1\right) = \frac{1}{\sqrt{N}}\left[\frac{N-2}{N}+\frac{2(N-1)}{N}\right]$$
$$= \frac{1}{\sqrt{N}}\frac{N-2}{N}+\frac{\sqrt{N-1}}{\sqrt{N}}\frac{2\sqrt{N-1}}{N}$$
$$= \sin\theta\cos(2\theta) + \cos\theta\sin(2\theta) = \sin(3\theta). \tag{5.83}$$

- The coefficients in front of $|i\rangle$ in $|\psi_1\rangle$ are likewise

$$\frac{1}{\sqrt{N}}\left(\frac{2N-4}{N}-1\right) = \frac{1}{\sqrt{N-1}}\left(\frac{\sqrt{N-1}}{\sqrt{N}}\frac{N-1}{N}-\frac{2\sqrt{N-1}}{\sqrt{N}N}\right)$$
$$= \frac{1}{\sqrt{N-1}}[\cos\theta\cos(2\theta) - \sin\theta\sin(2\theta)] = \frac{\cos(3\theta)}{\sqrt{N-1}}. \tag{5.84}$$

- Hence,

$$|\psi_1\rangle = \sum_{i=0,i\neq z}^{N-1} \frac{\cos(3\theta)}{\sqrt{N-1}}|i\rangle + \sin(3\theta)|z\rangle. \tag{5.85}$$

- The analysis above motivates us to define the unitary transformation $U_f := AR_f$. Then, we have

$$|\psi_1\rangle = A_f|\psi_0\rangle = \sum_{i=0,i\neq z}^{N-1} \frac{\cos(3\theta)}{\sqrt{N-1}}|i\rangle + \sin(3\theta)|z\rangle. \tag{5.86}$$

- Comparing $|\psi_1\rangle$ in (5.86) and $|\psi_0\rangle$ with the new variable θ,

$$|\psi_0\rangle = \sum_{i=0, i\neq z}^{N-1} \frac{\cos\theta}{\sqrt{N-1}} |i\rangle + \sin\theta |z\rangle ; \tag{5.87}$$

if we iterate the action of U_f k times on $|\psi_0\rangle$, we arrive at

$$|\psi_k\rangle = U_f^k |\psi_0\rangle = \sum_{i=0, i\neq z}^{N-1} \frac{\cos[(2k+1)\theta]}{\sqrt{N-1}} |i\rangle + \sin[(2k+1)\theta] |z\rangle. \tag{5.88}$$

Exercise 5.8 *What's the physical/geometric meaning of the unitary operator A_f in view of the angular variable θ defined above?*

Exercise 5.9 *Prove (5.88). Can you answer Exercise 5.8 now if you couldn't before?*

Exercise 5.10 *For any given N, is it possible to increase the probability of $|z\rangle$ to an optimal value while simultaneously decreasing the probabilities of the other basis states, without any iteration? Explain why the iterations in Grover's algorithm is necessary universally for general N.*

- The probability of finding the outcome z in $|\psi_k\rangle$ is clearly

$$p_{z,k} = \sin^2[(2k+1)\theta]. \tag{5.89}$$

Our job now is determining the iteration number k such that for $N \geq 1$, $p_{z,k} \geq 1 - 1/N$, which is good enough for us to locate the z in s_n. That is,

$$\sin^2[(2k+1)\theta] \geq 1 - \frac{1}{N} = 1 - \sin^2\theta$$
$$\Longrightarrow \cos^2[(2k+1)\theta] \leq \sin^2\theta$$
$$\Longrightarrow \sin^2\left[(2k+1)\theta - \tfrac{\pi}{2}\right] \leq \sin^2\theta$$
$$\Longrightarrow \sin\left|(2k+1)\theta - \tfrac{\pi}{2}\right| \leq \sin\theta$$
$$\Longrightarrow -\theta \leq (2k+1)\theta - \tfrac{\pi}{2} \leq \theta$$
$$\Longrightarrow \frac{\pi}{4\theta} - 1 \leq k \leq \frac{\pi}{4\theta}$$
$$\Longrightarrow \frac{\pi\sqrt{N}}{4} - 1 \leq k \leq \frac{\pi\sqrt{N}}{4}, \tag{5.90}$$

which leads to

$$k_{\max} = \left\lfloor \frac{\pi\sqrt{N}}{4} \right\rfloor = \mathcal{O}(\sqrt{N}).$$

This is a quadratic reduction of the classical complexity $\mathcal{O}(N)$.

Implementing Grover's Algorithm

Given that Grover's algorithm offers only a quadratic speedup for unstructured search problems, it's essential to note that quantum algorithms may not always achieve an exponential acceleration over their classical counterparts. Nonetheless, it's crucial to emphasize that, as of now, there is no rigorous proof or definitive no-go theorem prohibiting the existence of a quantum algorithm more efficient than Grover's. The field remains open to innovations and potential breakthroughs.

Example 5.4

Consider the case where $n = 6$. Then, $N = 2^6 = 64$ and $k_{max} = 6$. Suppose $z = 28$. Figure 5.11 depicts the probabilities of finding the correct $i = z$ for different k's. One can see that the probability peaks at about $k_{max} = 6$.

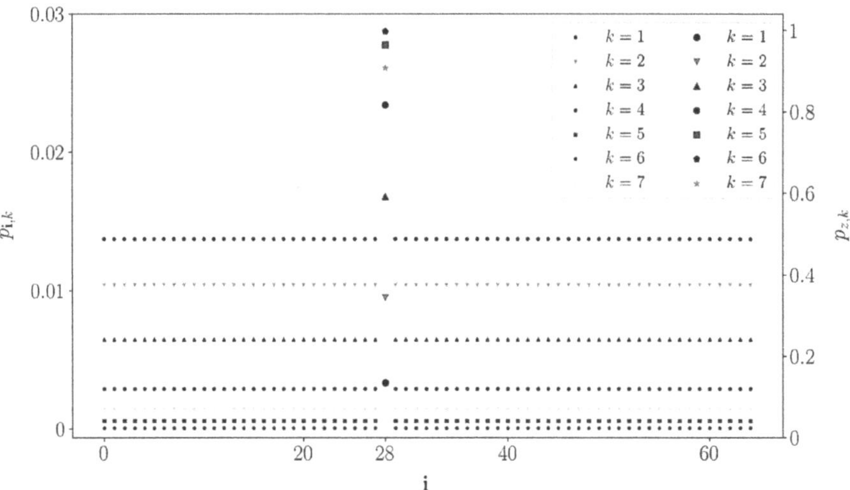

Figure 5.11 Probability $p_{i,k}$ at $k = 1, 2, \ldots, , 7$. It peaks at $k = 6$ and $z = 28$. Probabilities for $i \neq z$ substantially suppressed for $k > 1$. Vertical axis is logarithmically scaled in the region below 0.0137 for better visibility of the suppressed probabilities.

This example shows that in practice, for large N, it's unnecessary to iterate the unitary U_f k_{max} times because for sufficiently large $k < k_{max}$, the probability for $i \neq z$ has already been suppressed to be almost zero.

To implement Grover's algorithm on a quantum circuit, we have to implement the function f as an oracle, for which another n-qubit quantum register is needed to host the n ancillary qubits. Recall the oracle definition (5.4). Since in Grover's algorithm, the function is called upon when the SPRT R_f acts, the oracle and R_f can be combined together. We first draw the circuit for the unitary transformation $U_f = AR_f$. Putting all the pieces together, Figure 5.12 shows the quantum circuit of Grover's algorithm.

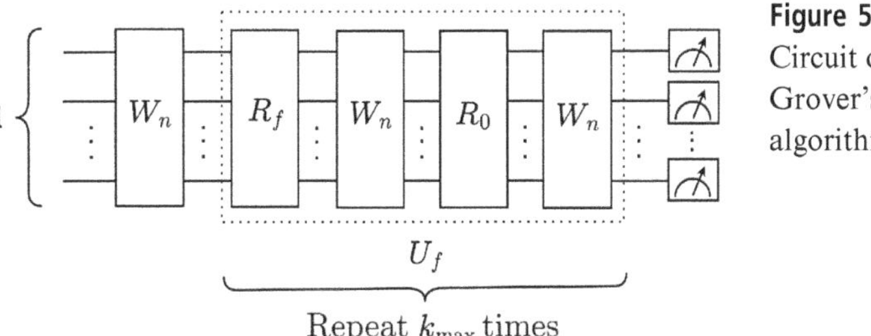

Figure 5.12 Circuit of Grover's algorithm.

> **Exercise 5.11** *Extending the Grover's algorithm for finding a single file to finding multiple files in a pile of unordered files is straightforward. Try to write down the Grover's algorithm for finding m files in a pile of $N = 2^n$ files.*

Conclusion

This chapter introduced foundational quantum algorithms, highlighting their essential roles and advantages in quantum computation. We began by exploring the Deutsch and Deutsch–Jozsa algorithms, which use quantum superposition to distinguish between constant and balanced functions with minimal evaluations, showcasing a quantum advantage over classical methods. These early algorithms set the stage for more advanced techniques by illustrating the power of quantum parallelism.

The chapter then examined the quantum Fourier transform, a critical tool in quantum algorithms for identifying periodicities in data, which serves as the backbone of period-finding algorithms. We saw this principle applied in Shor's algorithm, demonstrating its efficiency in integer factorization – a breakthrough with significant implications for cryptography due to its potential to challenge classical encryption methods.

Finally, Grover's search algorithm illustrated the application of quantum mechanics to unstructured search problems, achieving a quadratic speedup over classical search algorithms. Together, these algorithms underscore the computational benefits of quantum mechanics, particularly in tasks where superposition, entanglement, and phase manipulation enable efficiencies unattainable by classical approaches.

The next chapter will explore quantum decoherence, a phenomenon that poses challenges for maintaining quantum information over time. Understanding decoherence is essential for developing strategies to preserve quantum states, enabling more robust quantum computations in real-world conditions.

6   Quantum Decoherence

In the preceding chapter, the realm of quantum algorithms painted a promising picture, harnessing the profound intricacies of quantum mechanics to solve problems beyond the reach of classical computation. In this chapter, however, we venture into the Achilles' heel of quantum computing: quantum decoherence. While we've touched upon it in our earlier discussions in Chapter 3, a comprehensive understanding requires us to dive deeply into this phenomenon. Quantum decoherence, resulting from unintentional interactions between a quantum system and its environment, curtails the very quantum characteristics that make quantum computation powerful. This chapter seeks to unravel the mechanisms behind decoherence, its implications for quantum computing, and potential strategies to mitigate its effects. Our exploration aims to highlight the intricate dance between leveraging the marvels of quantum mechanics and managing its inherent susceptibilities.

6.1   A Rough Picture

To jog your memory, for a system in a pure state, its coherence is evidenced by the interference patterns between different basis states, as shown in Equation (3.6). Analogously, for a system in a mixed state, its coherence manifests as off-diagonal entries in the matrix as described by Equation (3.49). When these interference terms or off-diagonal matrix elements diminish, we say the system has undergone quantum decoherence, resulting in a state that more closely resembles a classical mixture with minimal quantum attributes.

This phenomenon of quantum decoherence is primarily observed in open quantum systems, meaning those that interact with an external environment. In the realm of quantum computing, our qubit systems are inherently open. They are influenced by various external factors such as their substrate, controlling units, and the quantum gates they pass through. If an environment saps a qubit system's quantum coherence, it no longer serves as an effective medium for quantum computation.

Yet, the inescapable truth is that we cannot entirely shield qubits from decoherence. This reality presents us with two primary avenues. First, we can engineer qubit systems with extended coherence periods, or coherence times, during which a significant number of quantum gate operations can take place. For instance, NMR (nuclear magnetic resonance)-based qubits boast coherence times spanning several seconds, accommodating roughly a thousand quantum gate operations. This avenue, deeply rooted in engineering, is beyond our current scope. Second, the field of quantum error correction proposes strategies to efficiently rectify qubits that might falter due to decoherence. We will study this subject in Chapter 7. Moreover, breaking free from the traditional qubit paradigm might offer alternative solutions to the decoherence challenge, such as topological quantum computation, which we'll introduce in Chapter 9.

In what follows, we'll use an illustrative example to provide a rudimentary understanding of quantum decoherence, laying the groundwork for developing your intuitive grasp of the topic.

Example 6.1

Let's consider the familiar CNOT gate, while regarding it as coupling a system – the control qubit – with an environment – the data qubit:

$$\alpha\,|0\rangle + \beta\,|1\rangle \quad\longrightarrow\quad ? \tag{6.1}$$

$$|0\rangle$$

- Input states:
  - System: $\alpha\,|0\rangle + \beta\,|1\rangle$, an arbitrary pure state, whose density matrix is

$$\rho = \begin{pmatrix} |\alpha|^2 & \alpha\beta* \\ \alpha^*\beta & |\beta|^2 \end{pmatrix}, \tag{6.2}$$

    where quantum coherence is evident.
  - Environment: $|0\rangle$.
  - Total: $|\psi\rangle = (\alpha\,|0\rangle + \beta\,|1\rangle) \otimes |0\rangle$.
- Output states:
  - Total: $|\psi'\rangle = \mathrm{CNOT}(\alpha\,|0\rangle + \beta\,|1\rangle) \otimes |0\rangle = \alpha\,|00\rangle + \beta\,|11\rangle$, entangled.
  - System: Since $|\psi'\rangle$ is entangled, the system's state is obtained by partial trace as

$$\rho' = \begin{pmatrix} |\alpha|^2 & 0 \\ 0 & |\beta|^2 \end{pmatrix}, \tag{6.3}$$

  which is completely decohered.

The result is strikingly clear: The quantum coherence inherent in the input state of the control qubit is wholly sapped by the CNOT gate, resulting in an output state that embodies a classical mixture.

What does this reveal? At its core, the CNOT gate facilitates a one-way flow of information. It transfers data from the control qubit (the system) to the data qubit (the environment). Consequently, when the system's state is extracted from the total state, the information that has flowed into the environment is irretrievably lost, as illustrated in Figure 6.1.

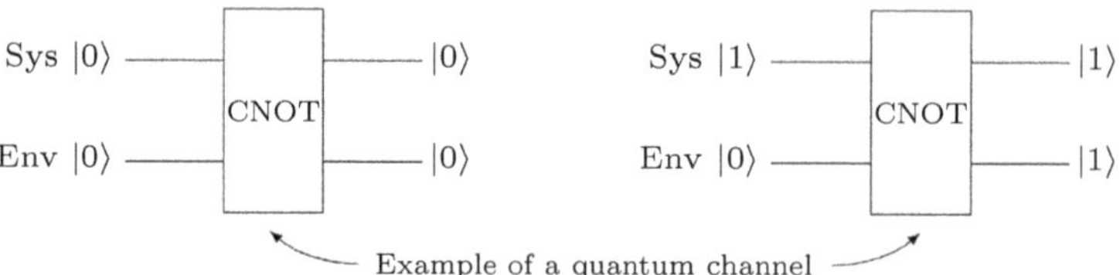

Figure 6.1 The CNOT gate relays information from the system (control qubit) exclusively to the environment (data qubit).

Looking at Figure 6.1, it's evident that the data qubit mirrors the state of the control qubit, irrespective of the basis state the latter occupies. Nevertheless, when faced with an arbitrary state like $\alpha\,|0\rangle + \beta\,|1\rangle$ of the control qubit, the CNOT gate cannot generate an exact duplicate. Instead, it outputs an entangled state. That said, after the CNOT gate's operation the initial data of α and β – in essence, the relative phase within the system's state – finds itself encoded within the entangled state of the combined system. Crucially, the environment lacks memory of the information's origin, a characteristic indicative of a **Markovian process** [191]. Consequently, the environment must be disregarded to ascertain the system's state, and the information transferred to the environment becomes irretrievable,[1] leading to the system's decoherence.

This depiction of the CNOT gate serves as a rudimentary example of what's known as a noisy quantum channel. This concept pertains to a quantum process within a system that encompasses an environment and describes its interplay with the system. From the system's perspective, a noisy quantum channel emerges as a nonunitary transformation because unitary transformations cannot turn a pure state into a mixed state or vice versa.

The illustration provided by this example lays a foundation for our upcoming in-depth exploration of open quantum systems and quantum decoherence. This foundational understanding will be instrumental in grasping the intricacies of the subject matter.

6.2 Open Quantum Systems

An open quantum system comprises a (principal) system interacting with an environment, through the Hamiltonian

$$H_T = H_S + H_E + H_I, \tag{6.4}$$

where H_T, H_S, H_E, and H_I denote the Hamiltonians of the total system, the principal system, the environment, and the interaction between them, respectively. To elucidate the dynamics of an open quantum system, it's essential to introduce the concept of a **superoperator**, also termed a **quantum operation**.[2]

---

[1] Recall the discussion on information loss in Chapter 3.

[2] A note on terminology: In earlier chapters, we loosely referred to a quantum operation as a unitary transformation or a sequence of quantum gates. Moving forward, "quantum operation" will be synonymous with "superoperator."

6.2.1 Superoperators

Quantum decoherence affects both pure and mixed states, leading us to work predominantly with density matrices. A quantum channel turns one density matrix to another, and this transformative process is encapsulated by the concept of a superoperator. To capture this transformation, it's crucial to expand our understanding beyond the standard quantum operator that acts within a Hilbert space. This expansion is necessitated by the fact that density matrices do not reside in but act on Hilbert spaces. We shall first present a formal definition of a superoperator and then further elucidate its nature through Example 6.2.

Definition 6.1. *Let $\mathcal{S}(\mathcal{H}_S)$ be the space of the density matrices acting on the Hilbert space of a quantum system S. A superoperator $\mathcal{E}$, also called a quantum operation, is a linear map*

$$\mathcal{E}: \mathcal{S}(\mathcal{H}_S) \to \mathcal{S}(\mathcal{H}_S). \tag{6.5}$$

Example 6.2

For a density matrix ρ_S of a closed system, any unitary time evolution operator $U(t)$ is a superoperator:

$$\mathcal{E}: \rho_S(t=0) \mapsto \rho_S(t) = U(t)\rho_S(t=0)U^\dagger(t). \tag{6.6}$$

For an open system, let's find out the superoperator corresponding to the CNOT gate.

$$
\begin{aligned}
\rho_S' &= \mathrm{Tr}_E |\psi'\rangle \langle\psi'| \\
&= \mathrm{Tr}_E(\mathrm{CNOT} |\psi\rangle |0\rangle \langle\psi|\langle0| \mathrm{CNOT}) \\
&= \sum_a \langle\varepsilon_a| \mathrm{CNOT} \underbrace{|\psi\rangle \langle\psi|}_{\rho_S} \otimes |0\rangle \langle0| \mathrm{CNOT} |\varepsilon_a\rangle \\
&= \sum_a \langle\varepsilon_a| \mathrm{CNOT} \underbrace{|0\rangle \rho_S \langle0|}_{\mathbb{1}_S\otimes|0\rangle\rho_S\mathbb{1}_S\otimes\langle0|} \mathrm{CNOT} |\varepsilon_a\rangle \\
&= \sum_a \underbrace{\langle\varepsilon_a| \mathrm{CNOT} |0\rangle}_{=:E_a} \rho_S \underbrace{\langle0| \mathrm{CNOT} |\varepsilon_a\rangle}_{=:E_a^\dagger} \\
&= \sum_a E_a \rho_S E_a^\dagger,
\end{aligned}
\tag{6.7}
$$

where $|\psi\rangle$ and $|\psi'\rangle$ are defined in Example 6.1, and $\{\varepsilon_a\}$ is an orthonormal basis of the environment Here, we also defined the Kraus operators E_a, which act on the system and by definition satisfy the condition

$$\sum_a E_a^\dagger E_a = \langle0| \mathrm{CNOT}^2 |0\rangle = \mathbb{1}_S, \tag{6.8}$$

where $\mathbb{1}_S$ is the identity operator on the system. Do these Kraus operators ring a bell to you? Recall generalized measurements defined in Equations (3.112) and (3.114).

> **Exercise 6.1** *Convince yourself that in the derivation above,*
>
> $$\rho_S \otimes |0\rangle \langle 0| = \mathbb{1}_S \otimes |0\rangle \rho_S \mathbb{1}_S \otimes \langle 0|. \tag{6.9}$$
>
> **Exercise 6.2** *Convince yourself that in the derivation above,*
>
> $$\langle \varepsilon_a | \, \text{CNOT} \, \mathbb{1}_S \otimes |0\rangle) = \langle \varepsilon_a | \, \text{CNOT} \, |0\rangle, \tag{6.10}$$
>
> *which makes sense of our definition of the Kraus operator E_a. Hint: Recall Equation (4.87).*

This example offers insight into the diverse manifestations of superoperators, which can vary based on the situation at hand. Naturally, one might wonder about the general representation of a superoperator. After all, in physics, we often place greater emphasis on an operator's representations rather than the operator in isolation. The Kraus operators introduced in Example 6.2 pave the way for the subsequent presentation of superoperator representations.

6.2.2 Kraus Representation

A superoperator is often cast in its Kraus representation [192], as illustrated in Example 6.2. To formally define the Kraus representation, we first need to set the scene.

We always assume that at $t = 0$, the system and its environment are uncorrelated, so the initial state of the total composite system at $t = 0$ takes the form

$$\rho_S \otimes \rho_E, \tag{6.11}$$

where ρ_S and ρ_E are the initial states of the system and its environment. As per Section 3.6, since any mixed state can be purified by augmenting its Hilbert space with a fictious Hilbert space of commensurate dimensions, we can simply assume that the environment's initial state is purified:

$$\rho_E = |\varepsilon_0\rangle \langle \varepsilon_0|. \tag{6.12}$$

Here, $|\varepsilon_0\rangle$ is a basis state in $\mathcal{H}_E = \text{span}\{|\varepsilon_a\rangle\}$, the possibly augmented Hilbert space of the environment. Hence, the initial total state is

$$\rho(0) = \rho_S \otimes |\varepsilon_0\rangle \langle \varepsilon_0|. \tag{6.13}$$

The total state still evolves unitarily:

$$\rho(t) = U(t)\rho(0)U^\dagger(t) = U(t)\rho_S \otimes |\varepsilon_0\rangle \langle \varepsilon_0|U^\dagger(t). \tag{6.14}$$

This next derivation yields the general definition of Kraus operators:

$$\rho_S(t) = \text{Tr}_E \rho(t)$$
$$= \text{Tr}_E \left(U(t)\rho_S \otimes |\varepsilon_0\rangle \langle \varepsilon_0|U^\dagger(t) \right)$$

$$
= \sum_a \langle \varepsilon_a | U(t) \underbrace{\rho_S \otimes |\varepsilon_0\rangle \langle \varepsilon_0|}_{\mathbb{1}_S \otimes |\varepsilon_0\rangle \rho_S \mathbb{1}_S \otimes \langle \varepsilon_0|} U^\dagger(t) |\varepsilon_a\rangle
$$

$$
= \sum_a \underbrace{\langle \varepsilon_a | U(t) |\varepsilon_0\rangle}_{E_a} \rho_S \underbrace{\langle \varepsilon_0 | U^\dagger(t) |\varepsilon_a\rangle}_{E_a^\dagger}
$$

$$
= \sum_a E_a(t) \rho_S E_a^\dagger(t), \tag{6.15}
$$

where we define the **Kraus operator** E_a by

$$
E_a := \langle \varepsilon_a | U(t) |\varepsilon_0\rangle. \tag{6.16}
$$

The expression

$$
\sum_a E_a(t) \rho_S E_a^\dagger(t) \tag{6.17}
$$

is termed the **Kraus representation** of the superoperator

$$
\mathcal{E} : \rho_S \mapsto \mathcal{E}(\rho_S) := \rho_S(t) = \sum_a E_a(t) \rho_S E_a^\dagger(t). \tag{6.18}
$$

This representation is also referred to as the **operator sum representation (OSR)**.

The Kraus operators thus defined adhere to the *completeness condition*:

$$
\sum_a E_a^\dagger E_a = \sum_a \langle \varepsilon_0 | U^\dagger(t) |\varepsilon_a\rangle \langle \varepsilon_a | U(t) |\varepsilon_0\rangle = \langle \varepsilon_0 | U^\dagger(t) U(t) |\varepsilon_0\rangle = \mathbb{1}_S. \tag{6.19}
$$

We reiterate that $U(t)$ might not exclusively denote a time evolution operator; it can represent a quantum gate or an unspecified unitary transformation.

In Example 6.2, the CNOT gate's Kraus representation was derived. For reference, a generalized measurement, as defined in Equation (3.112), can employ operators recognizable as Kraus operators.

Exercise 6.3 *Show that a superoperator indeed maps a density matrix to another density matrix.*

Exercise 6.4 *Prove that a superoperator can have multiple Kraus representations or OSRs.*

6.2.3 Kraus Representation Theorem

The importance of Kraus representations is captured by the Kraus representation theorem. To understand the theorem, we need to lay the groundwork first.

Noisy Quantum Channel: Definition

The Kraus representation or OSR of a superoperator can be interpreted as a **noisy quantum channel (NQC)**. Given an NQC, one typical way to describe it is through a probabilistic mixture of unitary transformations, formulated as in Equation (6.20).

Here, each unitary U_a acts on the system with a probability p_a, and the summation is over all such unitaries, with the constraint $\sum_a p_a = 1$:

$$\mathcal{E}(\rho) = \sum_a p_a U_a \rho U_a^\dagger. \tag{6.20}$$

To obtain the Kraus representation from this unitary set, one can view each term $\sqrt{p_a}U_a$ as a Kraus operator E_a, such that

$$\mathcal{E}(\rho) = \sum_a E_a \rho E_a^\dagger, \tag{6.21}$$

where $\mathcal{E}$ is the superoperator corresponding to the NQC, and $E_a = \sqrt{p_a}U_a$. It's evident from this relation that the probabilities p_a embedded in the unitary description manifest in the Kraus representation as factors in the Kraus operators. Moreover, the sum over all $E_a \rho E_a^\dagger$ terms in the Kraus representation corresponds to the average or expectation value of the system's state after undergoing the effects of the NQC. Under each unitary U_a, the system's density matrix undergoes the transformation

$$\rho_S \xrightarrow{U_a} \frac{E_a \rho_S E_a^\dagger}{\mathrm{Tr}(E_a \rho_S E_a^\dagger)}. \tag{6.22}$$

It's worth noting that although an NQC consists of a set of unitaries, its overall effect on a density matrix of the system is nonunitary in general.

Unitary Representation

We have seen that a set of unitaries representing an NQC leads to a Kraus representation of the superoperator corresponding to the NQC. What about the other way around? That is, given a superoperator in a Kraus representation or OSR, can we find a unitary representation of the superoperator or NQC? You might be confused: As stressed on a couple of times, an NQC acts on a system nonunitarily in general, so how can we represent it by a unitary transformation? Indeed, we cannot; however, similar to the idea of purification, our precise question is:

Given a system going through an NQC represented by certain Kraus operators, can we augment the system by attaching to it an environment, such that the total composite system undergoes a unitary transformation that recovers the NQC on the original system upon tracing out the environment?

The answer is affirmative, as we now see. Consider any NQC $\{U_a, p_a | a = 1, 2, \ldots, m\}$, which is equivalent to a Kraus representation or OSR by Equations (6.20) and (6.21). We may introduce an environment with Hilbert space $\mathcal{H}_E = \mathrm{span}\{|\varepsilon_a\rangle \,|\, a = 1, 2, \ldots, m\}$ and let

$$U = \sum_a U_a \otimes |\varepsilon_a\rangle \langle \varepsilon_a|, \tag{6.23}$$

which is obviously unitary because $\{|\varepsilon_a\rangle\}$ is an orthonormal basis.

Formally, we define

$$\rho_E = \sum_a p_a \, |\varepsilon_a\rangle \langle \varepsilon_a|. \tag{6.24}$$

Then, we show that the unitary transformation (6.23) indeed reproduces the OSR of the given NQC:

$$\mathcal{E}(\rho_S) = \mathrm{Tr}_E \left(U \rho_S \otimes \rho_E U^\dagger \right)$$

$$= \mathrm{Tr}_E \left(\sum_a U_a \otimes |\varepsilon_a\rangle \langle \varepsilon_a| \, \rho_S \otimes \underbrace{\sum_b p_b \, |\varepsilon_b\rangle \langle \varepsilon_b|}_{\sum_b p_b \mathbb{1}_S \otimes |\varepsilon_b\rangle \rho_S \mathbb{1}_S \otimes \langle \varepsilon_b|} \sum_c U_c^\dagger \otimes |\varepsilon_c\rangle \langle \varepsilon_c| \right)$$

$$= \mathrm{Tr}_E \left(\sum_a p_a U_a \rho_S U_a^\dagger \otimes |\varepsilon_a\rangle \langle \varepsilon_a| \right)$$

$$= \sum_a p_a U_a \rho_S U_a^\dagger. \tag{6.25}$$

Therefore, the unitary transformation (6.23) is a well-defined unitary representation of the NQC pertaining to the Kraus representation of the NQC.

In the derivation above, the environment being added to the system has a Hilbert space whose dimension is equal to the number of Kraus operators representing the NQC. This equality is not a coincidence. In fact, Preskill proved this theorem [193]:

Theorem 6.1. *Given any superoperator on an N-dimensional system, its OSR takes at most N^2 Kraus operators, and to obtain a unitary representation of the NQC, it suffices to add an N^2-dimensional environment.*

Preskill's theorem naturally brings us to ponder the complexity of superoperators, especially in terms of their parameterization. Let's analyze how many parameters define a superoperator and the inherent constraints that shape them, bearing in mind that a density matrix is Hermitian by definition and thus has N^2 real parameters. As an operator mapping one density matrix to another, a superoperator must then have $N^2 \times N^2 = N^4$ real parameters. That said, the N^4 parameters are not all free but are subject to constraints, to wit the completeness condition of the Kraus operators,

$$\sum_a E_a^\dagger E_a = \mathbb{1}_S. \tag{6.26}$$

This condition clearly supplies N^2 constraints because $\mathbb{1}_S$ has N^2 entries. Thus, a generic superoperator has

$$N^4 - N^2 \tag{6.27}$$

independent real parameters.

A Semigroup Structure

Having seen the complexity of superoperators, we are not surprised to acknowledge the existence of numerous superoperators on a given system. A natural question arises: Is there a mathematical structure confining all possible superoperators on a given system? Indeed there is: The superoperators on a given system form a semigroup, as we now explain.

Given any two superoperators $\mathcal{E}_1, \mathcal{E}_2 \colon \mathcal{S}(\mathcal{H}_S) \to \mathcal{S}(\mathcal{H}_S)$, their compositions are again superoperators

$$\mathcal{E}_1 \circ \mathcal{E}_2, \mathcal{E}_2 \circ \mathcal{E}_1 \colon \mathcal{S}(\mathcal{H}_S) \to \mathcal{S}(\mathcal{H}_S). \tag{6.28}$$

The identity $\mathbb{1}_{N^2}$ is a trivial superoperator. Nevertheless, since a superoperator $\mathcal{E}$ is in general nonunitary, $\mathcal{E}^{-1}$ does not necessarily exist. Therefore, the set $\{\mathcal{E} \colon \mathcal{S}(\mathcal{H}_S) \to \mathcal{S}(\mathcal{H}_S)\}$ forms a *semigroup*.

Once again, an NQC described by a nonunitary superoperator depletes the information of the system going through the channel, causing the system to decohere, in accordance to the noninvertibility of the channel due to information loss. Superoperators thus serve as a general treatment of decoherence.

Complete Positive Maps

A superoperator maps a density matrix to another one. Since density matrices are nonnegative by definition, a superoperator must preserve this nonnegativity. Indeed, a linear map on $\mathcal{S}(\mathcal{H}_S)$ is termed *positive* if it maps a nonnegative operator on $\mathcal{H}_S$ to another nonnegative operator. Consequently, superoperators are inherently **positive maps**.

When seeking unitary representations of an NQC, we considered extending a superoperator acting on the system to a unitary operator operating on a larger composite system. This enlarged system combines the original system with an auxiliary environment. For a positive linear map $\mathcal{E}$ acting on the system, if its extended maps $\mathcal{E} \otimes \mathbb{1}_E$ retain their positivity across the composite system for any conceivable environment E, then $\mathcal{E}$ is termed a **completely positive map**. Specifically, for any nonnegative density matrix ρ_{SE} characterizing the composite system, the matrix $\rho'_{SE} = \mathcal{E} \otimes \mathbb{1}_E(\rho_{SE})$ remains nonnegative irrespective of the choice of E.

Exercise 6.5 *It is imperative to note that not every positive map qualifies as completely positive. For illustration, consider the state (6.29) describing a composite system:*

$$|\psi_{SE}\rangle = \frac{1}{\sqrt{2}}(|0_S\rangle\,|0_E\rangle + |1\rangle_S\,|1_E\rangle). \tag{6.29}$$

Let ρ_S signify the density matrix of the system and designate a superoperator via the transpose T_S:

$$T_S(\rho_S) = \rho_S^T. \tag{6.30}$$

This superoperator is undeniably positive. Nevertheless, demonstrate that

$$T_S \otimes \mathbb{1}_E \tag{6.31}$$

> *loses its positivity when assessed against the state $|\psi_{SE}\rangle$ specified above. How would you interpret your finding physically?*

Now, with groundwork firmly in place, we are poised to introduce the Kraus representation theorem.

The Theorem

In 1983, Kraus elucidated a profound theorem concerning superoperators.

Theorem 6.2. *A superoperator $\mathcal{E}\colon S(\mathcal{H}_S) \to S(\mathcal{H}_S)$ that preserves hermiticity, trace, and is completely positive possesses an OSR. Moreover, it can be represented unitarily on an extended Hilbert space.*

This unitary representation for a completely positive superoperator is widely recognized as the **Stinespring representation** [194]. The significance of the Kraus representation theorem cannot be overstated. This theorem provides a mathematical framework to describe the action of quantum channels on quantum states. Given the fundamentally nonclassical nature of quantum systems, understanding how information is transformed and processed is of paramount importance. The Kraus representation gives insight into the effects of noise, decoherence, and other environmental interactions on a quantum system. This understanding is vital for tasks such as quantum error correction, which aims to counteract the effects of noise and decoherence in quantum computations.

The Stinespring representation further deepens this understanding. It offers a unitary perspective on quantum channels by introducing an auxiliary system, often termed the "environment." This representation provides a more intuitive and visualizable depiction of quantum operations, allowing one to think of the action of a quantum channel as a unitary evolution on a larger system. In essence, it extends the idea of unitary evolution, a core principle of quantum mechanics, to encompass nonunitary processes. This has been instrumental in understanding quantum-to-classical transitions and developing tools and protocols in quantum information theory.

Moreover, the Stinespring representation paves the way for more advanced mathematical tools and concepts, such as operator space theory and the **Choi–Jamiolkowski isomorphism** [195, 196]. These tools, built upon the foundation laid by Stinespring, have proven invaluable in the study and advancement of quantum information processing.

6.3 Noisy Quantum Channel: Examples

Having explored the formal definition of NQCs and their Kraus representations, it's time to strengthen our understanding by means of concrete NQCs and their consequent quantum decoherence. In this section, we shall use quantum circuits to emulate typical NQCs through which a single qubit decoheres in various ways and how two entangled qubits may disentangle.

6.3.1 Bit-Flip Channel

As indicated by its name, a **bit-flip channel** has a chance p of flipping a passing-through qubit. This channel is described by a superoperator $\mathcal{E}_{bf}$ taking the Kraus representation

$$\rho_S' = \mathcal{E}_{bf}(\rho_S) = (1 - p)\rho_S + p\sigma_x\rho_S\sigma_x, \quad 0 \leq p \leq 1, \tag{6.32}$$

where the two Kraus operators[3] are

$$E_0 = \sqrt{1 - p}\mathbb{1}_S, \quad E_1 = \sqrt{p}\sigma_x. \tag{6.33}$$

This NQC can be easily emulated by a CNOT gate with the control qubit in a superposition, but this time the data (control) qubit is the system (environment). The circuit looks like this:

$$\tag{6.34}$$

Let's compute the ρ_S' in this circuit:

$$\rho_S' = \mathrm{Tr}_E\left\{ \underbrace{(|0\rangle\langle 0| \otimes \mathbb{1}_S + |1\rangle\langle 1| \otimes X)}_{\text{CNOT}} \left[(1 - p)|0\rangle\langle 0| + \sqrt{p(1 - p)}|0\rangle\langle 1| \right.\right.$$

$$\left.\left. + \sqrt{p(1 - p)}|1\rangle\langle 0| + p|1\rangle\langle 1| \right] \otimes \rho_S \underbrace{(|0\rangle\langle 0| \otimes \mathbb{1}_S + |1\rangle\langle 1| \otimes X)}_{\text{CNOT}} \right\}$$

$$= \mathrm{Tr}_E\left[(1-p)|0\rangle\langle 0| \otimes \rho_S + \sqrt{p(1-p)}|0\rangle\langle 1| + \sqrt{p(1-p)}|1\rangle\langle 0| \right.$$

$$\left. + p|1\rangle\langle 1| \otimes X\rho_S X \right]$$

$$= (1 - p)\rho_S + pX\rho_S X, \tag{6.35}$$

which agrees with Equation (6.32).

When we mention a "circuit emulation" of a noisy quantum channel (NQC), what's really behind that term? It's essentially what we've touched on in the previous section: the unitary representation of an NQC. This is achieved by incorporating an auxiliary environment into our system. Take the bit-flip channel's circuit emulation as an example. The CNOT gate is our desired unitary transformation, and the control qubit serves as the auxiliary environment, complementing our main component – the data qubit as the system.

---

[3] Here, the Pauli σ_x matrix is used to emphasize the operator form of the Kraus operator, while the X gate is used when the Kraus operator is emulated or realized in a concrete quantum circuit, as in Equation (6.34).

> **Exercise 6.6** *The choice of the input state of the control qubit in circuit (6.34) is nonunique, and even a mixed state may work too. Find an input mixed state of the control qubit, such that the circuit results in the same ρ'_S.*
>
> **Exercise 6.7** *Multiple circuits may realize the bit-flip channel. Find a circuit other than (6.34) to realize the bit-flip channel.*

Decoherence

Let's quantitatively analyze the effect of the bit-flip channel on single-qubit states. Since single-qubit states live on the Bloch sphere and inside the Bloch ball, we can visualize the analysis as well. Recall that a generic single-qubit density matrix can be expressed as

$$\rho_S = \frac{1}{2}(\mathbb{1}_S + x\sigma_x + y\sigma_y + z\sigma_z) = \frac{1}{2}\begin{pmatrix} 1+z & x-\hat{\imath}y \\ x+\hat{\imath}y & 1-z \end{pmatrix}, \tag{6.36}$$

which encompasses both pure and mixed states. Then, after going through the bit-flip channel (6.32), the state becomes

$$\rho'_S = (1-p)\rho_S + p\sigma_x\rho_S\sigma_x = \frac{1}{2}\begin{pmatrix} 1+(1-2p)z & x-\hat{\imath}(1-2p)y \\ x+\hat{\imath}(1-2p)y & 1-(1-2p)z \end{pmatrix}. \tag{6.37}$$

Evidently, the space ρ'_S delineates is an ellipsoid formed by shrinking the Bloch ball in both y and z directions with scale factor $(1-2p)$. See Figure 6.2 for an illustration with $p = 0.25$. If $p = 0.5$, the Bloch ball would be completely crushed in both y and z directions and become the line segment $[-1, 1]$ along the x-axis. Nevertheless, as seen in Equation (6.37), the bit-flip channel cannot completely decohere the single-qubit system because the off-diagonal parameter x is intact through the channel.

6.3.2 Phase-Flip Channel

As its name implies, the phase-flip channel may flip the relative phase of a single qubit. Analogously to the bit-flip channel, the **phase-flip channel** $\mathcal{E}_{pf}$ is represented

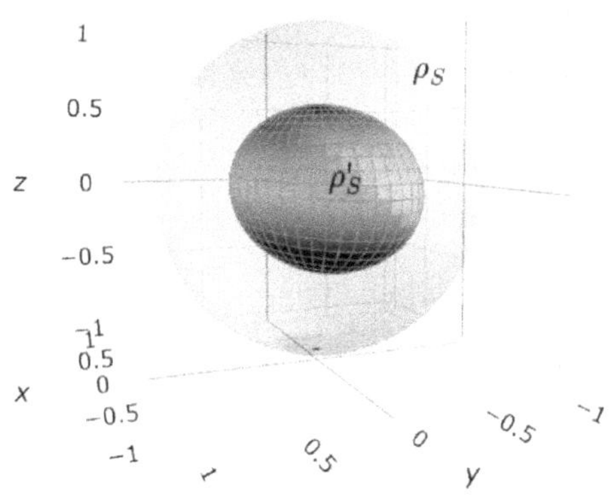

Figure 6.2 The original Bloch sphere/ball and the ellipsoid with $p = 0.25$ where ρ'_S resides.

by the Kraus operators

$$E_0 = \sqrt{1-p}\,\mathbb{1}_S, \quad E_1 = \sqrt{p}\,\sigma_z, \tag{6.38}$$

yielding

$$\rho_S' = \mathcal{E}_{\mathrm{pf}}(\rho_S) = (1-p)\rho_S + p\sigma_z\rho_S\sigma_z, \quad 0 \le p \le 1. \tag{6.39}$$

Exercise 6.8 *Recall that a superoperator can have multiple Kraus representations. Show that the phase-flip channel can also be represented by the Kraus operators*

$$E_0' = \begin{pmatrix} 1 & 0 \\ 0 & \cos\theta \end{pmatrix}, \quad E_1' = \begin{pmatrix} 0 & 0 \\ 0 & \sin\theta \end{pmatrix}, \tag{6.40}$$

provided with a specific relationship between θ and p. Can you derive the relationship between these Kraus operators and those defined in Equation (6.38)?

The phase-flip channel can be emulated using a CZ (controlled Z) gate:

$$\tag{6.41}$$

Evaluating the circuit, we deduce that

$$
\begin{aligned}
\rho_S' &= \mathrm{Tr}_E\Big\{(|0\rangle\langle 0| \otimes \mathbb{1}_S + |1\rangle\langle 1| \otimes Z)\big[(1-p)|0\rangle\langle 0| + p|1\rangle\langle 1|\big] \\
&\quad \otimes \rho_S(|0\rangle\langle 0| \otimes \mathbb{1}_S + |1\rangle\langle 1| \otimes Z)\Big\} \\
&= \mathrm{Tr}_E\Big[(1-p)\rho_S \otimes |0\rangle\langle 0| + pZ\rho_S Z \otimes |1\rangle\langle 1|\Big] \\
&= (1-p)\rho_S + pZ\rho_S Z,
\end{aligned}
\tag{6.42}
$$

which complies with Equation (6.39).

Decoherence

The post-channel density matrix, ρ_S', adopts the form

$$\rho_S' = \frac{1}{2}\begin{pmatrix} 1+z & (1-2p)(x-iy) \\ (1-2p)(x+iy) & 1-z \end{pmatrix}. \tag{6.43}$$

The state space of ρ_S' is another ellipsoid. This time, it results from the Bloch ball shrinking along the x and y axes by a scaling factor of $1-2p$. Refer to Figure 6.3 for a depiction.

For $p = 0.5$, the Bloch ball collapses entirely along the x and y axes, resulting in a line segment spanning $[-1, 1]$ on the z-axis. Notably, when $p = 0.5$, the quantum coherence of the system qubit disappears entirely.

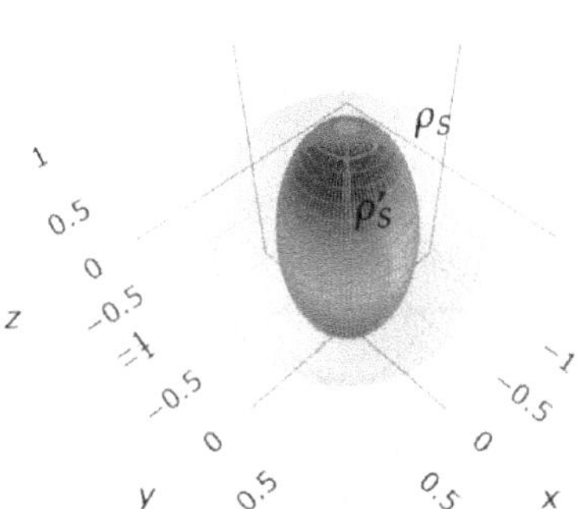

Figure 6.3 The original Bloch sphere/ball and the ellipsoid for $p = 0.25$ where ρ'_S is situated.

The phase-flip channel is also labeled as a *phase relaxation process*. In this process, a qubit's relative phase angle shifts from $\varphi = \arctan \frac{y}{x}$ to a statistical mixture of φ and $\varphi + \pi$. When a qubit's relative phase becomes a statistical mixture, it loses its computational utility. Therefore, a longer **phase relaxation time**, or T_2 **time**, is desirable. This metric is crucial for certain quantum computer types, including NMR, quantum dot, and superconducting quantum computers.

Exercise 6.9 *Consider a quantum channel that integrates the effects of both the bit-flip and phase-flip channels. Your task:*

1. *Construct a unified channel that can flip the state of the qubit and its relative phase concurrently.*
2. *Derive the Kraus representation for this integrated channel using precisely two Kraus operators.*
3. *Illustrate the emulation of this channel through a quantum circuit.*
4. *Investigate the quantum decoherence imparted by this combined channel on the qubit's state.*

Provide a comprehensive analysis and be sure to account for the interplay between the bit and phase flips.

6.3.3 Phase-Damping Channel

Unlike the phase-flip channel, which inverts a qubit's relative phase instantaneously, a more realistic decoherence mechanism can be depicted as a rotation

$$R_z(\theta) = \begin{pmatrix} e^{-i\frac{\theta}{2}} & 0 \\ 0 & e^{i\frac{\theta}{2}} \end{pmatrix}, \tag{6.44}$$

where θ is a stochastic variable obeying the Gaussian distribution

$$p(\theta) = \frac{1}{\sqrt{4\pi\lambda}} e^{-\frac{\theta^2}{4\lambda}}. \tag{6.45}$$

In this scenario, λ is a nonnegative real constant, the significance of which will be made clear momentarily.

Upon traversing this channel, the density matrix of a qubit transforms to

$$\rho_S' = \int_{-\infty}^{\infty} d\theta\, p(\theta) R_z(\theta) \rho_S R_z(-\theta) = \frac{1}{2}\begin{pmatrix} 1+z & e^{-\lambda}(x - \hat{\imath}y) \\ e^{-\lambda}(x + \hat{\imath}y) & 1-z \end{pmatrix}. \tag{6.46}$$

Exercise 6.10 *Verify Equation (6.46).*

Should the qubit navigate through n identical such channels in sequence, each having a time span Δt, and given that $t = n\Delta t$, the resultant post-channel state becomes

$$\rho_S(t) = \rho_S^{(n)} = \frac{1}{2}\begin{pmatrix} 1+z & e^{-n\lambda}(x - \hat{\imath}y) \\ e^{-n\lambda}(x + \hat{\imath}y) & 1-z \end{pmatrix} = \frac{1}{2}\begin{pmatrix} 1+z & e^{-\Gamma t}(x - \hat{\imath}y) \\ e^{-\Gamma t}(x + \hat{\imath}y) & 1-z \end{pmatrix}. \tag{6.47}$$

In the latter equality, we introduce the **decoherence rate**

$$\Gamma := \frac{\lambda}{\Delta t}, \tag{6.48}$$

and approach the limit where $\Delta t \to 0$. As this limit is reached, the n channels seamlessly merge into a singular **phase-damping channel**, causing the qubit to decohere at the rate Γ. For large values of either Γ or t, the qubit is effectively decohered. The state space of $\rho_S(t)$ mirrors an ellipsoid, analogous to that shown in Figure 6.3, but scaled by a factor of $\exp(-\Gamma t)$. This process of decoherence is known as a T_2 **process** – which a quantum computer should try to avoid, as when it occurs, the qubit state becomes a statistical mixture and is no longer qualified as a qubit. Therefore, the inverse $1/\Gamma$, which is the T_2 or phase relaxation time, is a key metric for quantum computers. Quantum computing scientists and engineers are striving to increase the T_2 times of their qubit systems.

6.3.4 Depolarizing Channel

The **depolarizing channel** is generally described by the superoperator

$$\rho_S' = \mathcal{E}_{\text{dp}}(\rho_S) = (1-p)\rho_S + \frac{p}{2}\mathbb{1}_S, \tag{6.49}$$

which, at first glance, does not seem to possess a straightforward Kraus representation. To unveil this representation, let's focus on the equation above. The first term on the right-hand side encodes a Kraus operator, $\propto \mathbb{1}_S$. Our goal is to recast the second term, $p\mathbb{1}_S/2$, in terms of Kraus operators acting on ρ_S. Since any single-qubit density matrix can be expressed as a combination of the identity matrix and the Pauli matrices, considering Equations (4.34) and (4.9), we obtain the relations

$$\rho_S = \frac{1}{2}(\mathbb{1}_S + x\sigma_x + y\sigma_y + z\sigma_z) \tag{6.50}$$

$$\sigma_x \rho_S \sigma_x = \frac{1}{2}(\mathbb{1}_S + x\sigma_x - y\sigma_y - z\sigma_z) \tag{6.51}$$

$$\sigma_y \rho_S \sigma_y = \frac{1}{2}(\mathbb{1}_S - x\sigma_x + y\sigma_y - z\sigma_z) \tag{6.52}$$

$$\sigma_z \rho_S \sigma_z = \frac{1}{2}(\mathbb{1}_S - x\sigma_x - y\sigma_y + z\sigma_z) \tag{6.53}$$

$$\rho_S + \sum_{i=x,y,z} \sigma_i \rho_S \sigma_i = 2\mathbb{1}_S. \tag{6.54}$$

Plugging these relations into Equation (6.49), we obtain

$$\rho_S' = \left(1 - \tfrac{3}{4}p\right)\rho_S + \tfrac{p}{4} \sum_{i=x,y,z} \sigma_i \rho_S \sigma_i, \tag{6.55}$$

implying the four Kraus operators

$$E_0 = \sqrt{1 - \tfrac{3}{4}p}\,\mathbb{1}_S, \quad E_i = \tfrac{\sqrt{p}}{2}\sigma_i, \quad i = x, y, z. \tag{6.56}$$

This is the first time we see more than two Kraus operators needed to represent an NQC. The number of Kraus operators determines the dimension of the auxiliary environment to be added to the system so that the NQC can be represented by a unitary transformation acting on the total composite system. As aforementioned, this unitary transformation can be emulated by a quantum circuit. Now that the depolarizing channel demands four Kraus operators, by Theorem 6.1 we would need an auxiliary environment that has a four-dimensional Hilbert space, namely a two-qubit environment. Including the system qubit, the circuit emulation should consists of three qubits. It turns out that a quantum Fredkin gate or CSWAP gate does the job:

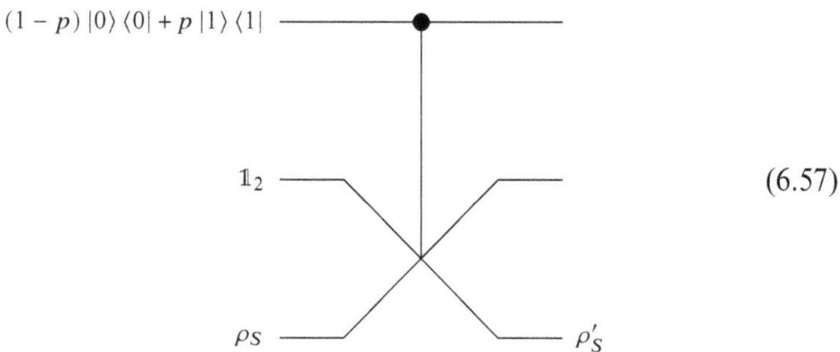

$$\tag{6.57}$$

In this circuit setting, the upper two qubits span the environment, regardless of whether the states of the lower two qubits are exchanged or not. The CSWAP gate is expressed as the unitary operator

$$|0\rangle\langle 0| \otimes \mathbb{1}_4 + |1\rangle\langle 1| \otimes \mathrm{SWAP}. \tag{6.58}$$

To proceed, you may need the result of Exercise 6.11.

Exercise 6.11 *Given two single-qubits in states ρ_1 and ρ_2, show that*

$$\mathrm{SWAP}(\rho_1 \otimes \rho_2)\,\mathrm{SWAP} = \rho_2 \otimes \rho_1. \tag{6.59}$$

Let's first evaluate the total output state of the circuit given the input states labeled in the circuit. By Equations (6.58) and (6.59),

$$\left(|0\rangle\langle 0| \otimes \mathbb{1}_4 + |1\rangle\langle 1| \otimes \mathrm{SWAP}\right)\left\{\left[(1-p)|0\rangle\langle 0| + p|1\rangle\langle 1|\right]\right.$$

$$\left.\otimes \frac{\mathbb{1}_2}{2} \otimes \rho_S\right\}\left(|0\rangle\langle 0| \otimes \mathbb{1}_4 + |1\rangle\langle 1| \otimes \mathrm{SWAP}\right) \tag{6.60}$$

$$= (1-p)|0\rangle\langle 0| \otimes \frac{\mathbb{1}_2}{2} \otimes \rho_S + p|1\rangle\langle 1| \otimes \rho_S \otimes \frac{\mathbb{1}_2}{2}.$$

We then partial-trace the state above over the environment to obtain ρ_S', keeping in mind that the upper two qubits are the environment:

$$\rho_S' = \mathrm{Tr}_E\left[(1-p)|0\rangle\langle 0| \otimes \frac{\mathbb{1}_2}{2} \otimes \rho_S + p|1\rangle\langle 1| \otimes \rho_S \otimes \frac{\mathbb{1}_2}{2}\right]$$

$$= \sum_{i,j=0}^{1}\left[\langle i|\langle j|\left(|0\rangle\langle 0| \otimes \frac{\mathbb{1}_2}{2}|i\rangle|j\rangle\right)(1-p)\rho_S + \langle i|\langle j|(|1\rangle\langle 1| \otimes \rho_S |i\rangle|j\rangle)\frac{p}{2}\mathbb{1}_2\right]$$

$$= \sum_{i,j=0}^{1}\left[\langle i|0\rangle\langle 0|i\rangle\langle j|\frac{\mathbb{1}_2}{2}|j\rangle(1-p)\rho_S + \langle i|1\rangle\langle 1|i\rangle\langle j|\rho_S|j\rangle\frac{p}{2}\mathbb{1}_2\right]$$

$$= (1-p)\rho_S + (\mathrm{Tr}\rho_S)\frac{p}{2}\mathbb{1}_2$$

$$= (1-p)\rho_S + \frac{p}{2}\mathbb{1}_S, \tag{6.61}$$

aligning perfectly with Equation (6.49).

Decoherence

The resultant state ρ_S' in matrix form is

$$\rho_S' = \frac{1}{2}\begin{pmatrix} 1 + (1-p)z & (1-p)(x - iy) \\ (1-p)(x + iy) & 1 - (1-p)z \end{pmatrix}, \tag{6.62}$$

depicting a Bloch ball that has been shrunk uniformly in all three dimensions by a factor of $1 - p$. This results in a smaller ball centered within the Bloch sphere, as showcased for $p = 0.5$ in Figure 6.4.

Figure 6.4 Comparison between the original Bloch sphere and the shrunken ball for $p = 0.5$, representing the state ρ_S' in Equation (6.62).

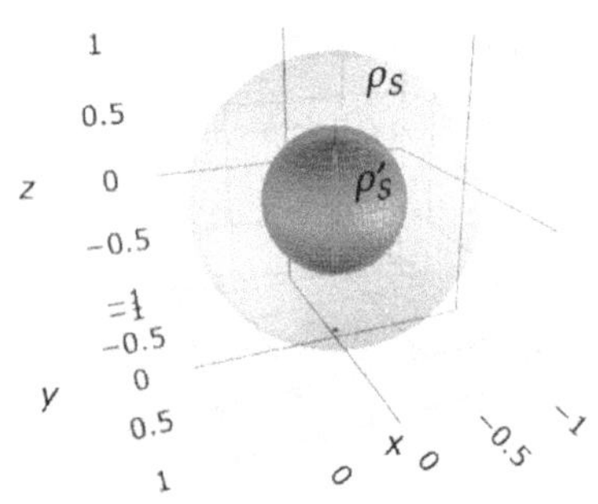

In essence, the depolarizing channel not only reduces the coherence entirely when $p = 1$, but it also collapses the entire Bloch ball to its center, signifying a state of total depolarization.

6.3.5 Amplitude-Damping Channel

Another noteworthy NQC is the **amplitude-damping channel**, which in its Kraus representation reads

$$\rho_S' = \mathcal{E}_{\mathrm{ad}}(\rho_S) = E_0 \rho_S E_0^\dagger + E_1 \rho_S E_1^\dagger, \tag{6.63}$$

where the Kraus operators are given by

$$E_0 = \begin{pmatrix} 1 & 0 \\ 0 & \sqrt{1-p} \end{pmatrix}, \quad E_1 = \sqrt{p} \begin{pmatrix} 0 & 1 \\ 0 & 0 \end{pmatrix}. \tag{6.64}$$

The circuit in Equation (6.65) serves as a unitary emulation of this particular NQC:

$$\tag{6.65}$$

Exercise 6.12 *Evaluate the circuit (6.65) to confirm that the output is indeed* ρ_S' *as described in Equation* (6.63).

Rather than emulating the amplitude-damping channel via a quantum circuit, Remark 6.1 offers an alternative physical representation of the channel through a *spin-chain model*.

Remark 6.1 A Spin-Chain Model Emulating the Amplitude-Damping Channel

In the theory of quantum communication, an amplitude-damping channel can be emulated using the **Heisenberg spin-chain model**. This model, governed by a time-independent Hamiltonian, allows for the transmission of quantum states from Alice to Bob, who are the sender and receiver respectively located at the left and right ends of the spin chain.

Consider a system of a spin-chain with N coupled spin-$\frac{1}{2}$ states. The Hamiltonian for this Heisenberg model is

$$H = -\sum_{i=1}^{N-1} J_i \left(\sigma_x^i \sigma_x^{i+1} + \sigma_y^i \sigma_y^{i+1} + \gamma \sigma_z^i \sigma_z^{i+1} \right) - \sum_{i=1}^{N} B_i \sigma_z^i, \qquad J_i > 0, \ B_i > 0. \tag{6.66}$$

Alice controls the first k spins and wants to send a quantum message to Bob, who can measure the last k spins; here $k < N$.

> **Remark 6.1, continued**
>
> Alice encodes her quantum message into the first k spins. She defines her $|0\rangle_s$ as the ground state of the system, characterized by all spins of the chain pointing downward:
>
> $$|0\rangle_s := |\downarrow\downarrow \cdots \downarrow\rangle . \tag{6.67}$$
>
> Her $|1\rangle_s$ state is defined as
>
> $$|1\rangle_s := \frac{1}{\sqrt{k}} \sum_{j=1}^{k} |j\rangle , \tag{6.68}$$
>
> where $|j\rangle$ is the excited state where only the j-th spin points upward:
>
> $$|j\rangle := \left|\downarrow^{j-1} \uparrow \downarrow^{N-j}\right\rangle . \tag{6.69}$$
>
> Alice's quantum message is thus
>
> $$|\Psi_0\rangle := \alpha\,|0\rangle_s + \beta\,|1\rangle_s , \tag{6.70}$$
>
> where α and β are complex amplitudes satisfying $|\alpha|^2 + |\beta|^2 = 1$.
> From $|\Psi_0\rangle$, the state of the entire chain at time t is
>
> $$|\Psi_t\rangle = \alpha\,|0\rangle + \beta \sum_{i=1}^{N} \sum_{j=1}^{k} f_{ij}(t)\,|i\rangle , \quad f_{ij}(t) := \langle i|e^{-iHt}|j\rangle . \tag{6.71}$$
>
> Bob decodes Alice's message by measuring the last k spins. The first $N-k$ spins, which are not accessible to Bob, are therefore traced out:
>
> $$\rho_B(t) := \mathrm{Tr}_{\mathrm{spin},\,j \leq N-k}\,|\Psi_t\rangle\langle\Psi_t| . \tag{6.72}$$
>
> **Exercise 6.13** *Answer questions 1 through 3:*
>
> 1. *Compute $\rho_B(t)$.*
> 2. *Define Bob's $|0\rangle_r$ and $|1\rangle_r$ states based on the last k spins and find the normalized $|1\rangle_r$.*
> 3. *Show that ρ_B can be obtained by applying amplitude damping on Alice's sent state $|\Psi_0\rangle = \alpha\,|0\rangle + \beta\,|1\rangle$:*
>
> $$\rho_B = E_0\,|\Psi_0\rangle\langle\Psi_0|\,E_0^{\dagger} + E_1\,|\Psi_0\rangle\langle\Psi_0|\,E_1^{\dagger}, \tag{6.73}$$
>
> *where E_0 and E_1 are the Kraus operators for amplitude damping:*
>
> $$E_0 = |0\rangle\langle0| + \sqrt{\eta}\,|1\rangle\langle1|, \qquad E_1 = \sqrt{1-\eta}\,|0\rangle\langle1|, \tag{6.74}$$
>
> *and find the **efficiency** η.*

Decoherence

The matrix form of ρ_S' immediately follows from Equation (6.63):

$$\rho_S' = \frac{1}{2}\begin{pmatrix} 1 + [p + (1-p)z] & \sqrt{1-p}(x - iy) \\ \sqrt{1-p}(x + iy) & 1 - [p + (1-p)z] \end{pmatrix} . \tag{6.75}$$

Note that since $-1 \leq z \leq 1$, $p + (1-p)z \leq 1$, the state space is an ellipsoid centered at $(0, 0, p)$, as depicted in Figure 6.5.

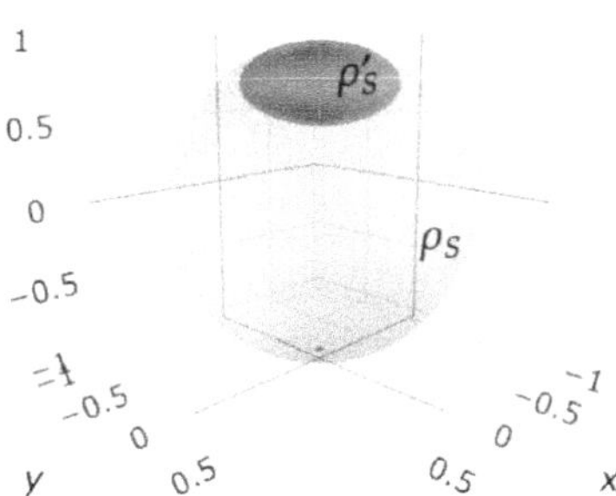

Figure 6.5 Comparison between the original Bloch sphere and the shrunken ellipsoid $p = 0.75$, representing the state ρ'_S in Equation (6.75).

In the extreme case, where $p = 1$, the ellipsoid collapses into a single point $(0, 0, 1)$, corresponding to the basis state $|0\rangle$. What's the physical meaning of p? Suppose $\rho_s = |1\rangle \langle 1|$, then by Equation (6.75), $\rho'_S = p |0\rangle \langle 0| + (1-p) |1\rangle \langle 1|$. Hence, p is the probability of the transition $|1\rangle \to |0\rangle$. This probability is termed the **damping probability**, which explains the name of the channel.

> **Exercise 6.14** *Describe what happens if initially $\rho_s = |0\rangle \langle 0|$.*

Similar to the case of the phase-damping channel, it's more physically intuitive to consider a time-dependent amplitude damping channel instead of the instantaneous one just discussed. For this purpose, we again consider repeatedly applying the channel (6.63) n times to a qubit, whose density matrix entries post-channel are

$$\left[\rho_s^{(n)}\right]_{11} = (1-p)^n \underbrace{(1-z)}_{[\rho_s]_{11}} = (1-p)^n [\rho_s]_{11} = e^{n \overbrace{\ln(1-p)}^{<0}} [\rho_s]_{11} \xrightarrow{n \to \infty} 0, \quad (6.76)$$

$$\underbrace{}_{\text{Initial probability of the qubit in } |1\rangle}$$

$$\left[\rho_s^{(n)}\right]_{01} = (\sqrt{1-p})^n (x - iy) = e^{\frac{n}{2} \ln(1-p)} [\rho_s]_{01} \xrightarrow{n \to \infty} 0, \quad (6.77)$$

$$\left[\rho_s^{(n)}\right]_{00} = 1 - \left[\rho_s^{(n)}\right]_{11} \xrightarrow{n \to \infty} 1. \quad (6.78)$$

That is,

$$\rho_S^{(n)} \xrightarrow{n \to \infty} |0\rangle \langle 0|. \quad (6.79)$$

Now, assume each repetition of the channel takes time Δt. Defining

$$\Gamma = -\frac{\ln(1-p)}{\Delta t}, \quad (6.80)$$

letting $t = n\Delta t$, and taking the limit $\Delta t \to 0$, we obtain

$$[\rho_s]_{11}(t) = \left[\rho_s^{(n)}\right]_{11} = e^{-n\Delta t \frac{-\ln(1-p)}{\Delta t}} [\rho_s]_{11} = e^{-\Gamma t} [\rho_s]_{11}(t = 0). \quad (6.81)$$

Thus, Γ is the transition rate of $|1\rangle \to |0\rangle$. This transition is known as a T_1 **process** – one that a quantum computer should try to avoid, as when it occurs, the computational basis collapses and the qubit is no longer qualified as a qubit. Therefore, the inverse $1/\Gamma$, referred to as the T_1 **time**, is a crucial metric for quantum computers. Quantum computing scientists and engineers also strive to increase the T_1 times of their qubit systems.

Figure 6.6 Quantum circuit reducing the entanglement of the top two qubits.

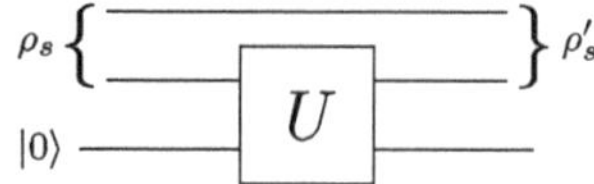

6.3.6 Disentangling Channel

As opposed to the quantum gates that generate entanglement among otherwise unentangled qubits, an NQC may be detrimental to entanglement, as decoherence due to the channel can reduce the entanglement among the qubits through the channel. To elucidate, consider the three-qubit quantum circuit in Figure 6.6, where the bottom two qubits are acted on by a unitary gate U that emulates a certain NQC, whereas the top two qubits (the system) are initially entangled.

To see how this circuit disentangles the two system qubits, we consider their initial state being the Bell state:

$$|\psi^+\rangle = \frac{1}{\sqrt{2}}(|01\rangle + |10\rangle). \tag{6.82}$$

So,

$$\rho_S = |\psi^+\rangle\langle\psi^+| = \frac{1}{2}(|01\rangle\langle01| + |01\rangle\langle10| + |10\rangle\langle01| + |10\rangle\langle10|) = \frac{1}{2}\begin{pmatrix} 0 & 0 & 0 & 0 \\ 0 & 1 & 1 & 0 \\ 0 & 1 & 1 & 0 \\ 0 & 0 & 0 & 0 \end{pmatrix}. \tag{6.83}$$

Multiple choices for the two-qubit gate, or the NQC it represents, can serve our purpose. We shall consider Example 6.3.

Example 6.3

Suppose U emulates the depolarizing channel, whose Kraus operators take the form

$$E_0' = \sqrt{1 - \frac{3}{4}p}\,\mathbb{1}_S, \quad E_i' = \frac{\sqrt{p}}{2}\sigma_i, \quad i = x, y, z, \tag{6.84}$$

as defined in Equation (6.56).

These four Kraus operators are considered to act on the middle qubit in the circuit in Figure 6.6. So, to examine their effect on the system consisting of the top and middle qubits, we need to augment them by a trivial action on the top qubit:

$$\widetilde{E_0'} := \sqrt{1 - \frac{3}{4}p}\,\mathbb{1}_2 \otimes \mathbb{1}_S, \quad \widetilde{E_1'} := \mathbb{1}_2 \otimes E_i' = \frac{\sqrt{p}}{2}\mathbb{1}_2 \otimes \sigma_i, \quad i = x, y, z. \tag{6.85}$$

Hence,

$$\rho_S' = \sum_{i=0}^{3} \widetilde{E_i'}\rho_S \widetilde{E_i'}^\dagger = \begin{pmatrix} \frac{p}{4} & 0 & 0 & 0 \\ 0 & \frac{2-p}{4} & \frac{1-p}{2} & 0 \\ 0 & \frac{1-p}{2} & \frac{2-p}{4} & 0 \\ 0 & 0 & 0 & \frac{p}{4} \end{pmatrix}. \tag{6.86}$$

Complete quantum decoherence of ρ_S' arises when $p = 1$. In this case, the output density matrix becomes a classical mixture:

$$\rho_S'(p = 1) = \frac{1}{4}(|00\rangle\langle00| + |01\rangle\langle01| + |10\rangle\langle10| + |11\rangle\langle11|)$$
$$= \frac{1}{2}(|0\rangle\langle0| + |1\rangle\langle1|) \otimes \frac{1}{2}(|0\rangle\langle0| + |1\rangle\langle1|). \tag{6.87}$$

Is this state separable or entangled? If you have done Exercise 3.16, you should find out that any bipartite density matrix admitting the decomposition

$$\rho = \sum_i \lambda_i \rho_i^{(1)} \otimes \rho_i^{(2)}, \tag{6.88}$$

where the superscripts label the two parties, describes a separable (mixed) state. As such, $\rho_S'(p = 1)$ describes a separable[4] mixed state, in which the two system qubits are unentangled, albeit they were entangled before going through the circuit.

> **Exercise 6.15** *Try other channels introduced in this section, and determine whether they can reduce the entanglement of the upper two qubits in Figure 6.6.*

Two insights can be gleaned from this example. First, as observed in the circuit depicted in Figure 6.6, the NQC influences only one of the two initially entangled system qubits. Nevertheless, the pair of system qubits astonishingly becomes disentangled after traversing the circuit, despite each qubit ending up in a mixed state. This phenomenon is characteristic of open systems. By contrast, in a closed system, an entangling gate such as the CNOT gate must couple both qubits simultaneously to generate entanglement.

Second, one might encounter an apparent paradox: Performing a partial trace on either qubit in $\rho_S'(p = 1)$ yields a reduced density matrix indicative of a mixed state. Indeed, each system qubit ends up in a mixed state after passing through the channel. The von Neumann entropy of the reduced density matrix is nonzero. One might then ask: Doesn't this imply that the bipartite state should be entangled? No! The nonzero von Neumann entropy[5] here should not be interpreted as evidence of entanglement between the two system qubits. Instead, it represents the information entropy of each individual qubit, whose state is a statistical mixture. The bipartite state essentially decomposes into two independent statistical mixtures, eliminating any sense of paradox.

6.4   The Lindblad Equation

Having been exposed to numerous examples of NQCs, a natural consequence of open systems, we are compelled to investigate the time evolution of open quantum systems more systematically.

---

[4] This is an example of **simply separable states**. If there is more than one nonzero λ_i, the state is non–simply separable.

[5] Entanglement entropy is not defined for mixed bipartite states.

Unlike a closed quantum system, which is described by a density matrix ρ_S and whose time evolution is governed by the Liouville–von Neumann equation (3.41),

$$\frac{\partial \rho_S}{\partial t} = -\mathrm{i}[H, \rho_S], \tag{6.89}$$

the time evolution of an open quantum system is dictated by a more intricate equation – the Lindblad equation [197]. Multiple avenues lead to this equation; here, we opt for an approach within the context of superoperators (quantum operations), owing to its greater generality and fewer underlying assumptions.

Consider an open quantum system coupled to an environment. The infinitesimal time evolution of the system's state, denoted by ρ_S, in its Kraus representation is given by

$$\rho_S(t + \mathrm{d}t) = \mathcal{E}(t + \mathrm{d}t, t)\rho_S(t) = \sum_{k=0}^{M-1} E_k \rho_S(t) E_k^\dagger, \tag{6.90}$$

where $N = \dim \mathcal{H}_S$, and $M \leq N^2$ is the number of Kraus operators. It's important to note that we write the action of the superoperator on the state as $\mathcal{E}(t + \mathrm{d}t, t)\rho_S(t)$ to emphasize that $\mathcal{E}$ is a linear operator.[6] This notation is convenient when dealing with infinitesimal changes in the system's state. Moreover, the Kraus operators E_k in the equation above are infinitesimal. Let's clarify the infinitesimal time dependence in the E_k's through this redefinition:

$$E_0 = \mathbb{1}_S + \frac{1}{\hbar} \underbrace{(-\mathrm{i}\underset{\text{Hermitian}}{\tilde{H}} + K)}_{\text{non-Hermitian}} \mathrm{d}t,$$

$$E_k = L_k\underset{\text{non-Hermitian}}{\sqrt{\mathrm{d}t}}. \tag{6.91}$$

It's always possible to redefine in this manner. For $\mathrm{d}t = 0$, only $E_0 = \mathbb{1}_S$ remains, aligning with $\mathcal{E}(t, t) = \mathbb{1}_S$. The operators K and L_k are not arbitrary; they are bounded by the completeness condition

$$\sum_k E_k^\dagger E_k = \mathbb{1}_S, \tag{6.92}$$

resulting in

$$\left[\mathbb{1}_S + \frac{1}{\hbar}(\mathrm{i}\tilde{H} + K)\mathrm{d}t\right]\left[\mathbb{1}_S + \frac{1}{\hbar}(-\mathrm{i}\tilde{H} + K)\mathrm{d}t\right] + \sum_{k=1}^{M-1} L_k^\dagger L_k \mathrm{d}t = \mathbb{1}_S$$

$$\implies \frac{2}{\hbar}K\mathrm{d}t + \sum_{k=1}^{M-1} L_k^\dagger L_k \mathrm{d}t + \mathcal{O}(\mathrm{d}t^2) = 0$$

$$\implies K = -\frac{\hbar}{2}\sum_{k=1}^{M-1} L_k^\dagger L_k. \tag{6.93}$$

---

[6] Previously, we wrote $\rho_S' = \mathcal{E}(\rho_S)$ to stress that $\mathcal{E}$ is a linear map on $\mathcal{S}(\mathcal{H}_S)$, because we were focused not on the time evolution but on the final state ρ_S'.

It's noteworthy that $\tilde{H}$ remains unaffected by the completeness condition. You may have inferred why we opted to use $\tilde{H}$ to represent this operator. The reason will be unveiled shortly.

Now, by inserting Equation (6.91) into the RHS of Equation (6.90) and remembering to retain only terms linear in $\mathrm{d}t$, we derive a difference equation:

$$\rho_S(t + \mathrm{d}t) = \rho_S(t) - \frac{i}{\hbar}[\tilde{H}, \rho(t)]\mathrm{d}t$$
$$+ \sum_{k=1}^{M-1} \left(L_k \rho_S(t) L_k^\dagger - \frac{1}{2} L_k^\dagger L_k \rho_S(t) - \frac{1}{2} \rho_S(t) L_k^\dagger L_k \right) \mathrm{d}t + \mathcal{O}(\mathrm{d}t^2). \quad (6.94)$$

At this point, one might choose to stop, particularly if no additional assumptions can be introduced. Nevertheless, it is common to employ the **Born–Markov approximation** [197, 198] which states:

The state $\rho_S(t)$ evolves according to a first-order differential equation. This means the environment lacks memory, resulting in a one-way information flow from the system to the environment. Such an assumption holds when the effect of the system on the environment's memory exists over a time scale much shorter than any pertinent system dynamics time scale.

Consequently, we postulate that

$$\rho_S(t + \mathrm{d}t) = \rho_S(t) + \dot{\rho}_S(t)\mathrm{d}t + \mathcal{O}(\mathrm{d}t^2). \qquad (6.95)$$

Using this assumption in the difference equation (6.94), we get the **GKLS** (Gorini, Kossakowski, Lindblad, and Sudarshan) **master equation** [197, 198]:

$$\frac{\partial \rho_S(t)}{\partial t} = -\frac{i}{\hbar}[\tilde{H}, \rho(t)] + \sum_{k=1}^{M-1} \left(L_k \rho_S(t) L_k^\dagger - \frac{1}{2} L_k^\dagger L_k \rho_S(t) - \frac{1}{2} \rho_S(t) L_k^\dagger L_k \right), \quad (6.96)$$

which is more widely recognized as the **Lindblad equation**. Here, the operator L_k is called a **Lindblad operator**.

While it might seem that the operator $\tilde{H}$ should be the system's Hamiltonian, it generally is not. Nonetheless, when the system interacts weakly with its environment, $\tilde{H}$ can be substituted by the system's Hamiltonian H. We will illustrate this with examples. Typically, $\tilde{H}$ is an effective Hamiltonian that acknowledges the interaction between the system and its environment. Computing this effective Hamiltonian can be intricate, yet for weak coupling, employing the Born–Markov approximation along with first-order perturbation theory often proves useful, as shown in an upcoming example.

Notably, by rescaling the Lindblad operators L_k using $L_k \rightarrow \sqrt{\gamma_k} L_k$ and positive parameters γ_k, whose interpretation depends on the specific case, the Lindblad equation can be expressed in an alternative form:

$$\frac{\partial \rho_S(t)}{\partial t} = -\frac{i}{\hbar}[\tilde{H}, \rho(t)] + \sum_{k=1}^{M-1} \gamma_k \left(L_k \rho_S(t) L_k^\dagger - \frac{1}{2} L_k^\dagger L_k \rho_S(t) - \frac{1}{2} \rho_S(t) L_k^\dagger L_k \right). \quad (6.97)$$

> **Exercise 6.16** *Show that the Lindblad equation (6.97) remains invariant under the transformation*
>
> $$L_k \rightarrow L_k + c_k, \quad \tilde{H} \rightarrow \tilde{H} + \frac{\hbar}{2i} \sum_k \gamma_k (c_k^* L_k - c_k L_k^\dagger) + c_0, \tag{6.98}$$
>
> *where $c_0 \in \mathbb{R}$ and $c_k \in \mathbb{C}$. Further, prove that because of this invariance, L_k can always be rendered traceless.*

Next, we will explore Examples 6.4 and 6.5 where the system's interaction with its environment is sufficiently weak, allowing us to assume $\tilde{H} = H_S$. Then, in Section 6.4.1, we will encounter a case where $\tilde{H} \neq H_S$.

Example 6.4

In this example, we consider a single-qubit system captured by the Hamiltonian

$$H = -\frac{\omega_0}{2}\sigma_z, \tag{6.99}$$

defining two energy levels E_0 and E_1 with an energy difference given by $\Delta E = E_1 - E_0 = \omega_0$. Suppose the environment is able to induce a phase flip on the qubit. But we don't care about the environment, so its influence on the system is encapsulated in a singular Lindblad operator, delineated in the computational basis as

$$L = \sqrt{p}\begin{pmatrix} 0 & 0 \\ 0 & 1 \end{pmatrix}, \tag{6.100}$$

which mirrors part of the phase-flip channel (6.40). Here, the parameter p plays the role of the γ_k in Equation (6.97). In the weak coupling limit, it holds that $\tilde{H} = H$. As a result, the Lindblad equation (6.97) becomes

$$\frac{\partial \rho}{\partial t} = -i[H, \rho] + L\rho L^\dagger - \frac{1}{2}(L^\dagger L \rho + \rho L^\dagger L), \tag{6.101}$$

which can be represented in matrix form as

$$\frac{\partial}{\partial t}\begin{pmatrix} \rho_{00} & \rho_{01} \\ \rho_{10} & \rho_{11} \end{pmatrix} = i\omega_0 \begin{pmatrix} 0 & \rho_{01} \\ -\rho_{10} & 0 \end{pmatrix} + p\begin{pmatrix} 0 & -\frac{1}{2}\rho_{01} \\ -\frac{1}{2}\rho_{10} & 0 \end{pmatrix} \tag{6.102}$$

yielding the solution

$$\begin{pmatrix} \rho_{00}(0) & \rho_{01}(0)e^{(i\omega_0 - p/2)t} \\ \rho_{10}(0)e^{(-i\omega_0 - p/2)t} & \rho_{11}(0) \end{pmatrix}. \tag{6.103}$$

Accordingly, we can discern the T_2 (phase flip or damping) time:

$$T_2 = \frac{2}{p}, \tag{6.104}$$

over which the system will decohere by a factor of $1/e$. Compare this result with Equation (6.47).

Example 6.5

In this example, we consider again the one-qubit Hamiltonian (6.99), but now assume that the environment introduces three Lindblad operators:

$$L_+ = \sqrt{\Gamma_+}\begin{pmatrix} 0 & 1 \\ 0 & 0 \end{pmatrix}, \quad L_- = \sqrt{\Gamma_-}\begin{pmatrix} 0 & 0 \\ 1 & 0 \end{pmatrix}, \quad L_z = \sqrt{\Gamma_z}\begin{pmatrix} 1 & 0 \\ 0 & -1 \end{pmatrix}, \tag{6.105}$$

where L_+ again facilitates the emission transition $|1\rangle \rightarrow |0\rangle$, L_- initiates the absorption transition $|0\rangle \rightarrow |1\rangle$, and L_z results in dephasing or phase damping. We continue to assume $\tilde{H} = H$. Thus, the Lindblad equation (6.97) in its matrix form becomes

$$\begin{aligned}
\frac{\partial}{\partial t}\begin{pmatrix} \rho_{00} & \rho_{01} \\ \rho_{10} & \rho_{11} \end{pmatrix} &= i\omega_0 \begin{pmatrix} 0 & \rho_{01} \\ -\rho_{10} & 0 \end{pmatrix} + \Gamma_+ \begin{pmatrix} \rho_{11} & -\frac{1}{2}\rho_{01} \\ -\frac{1}{2}\rho_{10} & -\rho_{11} \end{pmatrix} \\
&+ \Gamma_- \begin{pmatrix} -\rho_{00} & -\frac{1}{2}\rho_{01} \\ -\frac{1}{2}\rho_{10} & \rho_{00} \end{pmatrix} + \Gamma_z \begin{pmatrix} 0 & -2\rho_{01} \\ -2\rho_{10} & 0 \end{pmatrix}.
\end{aligned} \tag{6.106}$$

To gain more insight into the solution of this equation, let's define the equilibrium value of the density matrix ρ^{eq} as the state satisfying $\partial\rho/\partial t = 0$. The equilibrium condition yields

$$\Gamma_+\rho_{11}^{\mathrm{eq}} - \Gamma_-\rho_{00}^{\mathrm{eq}} = 0 \implies \frac{\rho_{11}^{\mathrm{eq}}}{\rho_{00}^{\mathrm{eq}}} = \frac{\Gamma_-}{\Gamma_+}. \tag{6.107}$$

Given that Γ_+ and Γ_- represent the photon emission and absorption rates respectively, they can be expressed as

$$\Gamma_+ = \gamma[1 + n(\omega_0)], \quad \Gamma_- = \gamma n(\omega_0), \tag{6.108}$$

where γ denotes a rate constant and $n(\omega_0)$ signifies the Bose-Einstein distribution

$$n(\omega_0) = \frac{1}{e^{\omega_0/k_B T} - 1}. \tag{6.109}$$

Consequently,

$$\frac{\Gamma_-}{\Gamma_+} = e^{-\frac{\omega_0}{k_B T}}. \tag{6.110}$$

With these considerations, the solution to the Lindblad equation can be expressed as

$$\begin{pmatrix} \rho_{00}^{\mathrm{eq}} + (\rho_{00} - \rho_{00}^{\mathrm{eq}})e^{-t/T_1} & \rho_{01}e^{i\omega_0 t - t/T_2} \\ \rho_{10}e^{-i\omega_0 t - t/T_2} & \rho_{11}^{\mathrm{eq}} + (\rho_{11} - \rho_{11}^{\mathrm{eq}})e^{-t/T_1} \end{pmatrix}, \tag{6.111}$$

where the T_1 and T_2 times are given by

$$T_1^{-1} = \Gamma_+ + \Gamma_- = \gamma\coth(\omega_0/2k_B T), \tag{6.112}$$

$$T_2^{-1} = \frac{1}{2}(\Gamma_+ + \Gamma_-) + 2\Gamma_z = \frac{T_1^{-1}}{2} + 2\Gamma_z. \tag{6.113}$$

The qubit will undergo complete decoherence when $t \gg T_2$. This scenario closely depicts situations encountered in all prototypical quantum circuit model–based quantum computers, such as superconducting, ion-trap, and NMR quantum computers. This example is so simple and physical that it can be found in many textbooks on quantum computation, for example, in [21].

6.4.1 The Effective Hamiltonian in a One-Qubit System Coupled to a Harmonic Oscillator

In the two examples above, we take the weak coupling limit to set $\tilde{H} = H$. Now, let's consider a scenario where the coupling is not weak enough to discard the environment's impact on the system's Hamiltonian. In such a case, the system's Hamiltonian H will be modified to be an effective Hamiltonian $\tilde{H}$ that appears in the Lindbald equation of the system. To be concrete, we shall derive the form of the effective Hamiltonian $\tilde{H}$ for a one-qubit system weakly coupled to a harmonic oscillator. Let's proceed as follows:

1. **Define raising and lowering operators:**

$$\sigma_+ = \sigma_x + i\sigma_y, \quad \sigma_- = \sigma_x - i\sigma_y \tag{6.114}$$

2. **System Hamiltonian:**

$$H_S = \omega_0 \sigma_z \tag{6.115}$$

3. **Environment Hamiltonian:**

$$H_R = \omega a^\dagger a \tag{6.116}$$

4. **Interaction Hamiltonian:**

$$H_I = g(\sigma_+ a + \sigma_- a^\dagger) \tag{6.117}$$

The full Hamiltonian is then

$$H = H_S + H_R + H_I \tag{6.118}$$

5. **Liouville–von Neumann equation for the total system:**

$$\frac{\partial \rho}{\partial t} = -i[H, \rho] \tag{6.119}$$

6. **Reduced density matrix:**

$$\rho_S = \mathrm{Tr}_R[\rho] \tag{6.120}$$

7. **Master equation for ρ_S:** We assume that, under the Born–Markov approximations, the master equation can be written in the general Lindblad form, where $\tilde{H}$ is the effective Hamiltonian we seek.

8. **Perturbation analysis**: We treat H_I as a perturbation and expand the master equation in powers of g. To first order, we typically find $\tilde{H} = H_S + \Delta H$, where ΔH contains terms dependent on g.
9. **Effective Hamiltonian**:

$$\tilde{H} = \omega_0 \sigma_z + \gamma \sigma_+ + \gamma^* \sigma_- \tag{6.121}$$

Exercise 6.17 *Fill the gaps in the procedure above:*

1. *Derive the form of ΔH using first-order perturbation theory. What are the coefficients that γ depends on?*
2. *Show that the effective Hamiltonian $\tilde{H}$ reduces to H_S in the limit $g \to 0$. What does this imply about the system–environment coupling?*

Exercise 6.18 *Application to quantum decoherence:*

1. *Given the effective Hamiltonian*

$$\tilde{H} = \omega_0 \sigma_z + \gamma (\sigma_+ + \sigma_-), \tag{6.122}$$

discuss how it leads to T_1 processes (amplitude damping) in the two-qubit system. Specifically, identify terms in the effective Hamiltonian that contribute to transitions between the basis states of the qubits, which can be considered as a T_1 process.
2. *Relate the T_1 relaxation process to quantum information storage and computation. How would the presence of a nonzero γ affect the performance of a quantum computer or the fidelity of quantum information storage? Provide concrete examples.*

Exercise 6.19 *Extend the above discussion to a two-qubit system weakly coupled to a harmonic oscillator. We assume that the two qubits do not interact directly with each other but only through the harmonic oscillator.*

1. ***Total Hamiltonian**: Show that the total Hamiltonian for a two-qubit system coupled to a harmonic oscillator is given by*

$$H = \omega_{01}\sigma_z^{(1)} + \omega_{02}\sigma_z^{(2)} + \omega a^\dagger a + g_1(\sigma_+^{(1)}a + \sigma_-^{(1)}a^\dagger) + g_2(\sigma_+^{(2)}a + \sigma_-^{(2)}a^\dagger)$$
$$\tag{6.123}$$

2. ***Effective Hamiltonian**: Find the form of the effective Hamiltonian $\tilde{H}$ for this two-qubit system under the Born–Markov approximations.*
3. ***Entanglement effects**: Explore how the interaction with the harmonic oscillator affects an initially entangled state between the two qubits. Calculate the state of the system after a short time interval and discuss whether this interaction tends to enhance or diminish the initial entanglement.*

In closing, the study of open quantum systems, epitomized by the Lindblad equation, is not just a niche domain within quantum computation; it represents a vast and continuously evolving frontier that investigates the rich interplay between quantum systems and their environments. Throughout this section, we've come to appreciate that grasping these dynamics is not solely for the sake of mastery of quantum computation, but is essential for cutting-edge developments and applications in the field.

It is noteworthy that the effective Hamiltonian $\tilde{H}$ in the Lindblad equation is commonly assumed to be Hermitian. This is not always the case, however Non-Hermitian systems, a consequential branch stemming from open quantum systems, have emerged as a fascinating area of study. These systems, defined by their non-Hermitian Hamiltonians, arise due to certain energy dissipations and feature mechanisms that can't be captured by traditional Hermitian descriptions. Their study is vital, as they can shed light on intricate quantum phenomena and drive innovative techniques in quantum computation.

Recent years have witnessed an upsurge in research dedicated to open quantum systems, unveiling aspects once deemed abstruse, from the complexities of the non-Hermitian realm to the expansion of quantum computational capabilities. From quantum error correction to the development of noise-resilient quantum algorithms, the influence of this domain is pervasive and profound. As we venture deeper into the era of quantum computation, the Lindblad equation, non-Hermitian systems, and their associated frameworks will undeniably stand as cornerstones, shaping our quantum computational endeavors. With every discovery and development, the promise of quantum computation grows stronger. To the next generation of quantum computer scientists and researchers, the realm of open quantum systems offers a treasure trove of challenges, insights, and innovations waiting to be unlocked.

6.5   Weak Measurement

In Chapter 3, we delineated various types of quantum measurements, such as projective, generalized, and positive-operator-valued measurements. As evidenced in Chapter 5, these measurements are foundational to quantum computation. Yet, this only scratches the surface of quantum measurement paradigms. In fact, in our first contact with quantum decoherence in this chapter, we subtly encountered another measurement type known as the weak measurement [199]. To fully grasp the nuances and significance of weak measurements, it's beneficial to revisit our initial interaction with a noisy quantum channel: the CNOT gate. For clarity, we reproduce Figure 6.1 here, renumbered as Figure 6.7.

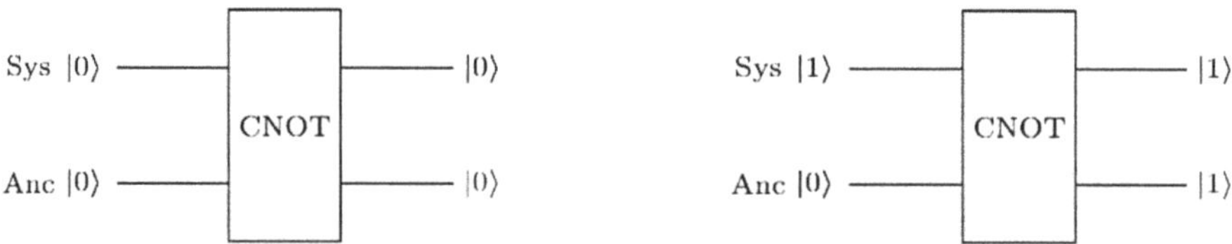

Figure 6.7 The CNOT gate as a premeasurement tool.

As the figure illustrates, when the system starts in a computational basis state, it remains undisturbed throughout the CNOT operation, and this state is mirrored by the ancillary qubit. This observation hints at the CNOT gate functioning as a "premeasurement," suggesting that the ancillary qubit, in a sense, "measures" the system qubit. Traditional measurements, particularly projective measurements, disrupt the quantum state being measured, leading to post-measurement collapse. In contexts like quantum computation and communication, there's often a desire to measure a qubit without changing its state. Unfortunately, this seems counterintuitive because measuring a qubit inherently involves an interaction between the qubit and the measuring apparatus, leading to potential changes in the qubit's state.

Yet, the behavior of the CNOT gate offers a glimpse into a potential workaround: Could we possibly link a qubit (or a group of qubits) to an ancilla (shorthand for ancillary qubit) and ascertain the state of the system through the ancilla measurement? This is a promising avenue, as demonstrated by the CNOT gate. But this method is not without its pitfalls. For instance, if the system qubit is in a non-standard state, and so not a computational basis state, the ancilla cannot simply replicate it but ends up entangled with it. When the ancilla is traced out, the system evolves into a mixed state, inheriting traces from the ancilla.

This brings us to some pivotal questions: How much information about the system's state can we glean solely by observing the ancilla? More importantly, is it possible to achieve a near-complete extraction of the system's state with minimal disruption? Weak measurements are designed to address these questions.

Instead of diving deeply into weak measurements, we will elucidate how they operate using the CNOT gate as a representative example.

Remember that our system qubit is the control qubit and the ancilla qubit serves as the data qubit. Their initial states are given by

$$|\psi_S\rangle = \alpha\,|0_S\rangle + \beta\,|1_S\rangle \tag{6.124}$$

$$|\psi_A\rangle = \cos\tfrac{\theta}{2}\,|0_A\rangle + \sin\tfrac{\theta}{2}\,|1_A\rangle, \tag{6.125}$$

where the subscripts S and A label the system and ancilla. As these states traverse the CNOT gate, their combined state becomes

$$\alpha\cos\tfrac{\theta}{2}\,|0_S0_A\rangle + \alpha\sin\tfrac{\theta}{2}\,|0_S1_A\rangle + \beta\sin\tfrac{\theta}{2}\,|1_S0_A\rangle + \beta\cos\tfrac{\theta}{2}\,|1_S1_A\rangle. \tag{6.126}$$

Upon taking a partial trace over the ancilla and the system qubits, respectively, we obtain

$$
\rho_S = \begin{pmatrix} |\alpha|^2 & \alpha\beta^*\sin\theta \\ \alpha^*\beta\sin\theta & |\beta|^2 \end{pmatrix},
$$

$$
\rho_A = \begin{pmatrix} |\alpha|^2\cos^2\tfrac{\theta}{2} + |\beta|^2\sin^2\tfrac{\theta}{2} & \tfrac{1}{2}\sin\theta \\ \tfrac{1}{2}\sin\theta & |\alpha|^2\sin^2\tfrac{\theta}{2} + |\beta|^2\cos^2\tfrac{\theta}{2} \end{pmatrix}. \tag{6.127}
$$

If we aim to measure the expectation value of σ_z in ρ_S, we get

$$\langle \sigma_z \rangle_S = |\alpha|^2 - |\beta|^2. \tag{6.128}$$

On the other hand,

$$\langle \sigma_z \rangle_A = (|\alpha|^2 - |\beta|^2) \cos \theta. \tag{6.129}$$

The essence of weak measurement is to read out $\langle \sigma_z \rangle_S$ by observing σ_z in the ancilla state. This is achievable as long as $\theta \neq \pi/2$, since

$$\langle \sigma_z \rangle_S = \frac{\langle \sigma_z \rangle_A}{\cos \theta}. \tag{6.130}$$

To ensure minimal disturbance to the system qubit, we require a θ value close to $\pi/2$. This kind of measurement is termed a **weak measurement**. Now, a small value of $\cos \theta$ causes $\langle \sigma_z \rangle_A$ to be exceedingly minute, complicating precise measurements. As such, in order to obtain a reliable outcome for $\langle \sigma_z \rangle_S$, it becomes necessary to repeat the weak measurement sufficiently many times to accomplish this.

Weak measurements have been a topic of considerable interest and active research in quantum mechanics over the past few decades [200]. They provide a way to gain insight into a quantum system without causing a full collapse of the system's state, as seen in traditional projective measurements. This delicate balance between obtaining information and minimally disturbing the system is what makes weak measurements a significant and versatile tool in quantum physics. Their applications range from fundamental quantum mechanics experiments, aiming to probe and understand the subtleties of quantum theory, to more practical scenarios in quantum computing and quantum communication where minimal disturbance is crucial. As our exploration of quantum systems deepens and the demands of quantum technologies grow, the techniques and principles of weak measurements are poised to play an ever-increasing role. While this chapter only scratches the surface, the reader is encouraged to explore deeper into the rich and intricate world of weak measurements to appreciate their full scope and potential.

Exercise 6.20 *Consider a data qubit prepared in the state*

$$|\psi_d\rangle = \alpha|0\rangle + \beta|1\rangle, \tag{6.131}$$

where $|0\rangle$ and $|1\rangle$ are the computational basis states, and $|\alpha|^2 + |\beta|^2 = 1$. An ancilla qubit is initialized in the state

$$|\psi_a\rangle = |0\rangle. \tag{6.132}$$

To perform a weak measurement of the σ_z observable on the data qubit, we let the data qubit and ancilla qubit interact via the Hamiltonian

$$H_{int} = g\,\sigma_z^{(d)} \otimes \sigma_y^{(a)}, \tag{6.133}$$

where

- $\sigma_z^{(d)}$ *acts on the data qubit,*
- $\sigma_y^{(a)}$ *acts on the ancilla qubit,*
- g *is a small coupling constant (ensuring a weak interaction).*

The system evolves for a short time δt, resulting in the evolution operator

$$U = e^{-iH_{int}\delta t} \approx I - iH_{int}\delta t, \tag{6.134}$$

where I is the identity operator and we set $\hbar = 1$ for simplicity. After the interaction, a projective measurement of σ_x is performed on the ancilla qubit. Complete tasks 1 through 5. (Assume $g\delta t \ll 1$ and neglect terms of order $(g\delta t)^2$ or higher.)

1. **State after interaction:**
   (a) *Compute the combined state $|\Psi_{after}\rangle$ of the data and ancilla qubits after the interaction and before the measurement.*
   (b) *Show that to first order in $g\delta t$,*

   $$|\Psi_{after}\rangle \approx |\psi_d\rangle \otimes |0\rangle - ig\delta t \left(\sigma_z^{(d)}|\psi_d\rangle\right) \otimes \left(\sigma_y^{(a)}|0\rangle\right), \tag{6.135}$$

2. **Measurement outcomes:**
   (a) *Calculate the probabilities P_+ and P_- of obtaining the measurement outcomes $|+\rangle_x$ and $|-\rangle_x$ when measuring σ_x on the ancilla qubit.*
   (b) *Express these probabilities in terms of α, β, and $g\delta t$.*

3. **Updated data qubit state:**
   (a) *Determine the normalized post-measurement state $|\psi_d^+\rangle$ of the data qubit, given the ancilla qubit is found in $|+\rangle_x$.*
   (b) *Repeat for the outcome $|-\rangle_x$, obtaining $|\psi_d^-\rangle$.*

4. **Analysis of disturbance:**
   (a) *Show that the disturbance to the data qubit's state is proportional to $g\delta t$.*
   (b) *Discuss how the limit $g\delta t \to 0$ minimizes the disturbance.*

5. **Information gain vs. disturbance:**
   (a) *Explain how measuring the ancilla qubit provides information about the σ_z observable of the data qubit.*
   (b) *Discuss the trade-off between the amount of information gained and the disturbance to the data qubit as $g\delta t$ varies.*

Additionally, answer the following questions.
- *How does the weak coupling $g\delta t$ affect the probabilities P_+ and P_-?*
- *How does the measurement of the ancilla qubit indirectly measure the data qubit?*
- *How is the concept of "weak measurement" reflected in the minimal disturbance to the data qubit's state?*

Conclusion

In this chapter, we explored the phenomenon of quantum decoherence, a fundamental challenge to sustaining quantum information. Beginning with an introduction to the concept of decoherence, we saw how the interaction between a quantum system and its environment leads to a loss of coherence, shifting the system's state toward a classical mixture. Through practical examples, such as the bit-flip, phase-flip, and depolarizing channels, we illustrated how specific interactions cause qubit states to decohere, each channel affecting the qubit's representation on the Bloch sphere in a unique way.

We examined how superoperators and the Kraus representation provide a mathematical framework for describing noisy quantum channels, emphasizing the Kraus operators' role in capturing the probabilistic effects of environmental interactions. The Lindblad equation was introduced as a model for the time evolution of open quantum systems, allowing us to quantify the rate of decoherence through metrics such as T_1 and T_2 times, critical parameters for qubit stability in quantum computing.

Chapter 7 will address strategies to mitigate the effects of decoherence. This includes examining quantum error correction codes, which are crucial for preserving quantum information in the presence of noise, thereby enabling practical, large-scale quantum computations.

7   Quantum Error Correction

In the realm of quantum computation, qubits are particularly susceptible to environmental interference, especially within the context of quantum circuit-based computation. As discussed in the preceding chapter, qubits can be compromised due to their traversal through noisy quantum channels, which can result in various types of errors. A bit-flip channel, for instance, might precipitate a transition $|0\rangle \leftrightarrow |1\rangle$, thus inducing a bit-flip error.

It's also worth noting that errors can arise due to measurements, where the act of observing a quantum state can alter it, leading to measurement errors. Nevertheless, in this chapter, our focus is primarily on errors arising from noisy quantum channels. Measurement errors, characterized by their non–trace preserving nature,[1] fall outside the scope of our current discussion on quantum error correction.

While technological advancements allow for the fabrication of qubit systems with progressively lower fault rates, achieving an impeccable qubit system remains physically impossible due to the presence of an environment, which is unavoidable as long as we need to control the qubits. In quantum computation theory, these inherently susceptible qubits are referred to as **physical qubits**. To perform reliable computations using such qubit systems, one must both monitor these physical qubits and correct any errors that may arise. Quantum error correction codes are a solution to this challenge. A **quantum error correction code** (**QECC**) encodes a solitary qubit of information into a composite state of multiple physical qubits, termed a logical qubit state. These multiple physical qubits collectively represent a single **logical qubit**.

This chapter will expound on various types of QECCs, starting with elementary three-qubit codes for bit-flip and phase-flip errors [201], progressing to Shor's comprehensive QECC [202], and culminating in a discussion of general properties essential for quantum error correction. This treatment sets the foundation for our subsequent expedition into stabilizer codes [203], which will in turn pave the way for an exploration of surface codes and surface code quantum computation in Chapter 8.

---

[1] Recall that the Kraus representation theorem 6.2 applies to trace-preserving superoperators only.

7.1   Classical Error Correction and Quantum No-Cloning Theorem

Error correction codes are by no means exclusive to the domain of quantum computation; they have a longstanding history in classical computation as well. Just like qubits, classical bits are also subject to unavoidable physical errors. Therefore, classical computing systems also employ strategies to encode a single logical bit of information across multiple physical bits. We shall begin by exploring the mechanisms behind the simplest classical error correction scheme, as it will serve as the conceptual springboard for understanding the simplest QECCs.

7.1.1   Classical Repetition Code

In any bit-based system, a physically determined error rate exists, characterized by a fixed probability p of bit-flipping for each bit. Provided that this probability is sufficiently low, $p \ll 1$, the system can be considered reliable. It follows that for any given string of bits, the likelihood of a single bit being flipped is much greater than that of multiple bits being flipped. This observation forms the basis for encoding a logical bit as multiple copies of itself, often termed the n-bit repetition code.

To elaborate, consider a bit of information, denoted $b \in \{0, 1\}$. We can replicate this bit n times to form a string $bb \cdots b$, which collectively serves as the logical bit b_L, defined as

$$0_L = \underbrace{00 \cdots 0}_{n}, \quad 1_L = \underbrace{11 \cdots 1}_{n}. \tag{7.1}$$

According to Figure 7.1, let's quantify the likelihood of either no bit or only a single bit being flipped within b_L, denoted as $p_{\leq 1}$. Given that the probability for each bit to remain unaltered during transmission is $1 - p$, we can express this as

$$p_{\leq 1} = (1 - p)^n + np(1 - p)^{n-1}. \tag{7.2}$$

Consequently, the probability that two or more bits are flipped is $p_{\geq 2} = 1 - p_{\leq 1}$.

It follows from $p \ll 1$ that $p_{\leq 1} \gg p_{\geq 2}$. For example, if $n = 3$ and $p = 0.05$, then

$$p_{\leq 1} = 0.99275, \quad p_{\geq 2} = 0.00725. \tag{7.3}$$

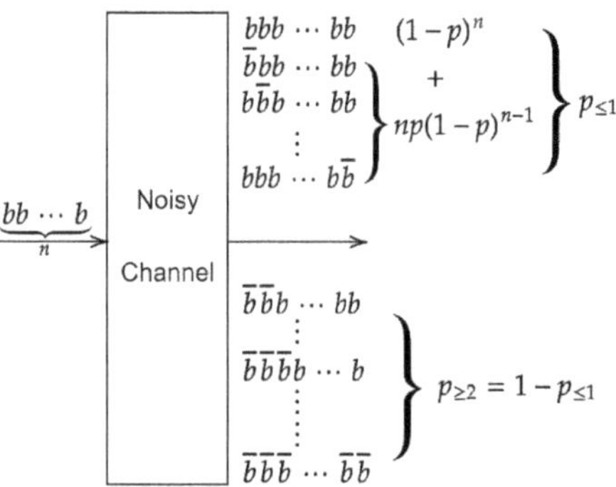

Figure 7.1 A classical logical bit encoded in n bits going through a noisy channel.

This makes it highly likely for the receiver to get either a correct logical bit or one with a single bit error. In the latter case, for a received bit string like $b\bar{b}b\cdots b$, the receiver can simply apply a NOT gate to flip $\bar{b}$ back to b, effectively correcting the error and decoding the correct logical bit b.

> **Exercise 7.1** *Discuss the case where $n = 2$.*

In summary, the classical repetition code teaches us that an error correction procedure typically involves four key steps:

1. Encoding: The logical bit is encoded across a predetermined number of physical bits.
2. Error syndrome detection: The receiver identifies the location and type of errors. It's worth noting that while a classical bit can only suffer a bit-flip error, a qubit is susceptible to both bit-flip and phase-flip errors.
3. Error correction: The receiver corrects the identified errors.
4. Decoding: The receiver decodes the corrected logical bit to obtain the original information.

These principles extend to quantum error correction as well. The looming question now is: Can this repetition code paradigm be implemented for quantum error correction? The answer is nuanced, requiring an understanding of the quantum no-cloning theorem.

7.1.2 The Quantum No-Cloning Theorem

According to quantum mechanics, an unknown state cannot be accurately replicated, or cloned, by any unitary process. This fact can be rigorously stated in Theorem 7.1.

Theorem 7.1. *No-Cloning Theorem [204]: Given a generic unknown state $|\psi\rangle$ of a quantum system, such as a qubit, there exists no unitary transformation U such that*

$$U |\psi\rangle |\phi\rangle = |\psi\rangle |\psi\rangle, \tag{7.4}$$

where $|\phi\rangle$ is any known state of another quantum system, for example another qubit, used to clone the state $|\psi\rangle$.

Proof We can prove this theorem by contradiction. Assume such a unitary transformation U that realizes Equation (7.4) exists. Consider two distinct states $|\psi_1\rangle$ and $|\psi_2\rangle$ to be cloned. That is, we can have

$$U |\psi_1\rangle |\phi\rangle = |\psi_1\rangle |\psi_1\rangle, \tag{7.5}$$

$$U |\psi_2\rangle |\phi\rangle = |\psi_2\rangle |\psi_2\rangle. \tag{7.6}$$

Since U is linear, it should be able to clone the state $|\psi\rangle = \alpha |\psi_1\rangle + \beta |\psi_2\rangle$ too, but this leads to two contradictory results:

$$U |\psi\rangle |\phi\rangle = |\psi\rangle |\psi\rangle = (\alpha |\psi_1\rangle + \beta |\psi_2\rangle) \otimes (\alpha |\psi_1\rangle + \beta |\psi_2\rangle) \tag{7.7}$$

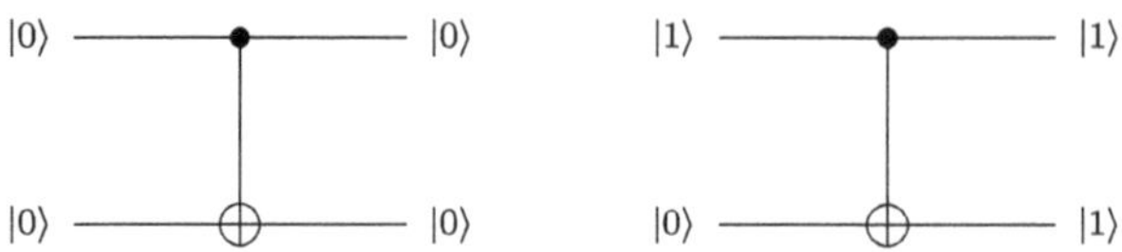

Figure 7.2 CNOT can copy the computational basis states $|0\rangle$ and $|1\rangle$.

and

$$U |\psi\rangle |\phi\rangle = \alpha U |\psi_1\rangle |\phi\rangle + \beta U |\psi_2\rangle |\phi\rangle = \alpha |\psi_1\rangle |\psi_1\rangle + \beta |\psi_2\rangle |\psi_2\rangle. \qquad (7.8)$$

Therefore, such a U cannot exist. $\qquad\qquad\qquad\qquad\qquad\qquad\qquad$ □

By the quantum no-cloning theorem, a blatant replication of an unknown quantum state is impossible.[2] Does this forbid a quantum repetition code for error correction? Not exactly. Can you see how to circumvent the conundrum?

7.1.3 Quantum Partial Copying

Astute readers may see a way out of the dilemma of being unable to clone an unknown quantum state. The key is the CNOT gate. We have already seen that the CNOT gate can copy the state of the control qubit onto the data qubit, provided that the control qubit is in one of its computational basis states and the data qubit is initially in the state $|0\rangle$. Figure 7.2 should serve as a reminder. Importantly, this state replication does not violate the quantum no-cloning theorem because an arbitrary state like $\alpha |0\rangle + \beta |1\rangle$ of the control qubit cannot be exactly duplicated onto the data qubit. Instead, an arbitrary control qubit state $\alpha |0\rangle + \beta |1\rangle$ will result in the entangled state $\alpha |00\rangle + \beta |11\rangle$ of the two qubits.

Hold on a moment. It appears that although an arbitrary state cannot be duplicated, its components can be [206, 207]! Could this provide a pathway to tackle our challenge of designing a quantum repetition code? Indeed, it does, as we will soon demonstrate. And stay tuned; the entanglement that just appeared will prove to be crucial.

7.2 Three-Qubit Bit-Flip and Phase-Flip QECCs

Let's first clarify the two types of errors a qubit may suffer from. Suppose a sequence of qubits is passing through an NQC. Let $|\psi\rangle = a |0\rangle + b |1\rangle$ represent the state of a single qubit. A *bit-flip error* is defined as an error that transforms $|\psi\rangle$ into

$$|\psi'\rangle = a |1\rangle + b |0\rangle, \qquad (7.9)$$

---

2 The quantum no-cloning theorem is a no-go theorem. Every no-go theorem holds under certain conditions, and violating those conditions can break it. In particular, the quantum no-cloning theorem requires copying an unknown pure state to another pure state while preserving the original. A recent study [205] circumvents the no-cloning theorem by employing a method called *encrypted cloning*. Interested readers are invited to consult the paper and determine which condition(s) of the no-cloning theorem are relaxed under encrypted cloning.

whereas a *phase-flip error* transforms it into

$$|\psi'\rangle = a\,|0\rangle - b\,|1\rangle. \tag{7.10}$$

We will explore how to address these two kinds of errors individually and then discuss how to handle their combination and possible variations.

7.2.1 Bit-Flip Error

Motivated by the classical repetition code and the partial copying capability of the CNOT gate, we aim to transmit a single qubit of information $|\psi\rangle = a\,|0\rangle + b\,|1\rangle$ through an NQC. Each basis state in $|\psi\rangle$ is replicated multiple times. Following the same analysis as in Section 7.1.1 and Exercise 7.1, the minimal number of replications is three. For illustrative purposes, we shall consider just three.

Encoding

We encode the qubit of information $|\psi\rangle$ to be transmitted into a logical qubit comprising three physical qubits:

$$|\psi\rangle \rightarrow |\psi_L\rangle = a\,|000\rangle + b\,|111\rangle. \tag{7.11}$$

Two CNOT gates and two additional physical qubits are required to construct the logical qubit (7.11), as implemented by this circuit:

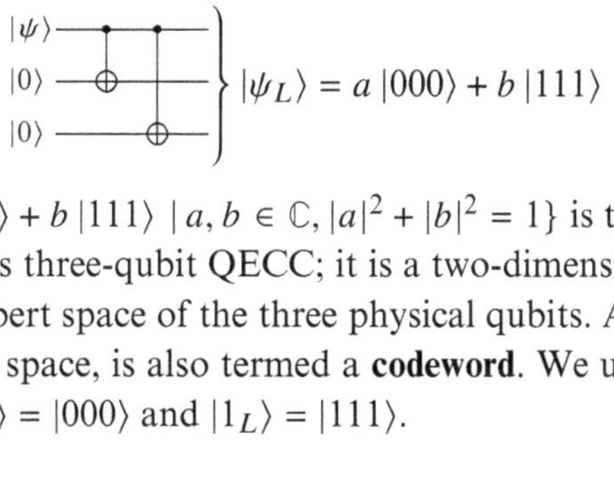

Figure 7.3 Circuit for implementing the qubit $|\psi\rangle$.

The space $C = \{a\,|000\rangle + b\,|111\rangle \mid a, b \in \mathbb{C}, |a|^2 + |b|^2 = 1\}$ is termed the **code space** or **logical space** for this three-qubit QECC; it is a two-dimensional subspace of the eight-dimensional Hilbert space of the three physical qubits. A logical state, meaning a state in the code space, is also termed a **codeword**. We usually define the two logical basis states $|0_L\rangle = |000\rangle$ and $|1_L\rangle = |111\rangle$.

Correctable Errors

Consider transmitting a logical qubit through three independent bit-flip NQCs, each denoted as $\mathcal{E}_{\mathrm{bf}}$. Each of the three physical qubits that compose this logical qubit may endure a bit-flip error with an independent probability p. Before exploring techniques to identify and rectify the flipped qubits, it is essential to examine the maximum number of bit-flip errors this three-qubit QECC can correct. This insight will be generalized in subsequent discussions.

Given the possibility that each physical qubit may flip, all eight three-qubit basis states of the logical qubit,

$$\{|i_2 i_1 i_0\rangle \mid i_2, i_1, i_0 = 0, 1\}, \tag{7.12}$$

are accessible via the NQCs. Adopting a geometric perspective, these eight basis states can be positioned at the corners of a cube [21]. Each edge of this cube signifies

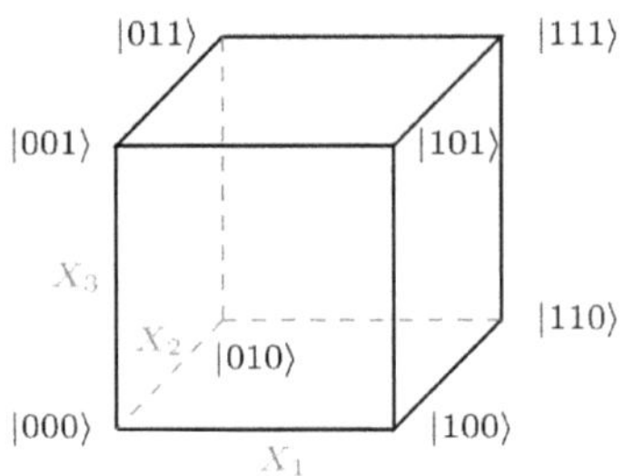

Figure 7.4 The cube of the eight three-qubit basis states. Each edge denotes a bit-flip error, labeled by X_1, X_2, or X_3, indicating which qubit is affected. Any two states connected by an edge are separated by exactly one bit-flip error.

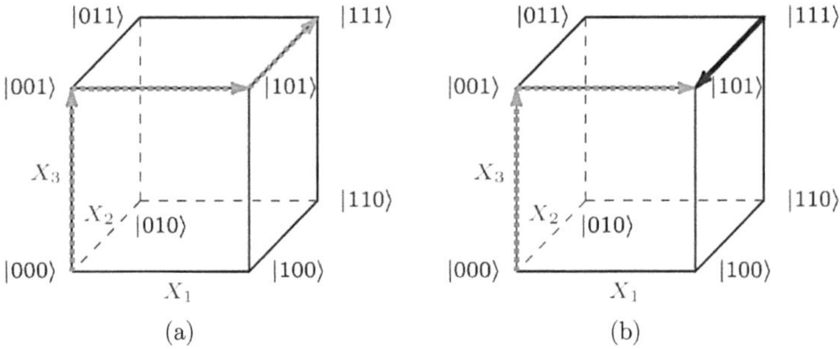

Figure 7.5 (a) Three single-qubit errors connect $|0_L\rangle$ and $|1_L\rangle$. (b) Two single-qubit errors on $|0_L\rangle$ and one on $|1_L\rangle$ map both to the same basis state.

a bit-flip error, connecting two states that differ by exactly one flipped qubit, as illustrated in Figure 7.4. Edges along the three orthogonal directions correspond to bit-flip errors on the respective qubits.

If the NQCs can swap the logical basis states $|0_L\rangle$ and $|1_L\rangle$ through bit-flip errors, the three-qubit QECC would be incapable of discerning whether errors have occurred, let alone correcting them. Discriminating between $a\,|000\rangle + b\,|111\rangle$ and $a\,|111\rangle + b\,|000\rangle$ would become unfeasible.

If all three physical qubits endure bit-flip errors, they could transform $|0_L\rangle$ and $|1_L\rangle$ into each other, as illustrated by a path comprising three orthogonal edges connecting these states on the cube. See the straight-arrow path in Figure 7.5(a) for an example. Alternatively, if either one or two physical qubits in a logical qubit may flip, both $|0_L\rangle$ and $|1_L\rangle$ can be converted to the same incorrect basis state, for example $|101\rangle$ in Figure 7.5(b). Although one knows that $|101\rangle$ is incorrect, it is impossible to tell whether it originated from $|0_L\rangle$ or $|1_L\rangle$.

Thus, the three-qubit QECC can only unambiguously detect and correct an error if a single physical qubit in the logical qubit is flipped. As shown in Figure 7.6, this unambiguity arises because the incorrect basis states reached from $|0_L\rangle$ are

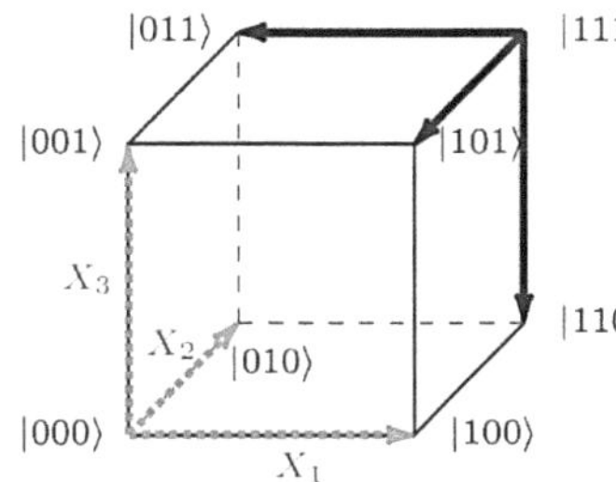

Figure 7.6 The incorrect basis states stemming from $|0_L\rangle$ are orthogonal to those stemming from $|1_L\rangle$.

orthogonal to those originating from $|1_L\rangle$, a pivotal concept for our broader exploration of quantum error correction.

In practical scenarios, similar to classical computation, the likelihood of encountering two or three bit-flip errors is significantly lower in a reliable qubit system. Therefore, the discussion will focus on the case where a logical qubit is susceptible to at most a single-qubit error. The pressing question then becomes: How can the receiver of a logical qubit identify and rectify a single bit-flip error?

Error Syndrome Detection

Focusing on single-qubit errors, a logical qubit traversing the bit-flip channels may emerge in one of four potential states, including the unaltered state:

$$
\begin{aligned}
a\,|000\rangle + b\,|111\rangle, \\
a\,|100\rangle + b\,|011\rangle, \\
a\,|010\rangle + b\,|101\rangle, \\
a\,|001\rangle + b\,|110\rangle.
\end{aligned}
\tag{7.13}
$$

A direct measurement could, of course, discern these states. Such an action would collapse the quantum state, however, negating its utility for future computations or communications. So, what is the solution?

The answer lies in an approach akin to weak measurement. Unlike traditional weak measurements, there's no need to determine the exact received state; identifying the flipped physical qubit suffices. To achieve this, the receiver entangles the logical qubit with a certain number of ancillary qubits. These ancilla qubits, known as **syndrome qubits**, interact in a way that correlates their states with the possible logical states. Since four logical states exist, precisely two syndrome qubits are required. Figure 7.7 shows the entangling quantum circuit.

Prior to measuring the syndrome qubits, the state of all five qubits in the circuit shown in Figure 7.7 forms an entangled superposition of the following states:

$$
\begin{aligned}
(a\,|000\rangle + b\,|111\rangle) \otimes |00\rangle, \\
(a\,|100\rangle + b\,|011\rangle) \otimes |11\rangle, \\
(a\,|010\rangle + b\,|101\rangle) \otimes |10\rangle, \\
(a\,|001\rangle + b\,|110\rangle) \otimes |01\rangle.
\end{aligned}
\tag{7.14}
$$

Evidently, a measurement on the syndrome qubits leaves the logical state untouched due to the entanglement. If the syndrome qubits yield $SS' = 00$, the logical qubit

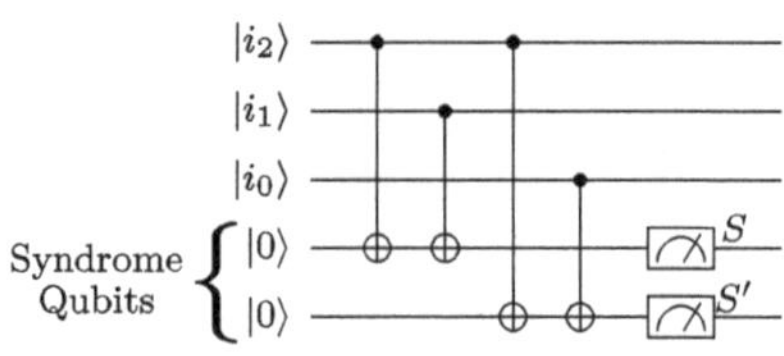

Figure 7.7 Circuit for detecting single-qubit bit-flip errors in a three-qubit QECC. The bottom two qubits are the syndrome qubits, and their measurement outcomes S and S' compose the error syndrome bit-string SS'.

Figure 7.8 Circuit designed to correct single bit-flip errors within a three-qubit QECC.

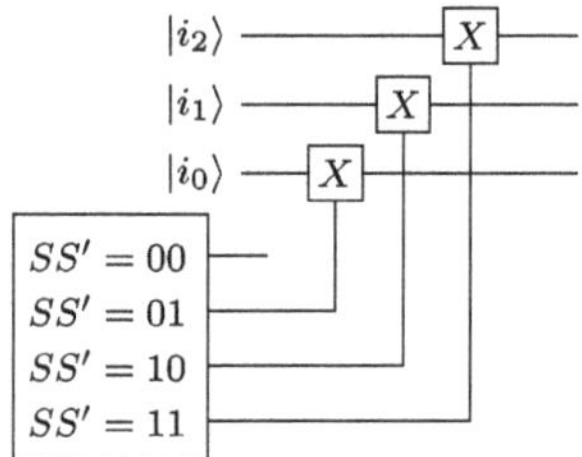

remains in its correct state. Otherwise, a single bit-flip error has occurred, which can be unambiguously identified via $SS' \neq 00$.

> **Exercise 7.2** *The circuit in Figure 7.7 is not the sole configuration that can achieve error syndrome detection for a three-qubit QECC. Design an alternative circuit that also performs unambiguous error syndrome detection. In fact, how many different circuits are there that achieve this goal?*

Error Correction and Fault Tolerance

Upon having detected the bit-flipped physical qubit within the logical qubit, the natural progression is to correct the error. This can be achieved by applying a NOT (X) gate to the misbehaving qubit. The specific circuit designed for this operation is illustrated in Figure 7.8.

A perceptive reader might question the necessity of measuring the syndrome qubits for error detection and correction. Intriguingly, it's not a strict requirement; it is entirely feasible to construct a circuit that corrects errors without any syndrome measurement.

> **Exercise 7.3** *Design a circuit capable of rectifying errors in a three-qubit QECC without resorting to the measurement of syndrome qubits. For this task, you are allowed to utilize C_2NOT gates without the need for decomposing them into their elementary gates.*

Should you attempt Exercise 7.3, you'll discern that such a circuit tends to involve a greater number of elementary gates. This intricacy amplifies the risk of introducing new errors. This realization guides us to a pivotal concept in the realm of QECCs: **fault tolerance**. Essentially, fault tolerance encapsulates the ability of a QECC to endure errors that might transpire during the error correction phase itself, in addition to the routine operations of quantum computation.

In designing fault-tolerant QECCs, it's imperative to ensure that errors introduced during the correction process can be effectively managed. This often necessitates encoding logical qubits in a manner that remains resilient to a designated set of errors and performing logical operations that uphold this resilience. A crucial method in this context is employing "transversal gates." These are logical actions that can be actualized as a sequence of physical gates, each acting autonomously on respective physical qubits from disparate code blocks.

Fault tolerance involves additional considerations, including the error rates of the quantum gates used in error correction and logical operations, as well as the architecture of the quantum computer itself. While this text won't provide a comprehensive discussion on fault tolerance, it will introduce Shor's nine-qubit QECC [202]. Though not inherently fault-tolerant, when embedded in a scheme of fault-tolerant quantum computation, it can operate in a fault-tolerant manner. Achieving this requires implementing its syndrome measurement and recovery operations with specific fault-tolerant protocols, such as using verified ancilla qubits, to prevent error propagation during the correction process. (The Steane seven-qubit QECC [208], another key QECC suitable for fault-tolerant quantum computing, will not be discussed. The Steane code is particularly suited for fault-tolerant quantum computing because its Clifford gates (H, S, CNOT) are transversal, meaning they act independently on physical qubits to naturally contain errors.)

Chapter 8 will provide a more exhaustive exploration of by far the most promising fault-tolerant QECC, known as the surface code.

Decoding

The state after error detection and correction is presumably the correct logical state, from which the receiver can extract the original information transmitted. This procedure of information extraction is referred to as decoding. But how to decode the three-qubit QECC?

The key to decoding lies in encoding. The logical qubit is encoded via the CNOT gates. Remember $\mathrm{CNOT}^2 = \mathbb{1}$? Hence, to decode, the receiver simply needs to reverse the encoding procedure:

$$\mathrm{CNOT}_{21}\,\mathrm{CNOT}_{20}(a\,|000\rangle + b\,|111\rangle) = (a\,|0\rangle + b\,|1\rangle) \otimes |00\rangle, \qquad (7.15)$$

where the subscript 2 of the CNOT gates labels the control qubit, while 1 and 0 label the two data qubits respectively.

Figure 7.9 combines the previously shown quantum circuits respectively for encoding, error syndrome detection, and error correction, as well as appending the circuit for decoding to construct the complete quantum circuit for three-qubit quantum error correction.

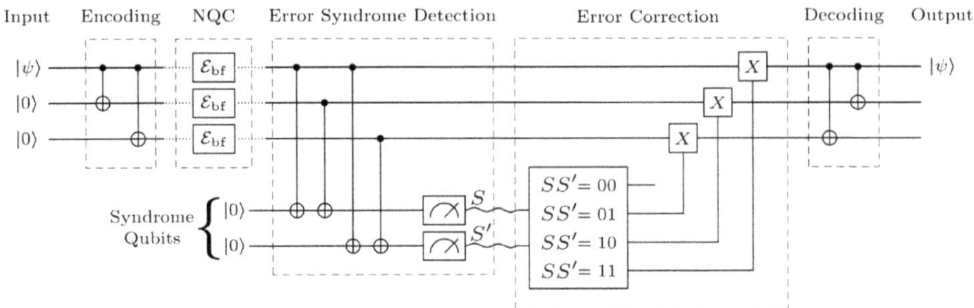

Figure 7.9 Quantum circuit of the three-qubit QECC for single bit-flip errors.

Continuous Qubit Rotation

In practical scenarios, a bit-flip channel might not be precisely represented by an X gate that performs a straightforward qubit flip. Instead, it might induce a continuous rotation of the qubit, generated by the operator X. This rotation can be described by the unitary

$$R_x(\theta) = e^{i\theta X} = (\cos\theta)\mathbb{1}_2 + i(\sin\theta)X, \tag{7.16}$$

resulting in the post-channel state of the qubit being

$$R_x(\theta)\,|\psi\rangle = \cos\theta\,|\psi\rangle + i\sin\theta X\,|\psi\rangle. \tag{7.17}$$

A pertinent question then arises: Can our three-qubit QECC address such errors? The answer is affirmative. To understand why, let's explore a specific case:

Assume that the leftmost (or topmost in the circuit) physical qubit – the one holding the original information – undergoes this imperfect bit-flip channel. The resulting logical state becomes

$$U_\theta \otimes \mathbb{1}_2 \otimes \mathbb{1}_2\,|\psi_L\rangle = \cos\theta\,|\psi_L\rangle + i\sin\theta(a\,|100\rangle + b\,|011\rangle), \tag{7.18}$$

which, when combined with the syndrome qubits prior to syndrome detection, evolves into

$$\cos\theta\,|\psi_L\rangle \otimes |00\rangle + i\sin\theta(a\,|100\rangle + b\,|011\rangle) \otimes |11\rangle. \tag{7.19}$$

From this state, syndrome measurement reveals two potential outcomes:

$$\begin{aligned} SS' = 00: &\quad |\psi_L\rangle, \\ SS' = 11: &\quad a\,|100\rangle + b\,|011\rangle, \end{aligned} \tag{7.20}$$

where the immaterial phase factors $\cos\theta$ and $i\sin\theta$ have been disregarded. Notably, these outcomes match those obtained from a perfect bit-flip channel. Thus, they can be corrected in the same manner. The situation is similar if either of the two ancilla qubits endure such an error. Consequently, the three-qubit QECC and the procedure outlined in Figure 7.9 remain effective against single-qubit errors due to imperfect bit-flip channels. The inherent resilience of QECCs in such situations underlines their importance, a theme we will revisit and generalize later.

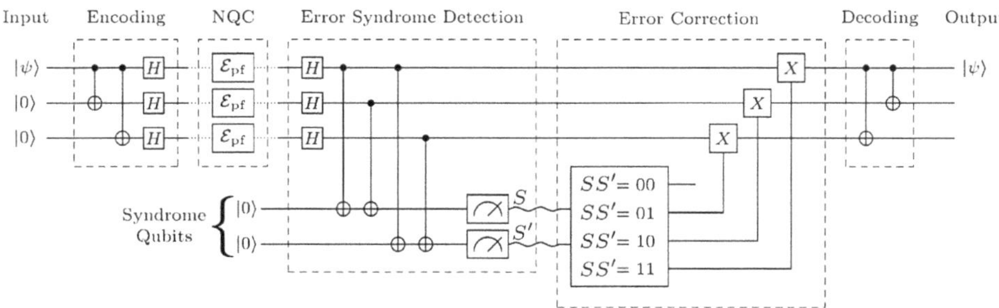

Figure 7.10 Quantum circuit of the three-qubit QECC for single phase-flip errors.

7.2.2 Phase-Flip Error

A phase-flip error, as depicted by Equation (7.10), can be succinctly expressed as $|i\rangle \to (-1)^i |i\rangle$. The NQC $\mathcal{E}_{\mathrm{pf}}$ that precipitates this error can be emulated by a Z gate because $Z|i\rangle = (-1)^i |i\rangle$. You might be thinking: Do we need to build a whole new QECC just for phase-flip errors?

The answer is, surprisingly, no. With a little tweak to our three-qubit QECC designed for single bit-flip errors, it's ready to correct single phase-flip errors as well. I bet you remember the useful conjugation relation $HZH = X$ from Exercise 4.15. With that in mind, the strategy is quite clever: Just sandwich each phase-flip channel between two Hadamard gates, effectively transforming it into a bit-flip channel. So, in our existing Figure 7.9, we simply swap out each NQC $\mathcal{E}_{\mathrm{bf}}$ with $H\mathcal{E}_{\mathrm{pf}}H$. The outcome is a modified quantum circuit (Figure 7.10) that's proficient in detecting and rectifying single phase-flip errors within our three-qubit QECC.

Exercise 7.4 *As with the bit-flip error, a phase-flip channel might not be impeccable. In some cases, it's more accurately depicted by a unitary phase rotation:*

$$R_z(\phi) = e^{\tfrac{i}{2}\phi Z}. \tag{7.21}$$

Show that the three-qubit QECC in Figure 7.10 remains resilient to single-qubit phase rotation errors.

7.3 Shor's QECC for Single-Qubit General Errors

Although our previously designed three-qubit QECCs for single bit-flip and phase-flip errors respectively are toy QECCs, they not only cultivate our intuition for designing QECCs but also serve as building blocks for more complicated and versatile QECCs. Imagine the scenario where a physical qubit may endure bit-flip, phase-flip, or both errors simultaneously. Can we construct a single QECC that is capable of detecting and correcting all such errors? There are in fact many such QECCs, and a simple and typical one is Shor's QECC, on which we shall expound in this section.

Recall the definitions of bit-flip and phase-flip errors from Equations (7.9) and (7.10). These errors can be emulated by the X and Z gates, respectively. If both errors concur on a qubit, the qubit endures a combined bit- and phase-flip error:

$$a\,|0\rangle + b\,|1\rangle \rightarrow -b\,|0\rangle + a\,|1\rangle. \tag{7.22}$$

Which unitary gate can simulate this combined error channel? Referring back to Equation (4.9) and Figure 4.7, the gate in question should be $Y = XZ$. A pivotal observation is that since $X, Y,$ and Z gates can generate all single-qubit unitary operations, any single-qubit error can be expressed in terms of these three fundamental error types.

Before diving straight into Shor's QECC, can you estimate how many physical qubits does the QECC need to encode a logical qubit? Well, we have designed a three-qubit QECC for single bit-flip errors and a similar one for single phase-flip errors. To detect and correct concurrent bit flip and phase flip on a single qubit, we may also need three qubits. So, this naive estimation says nine qubits. Although this estimation is correct, we will see that the actual encoding cannot be a naive combination of the previously constructed three-qubit QECCs because a qubit may suffer from both bit-flip and phase-flip errors simultaneously. Now, let's explore Shor's nine-qubit QECC step by step.

7.3.1 Encoding

To understand the encoding in Shor's nine-qubit QECC, let's first consider two orthogonal three-qubit states:

$$|+\rangle := \frac{1}{\sqrt{2}}(|000\rangle + |111\rangle), \quad |-\rangle := \frac{1}{\sqrt{2}}(|000\rangle - |111\rangle). \tag{7.23}$$

Suppose one, say for example the first one, of the three qubits had a bit-flip error; these two states would have become

$$|+\rangle \rightarrow \frac{1}{\sqrt{2}}(|100\rangle + |011\rangle), \quad |-\rangle \rightarrow \frac{1}{\sqrt{2}}(|100\rangle - |011\rangle). \tag{7.24}$$

If these two resultant states are regarded as two logical states of the three-qubit QECC for single bit-flip errors, one can simply apply the corresponding error syndrome detection and correction procedure to rectify the bit-flip error.

On the other hand, if one, for example the second one, of the three qubits in (7.23) were phase-flipped, we would have

$$|+\rangle \rightarrow \frac{1}{\sqrt{2}}(|000\rangle - |111\rangle) = |-\rangle, \quad |-\rangle \rightarrow \frac{1}{\sqrt{2}}(|000\rangle + |111\rangle) = |+\rangle. \tag{7.25}$$

That is, $|+\rangle$ and $|-\rangle$ are exchanged. This is true no matter which one of the three qubits endures a phase flip. Hence, if $|\pm\rangle$ are taken as the two basis states of a "block qubit" consisting of three qubits, a phase flip on any of the three qubits turns out to bit-flip the block qubit. As we know, we need a three-qubit QECC to detect and correct a single bit-flip error. Likewise, to detect and correct a bit-flip on a block qubit, we would need to triplicate the block qubit. This analysis suggests these logical qubit basis states:

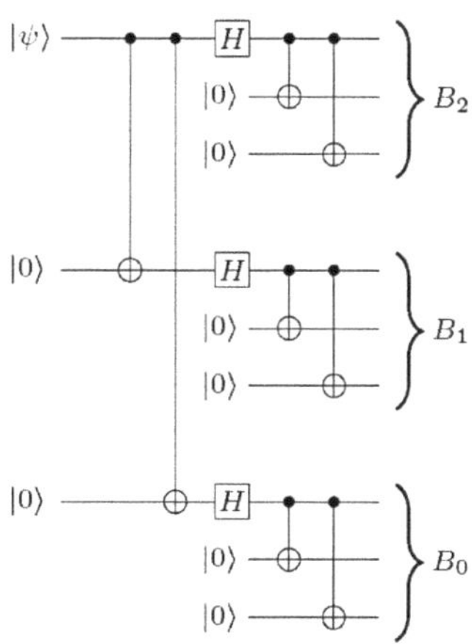

Figure 7.11 Encoding circuit for Shor's nine-qubit QECC.

$$|0_L\rangle := |+++\rangle = \underbrace{\frac{1}{\sqrt{2}}(|000\rangle + |111\rangle)}_{\text{Block 1}} \otimes \underbrace{\frac{1}{\sqrt{2}}(|000\rangle + |111\rangle)}_{\text{Block 2}} \otimes \underbrace{\frac{1}{\sqrt{2}}(|000\rangle + |111\rangle)}_{\text{Block 3}},$$

$$|1_L\rangle := |---\rangle = \frac{1}{\sqrt{2}}(|000\rangle - |111\rangle) \otimes \frac{1}{\sqrt{2}}(|000\rangle - |111\rangle) \otimes \frac{1}{\sqrt{2}}(|000\rangle - |111\rangle),$$

$$(7.26)$$

such that a logical qubit is represented as

$$|\psi_L\rangle = a\,|+++\rangle + b\,|---\rangle. \tag{7.27}$$

The idea is plain: Any single bit-flip error on a physical qubit can be detected and corrected within the block containing the physical qubit; any phase-flip error on a physical qubit bit-flips the corresponding block qubit and can thus be rectified by a three-block-qubit QECC. The circuit of this encoding is depicted in Figure 7.11.

7.3.2 Error Syndrome Detection and Correction

Now we send a logical qubit consisting of nine physical qubits through nine independent NQCs, each of which may induce all three fundamental types of errors. Given the low error probability p, for each logical qubit we'll focus on scenarios where at most one of the nine physical qubits becomes faulty post-NQC traversal.

To discern the different types of errors and locate the faulty physical qubits, we need to entangle the logical qubit with syndrome qubits. But how many? Recall that each three-qubit block, serving as a three-qubit QECC for single bit-flip errors on its own, necessitates two syndrome qubits to handle intra-block bit flips. That's six syndrome qubits for three blocks. To deal with phase-flip errors, our analysis in Section 7.3.2 taught us that a phase-flip error always manifests as a bit flip of the corresponding block qubit. Hence, to stain a faulty block qubit among three block qubits, two more syndrome qubits suffice. Therefore, Shor's nine-qubit QECC would require eight syndrome qubits in total.

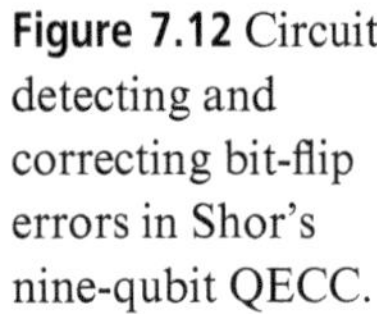

Figure 7.12 Circuit detecting and correcting bit-flip errors in Shor's nine-qubit QECC.

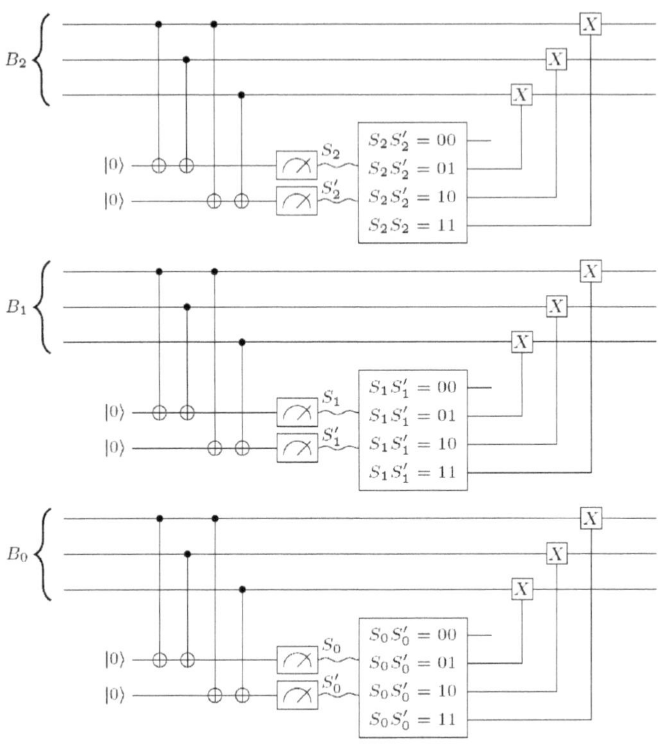

The process to append the six bit-flip syndrome qubits mirrors that in the three-qubit QECC for single bit-flip errors. That said, attaching the two phase-flip syndrome qubits is somewhat more intricate. So, let's discuss the two scenarios separately.

Bit-Flip Error

To append the six bit-flip syndrome qubits, we can simply triplicate the quantum circuit in Figure 7.7 in parallel and obtain the circuit diagram in Figure 7.12, including the error correction components.

This circuit should work as expected because any bit-flipped qubit in the m-th block qubit is only reflected in the corresponding syndrome outcome bit string $S_m S'_m$ and thus can be detected and corrected within the block, just like in a stand-alone three-qubit QECC.

Phase-Flip Error

A phase flip of any physical qubit within a block invariably bit-flips the block qubit. Yet, when referring to a bit flip, it's imperative to specify the basis of the quantum observable it is defined in relation to. The block qubit basis states, $|+\rangle$ and $|-\rangle$, as defined in Equation (7.23), are the eigenstates of the observable $X \otimes X \otimes X$:

$$X \otimes X \otimes X \,|\pm\rangle = \pm\,|\pm\rangle. \tag{7.28}$$

We can introduce $X_B := X \otimes X \otimes X$ and $Z_B := Z \otimes Z \otimes Z$ as the block X and Z gates. It's worth noting that the block qubit basis states $|\pm\rangle$ are not the eigenstates of Z_B. While we can employ the logic of the quantum circuit in Figure 7.7 to entangle the three block qubits with the two syndrome qubits, it's inconvenient to design a circuit to turn the eigenstates of X_B into those of Z_B. Instead, we can use the standard CNOT gates to accomplish the same outcome. To understand the mechanics, consider how $H \otimes H \otimes H$ influences the block qubit basis $|\pm\rangle$:

$$H \otimes H \otimes H\,|+\rangle \;\propto\; H \otimes H \otimes H(|000\rangle + |111\rangle) \propto |000\rangle + |011\rangle + |101\rangle + |110\rangle, \tag{7.29}$$

$$H \otimes H \otimes H\,|-\rangle \;\propto\; H \otimes H \otimes H(|000\rangle - |111\rangle) \propto |111\rangle + |100\rangle + |010\rangle + |001\rangle. \tag{7.30}$$

That is, $H \otimes H \otimes H\,|+\rangle$ $(H \otimes H \otimes H\,|-\rangle)$ results in a state where only an even (odd) number of qubits can be in state $|1\rangle$. Hence, if we exert $H \otimes H \otimes H$ on a block qubit, and then take the three physical qubits in the block as the respective control qubits of three CNOT gates to control a syndrome qubit, the state of the syndrome qubit after the action of the three CNOT gates can discern the block qubit states $|+\rangle$ and $|-\rangle$. Therefore, we can employ the logic of the quantum circuit in Figure 7.7 to entangle the three block qubits with the two syndrome qubits as in Figure 7.13.

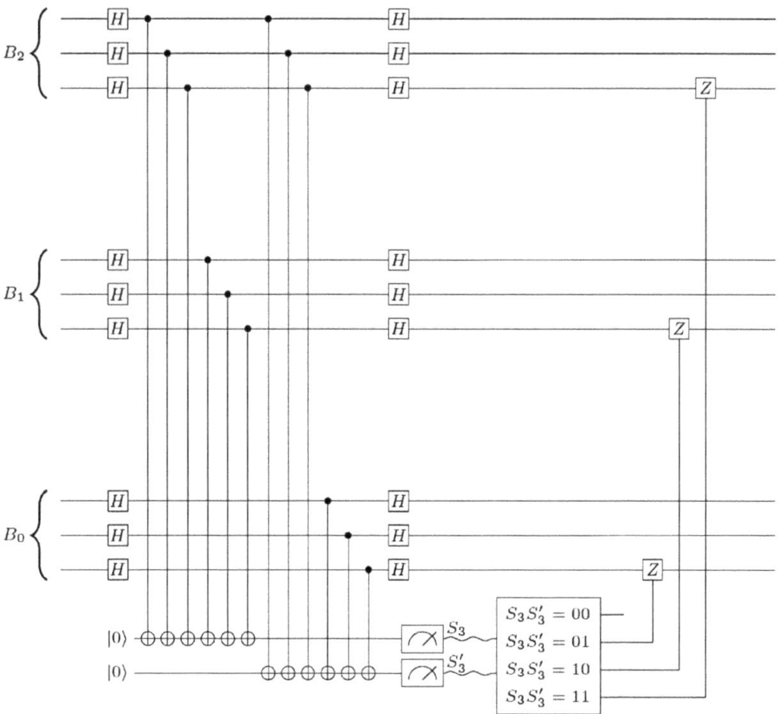

Figure 7.13 Circuit detecting and correcting phase-flip errors in Shor's nine-qubit QECC.

> **Exercise 7.5** *Verify that the circuit in Figure 7.13 truly operates as anticipated. You might choose to phase-flip any of the nine physical qubits, then ascertain whether the corresponding error is uniquely entangled with an error syndrome.*

It's vital to reapply the Hadamard gates to revert the block qubits to the basis states $|\pm\rangle$. In Figure 7.13, the circuit contains the correction stage for phase-flip errors too. Since a phase flip in any physical qubit always bit-flips the corresponding block qubit, to correct such an error, one needs only to apply a Z gate (phase-flip gate) to any physical qubit in the block qubit in which a phase flip occurred.

> **Exercise 7.6** *Design a quantum circuit that converts the block qubit basis states $|\pm\rangle$ to the basis states of Z_B. Use elementary quantum gates only.*

Decoding

The decoding process in Shor's nine-qubit QECC resembles that of the three-qubit QECC, with the primary differences being the greater number of physical qubits and the addition of Hadamard gates in the encoding procedure. Given that $H^2 = \mathbb{1}_2$, the decoding is a straightforward reversal of the encoding steps. Figure 7.14 presents the decoding sequence along with the comprehensive quantum circuit implementing Shor's nine-qubit QECC, joining the previously discussed components.

It's paramount in the circuit to address and correct the bit-flip errors before turning to the phase-flip errors. The order is vital because rectifying phase-flip errors requires that a block qubit remains in its basis states $|\pm\rangle$. On the other hand, a bit-flip error can shift a block qubit away from both $|+\rangle$ and $|-\rangle$, as indicated in Equation (7.24).

7.4 General Properties of Quantum Error Correction

While there exist numerous notable QECCs, such as the Steane seven-qubit QECC, diving into the intricacies of each can be time-consuming. Instead, in this section, we focus on the overarching properties of quantum error correction. Mastery of these properties should equip the reader to understand any specific QECC. Interestingly, these general properties pave the way for stabilizer codes – a formalism encompassing a vast array of QECCs, which we will cover in Section 7.5.

7.4.1 Error Generators

All errors on the qubits in an n-qubit system are generated by a finite set of generators. To nail down these generators, it's better to first recapitulate something we learned in Chapter 6, on quantum decoherence.

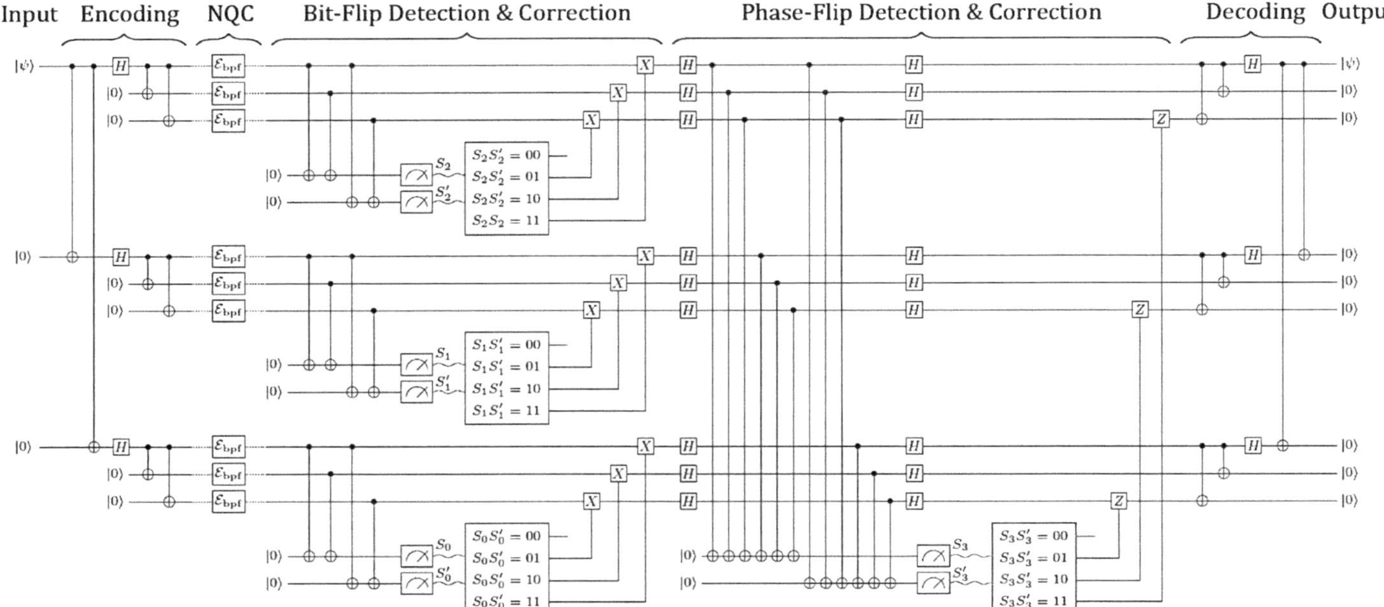

Figure 7.14 Full circuit for Shor's nine-qubit QECC.

Recall that an error on a qubit is caused by an NQC that couples the qubit with its environment. An NQC is represented by a unitary operator U acting on the composite system of the qubit and the environment. Using the same notations as those in Chapter 6, the action of U reads

$$
\begin{aligned}
U\,|\psi\rangle\,|\varepsilon_0\rangle &= \sum_{a=1}^{4}\sum_{n=0}^{1} |n\rangle\,|\varepsilon_a\rangle\,\langle n|\langle\varepsilon_a|U\,|\psi\rangle\,|\varepsilon_0\rangle \\
&= \sum_{a=1}^{4}\sum_{n=0}^{1} \langle n|\langle\varepsilon_a|U\,|\psi\rangle\,|\varepsilon_0\rangle\,|n\rangle\,|\varepsilon_a\rangle \\
&= \sum_{a=1}^{4}\sum_{n=0}^{1} \langle n|E_a\,|\psi\rangle\,|n\rangle\,|\varepsilon_a\rangle \\
&= \sum_{a=1}^{4}\sum_{n=0}^{1} |n\rangle\,\langle n|E_a\,|\psi\rangle\,|\varepsilon_a\rangle \\
&= \sum_{a=1}^{4} E_a\,|\psi\rangle\,|\varepsilon_a\rangle,
\end{aligned}
\tag{7.31}
$$

where $|\varepsilon_0\rangle$ is any initial state of the environment, $\{\varepsilon_a | a = 1, 2, 3, 4\}$ is an orthonormal basis of the environment, and $E_a := \langle\varepsilon_a|U\,|\varepsilon_0\rangle$ are the Kraus operators that represent the NQC. Remind yourself of Theorem 6.1 if you forget why the environment is four-dimensional, such that there are four Kraus operators.

What are these four Kraus operators? As alluded to in Chapter 6 and also observed in Equation (6.56), on a single qubit, any Kraus operator E_a can be expressed in terms of the operators in the set

$$
\{\mathbb{1}_2, \sigma_x, \sigma_y, \sigma_z\}.
\tag{7.32}
$$

In a conveniently chosen basis of the environment, we can identify the four Kraus operators E_a with the four operators in the set. The identity $\mathbb{1}_2$ represents no error, and the three Pauli matrices represent respectively bit-flip, phase-flip, and combined bit- and phase-flip errors.

Now let's move on to n qubits. To model an NQC on an n-qubit system, the commensurate environment should be 4^n-dimensional, resulting in 4^n Kraus operators. Since each single-qubit Kraus operator is one in the set (7.32), and since an n-qubit error is also qubit by qubit, n-qubit Kraus operators can be expressed as tensor products of these Pauli operators.[3] Namely,

$$
E_a \in \left\{ \bigotimes_{i=0}^{n-1} \sigma_i \,\middle|\, \sigma_i = \mathbb{1}_2, \sigma_x, \sigma_y, \sigma_z \right\}.
\tag{7.33}
$$

---

[3] While n-qubit Kraus operators can be expressed as tensor products of Pauli operators, it is important to note that this represents a practical simplification. In reality, the set of all possible unitary errors on n-qubits forms the group $U(2^n)$, which includes a much broader range of operations beyond the tensor products of the Pauli matrices. Each unitary operation can be represented as an exponential of a Hermitian operator, which can be expanded in terms of Pauli matrices. In fact, this expansion can be infinite. In practice, quantum error correction focuses on detecting and correcting errors that project onto the Pauli basis, as these errors are sufficient to handle the most probable and impactful error types. This approach simplifies the correction process and is feasible within the computational limits of current technology.

The number of Pauli matrices in a Kraus operator E_a is termed the *weight* of the error due to the Kraus operator.

Example 7.1

The three-qubit QECCs we have studied exemplify the abstract discussion above. For three physical qubits, $4^3 = 64$ Kraus operators are required, specifically,

$$E_a = \sigma_2 \otimes \sigma_1 \otimes \sigma_0, \quad a = 0, 1, \ldots, 63, \tag{7.34}$$

where each σ_i is independently chosen from the set (7.32). Nevertheless, our three-qubit QECCs are designed to only rectify single-qubit errors. Consequently, we consider only the 10 Kraus operators shown in Equation (7.36):

$$\begin{aligned}
&\mathbb{1}_2 \otimes \mathbb{1}_2 \otimes \mathbb{1}_2, \\
&\sigma_x \otimes \mathbb{1}_2 \otimes \mathbb{1}_2, \\
&\mathbb{1}_2 \otimes \sigma_x \otimes \mathbb{1}_2, \\
&\mathbb{1}_2 \otimes \mathbb{1}_2 \otimes \sigma_x, \\
&\sigma_y \otimes \mathbb{1}_2 \otimes \mathbb{1}_2, \\
&\quad \cdots \\
&\sigma_z \otimes \mathbb{1}_2 \otimes \mathbb{1}_2,
\end{aligned} \tag{7.35}$$

where each line represents an independent error on a single qubit, making these errors of weight 1.

7.4.2 Error Structure

After identifying the generators of all possible errors in an n-qubit system, we can now explore the mathematical structure of these errors.

The Group of Errors

For n qubits, the 4^n Kraus operators (7.33) generate all possible errors. Multiplying these operators together, the resultant set is closed under multiplication, thus forming a group (you should verify the other defining criteria of a group). This group of all n-qubit errors, represented by the Kraus operators and their products, is denoted by $\mathcal{G}_n$, termed the n-qubit **Pauli group**. The multiplication of Kraus operators does not generate any new elements outside this set, owing to the algebraic relations (4.9) of the Pauli matrices. These relationships yield possible sign factors ± 1 and $\pm i$, which are overall phase factors of n-qubit states caused by errors. We express this group formally as

$$\mathcal{G}_n = \{\pm 1, \pm i\} \times \{E_a | a = 0, 1, \ldots, n - 1\}, \tag{7.36}$$

where E_a is defined in Equation (7.33). The symbol $\times$ simply means that each of the phase factors can be applied to any E_a. Since the four overall phase factors are usually immaterial, they are separated out as a distinct group (forming the cyclic group $\mathbb{Z}_4$). The identity element of $\mathcal{G}_n$ is the $2^n \times 2^n$ identity matrix, with the inverse of an element being itself, adjusted by an appropriate overall sign, as Pauli matrices are both Hermitian and unitary.

Correctable Errors

Our examination of the three-qubit QECCs showed their capability to correct single-qubit bit-flip and phase-flip errors. The critical question for an n-qubit QECC is: What errors and how many concurrent single-qubit errors can it correct? To address this, we introduce the concept of correctable errors.

Let $\mathcal{E}_c$ be the set of correctable errors for a given QECC. To define this set, we revisit the case of the three-qubit QECC for single bit-flip errors for insight. This QECC can correct individual errors like $\sigma_x \otimes \mathbb{1}_2 \otimes \mathbb{1}_2$, but not the products of two such errors, which result in two-qubit errors. Hence, $\mathcal{E}_c \subset \mathcal{G}_n$ is not closed under multiplication and is merely a subset of $\mathcal{G}_n$, not a subgroup.

Figure 7.15 illustrates that single-qubit bit-flip errors are correctable, while multiqubit bit-flip errors are not. Mathematically, we contrast the two scenarios as follows:

$$\langle 000|\underbrace{(\sigma_x \otimes \mathbb{1}_2 \otimes \mathbb{1}_2)^\dagger}_{\text{Correctable}} \underbrace{(\mathbb{1}_2 \otimes \mathbb{1}_2 \otimes \sigma_x)}_{\text{Correctable}} |111\rangle = 0, \tag{7.37}$$

$$\langle 000|\underbrace{(\sigma_x \otimes \mathbb{1}_2 \otimes \sigma_x)^\dagger}_{\text{Uncorrectable}} \underbrace{(\mathbb{1}_2 \otimes \sigma_x \otimes \mathbb{1}_2)}_{\text{Correctable}} |111\rangle = 1. \tag{7.38}$$

These examples imply that if two errors $E_a, E_b \in \mathcal{G}_n$ are correctable, then their action on distinct codewords must yield orthogonal states. Otherwise, it would be impossible to discern the origin of the resultant faulty state, precluding any correction.

Thus, for an n-qubit QECC with code space C, we require that

$$\langle w_L | E_b^\dagger E_a | w_L' \rangle = 0, \tag{7.39}$$

for any two orthogonal codewords $|w_L\rangle, |w_L'\rangle \in C$, and for all $E_a, E_b \in \mathcal{E}_c$.

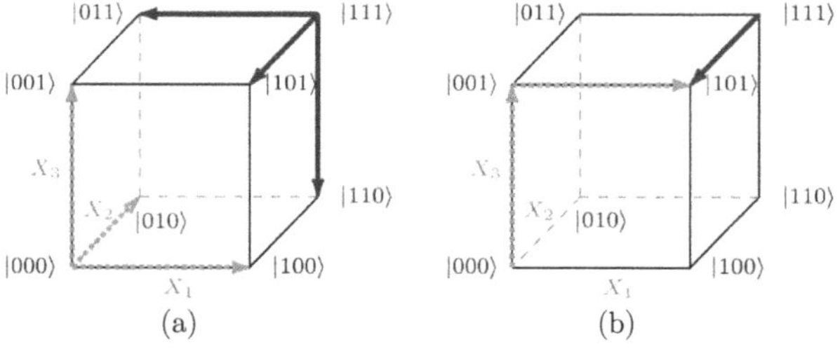

Figure 7.15 (a) Correctable errors and (b) uncorrectable errors for the three-qubit QECC for single bit-flip errors.

Throughout a quantum error correction process, the logical-qubit state should remain completely unknown to the error-syndrome detection and correction mechanisms; otherwise, the logical state would inevitably be disturbed. For example, consider a codeword $p\,|000\rangle + q\,|111\rangle$ ($p, q \in \mathbb{R}$ and $p^2 + q^2 = 1$) of the three-qubit QECC for bit-flip errors. Take a correctable error $\mathbb{1}_2 \otimes \mathbb{1}_2 \otimes \sigma_x$ and two uncorrectable ones, $\sigma_x \otimes \sigma_x \otimes \mathbb{1}_2$ and $\sigma_x \otimes \sigma_x \otimes \sigma_x$. We compute

$$(p\langle 000| + q\langle 111|)\underbrace{(\mathbb{1}_2 \otimes \mathbb{1}_2 \otimes \sigma_x)^\dagger}_{\text{Correctable}}\underbrace{(\mathbb{1}_2 \otimes \mathbb{1}_2 \otimes \sigma_x)}_{\text{Correctable}}(p\,|000\rangle + q\,|111\rangle) = 1, \qquad (7.40)$$

$$(p\langle 000| + q\langle 111|)\underbrace{(\sigma_x \otimes \sigma_x \otimes \mathbb{1}_2)^\dagger}_{\text{uncorrectable}}\underbrace{(\mathbb{1}_2 \otimes \mathbb{1}_2 \otimes \sigma_x)}_{\text{Correctable}}(p\,|000\rangle + q\,|111\rangle) = 2pq, \qquad (7.41)$$

$$(p\langle 000| + q\langle 111|)\underbrace{(\mathbb{1}_2 \otimes \mathbb{1}_2 \otimes \mathbb{1}_2)^\dagger}_{\text{No error}}\underbrace{(\sigma_x \otimes \sigma_x \otimes \sigma_x)}_{\text{uncorrectable}}(p\,|000\rangle + q\,|111\rangle) = 2pq. \qquad (7.42)$$

These computations show that when an uncorrectable error combines with a correctable error, the expectation value of the combined error in the original codeword may encode specific information, such as the $2pq$, about the original codeword. If this were the case, the syndrome measurement process could extract this information, thereby affecting the logical state, in a way similar to that in the scenario of weak measurement discussed in Section 6.5. In contrast, for correctable errors, such inner products are devoid of logical information; they are either 1 or 0 for the three-qubit QECC but may generally depend on the errors. Therefore, we can deduce the condition (7.43): For a QECC with code space C, $\forall E_a, E_b \in \mathcal{E}_c$ and $\forall w_L \in C$,

$$\langle w_L | E_a^\dagger E_b | w_L \rangle = \alpha_{ab}, \qquad (7.43)$$

where $\alpha_{ab} \in \mathbb{C}$ and $\alpha_{ab} = \alpha_{ba}^*$. For three-qubit QECCs, evidently, $\alpha_{ab} = \delta_{ab}$. These complex numbers α_{ab} depend only on the errors E_a and E_b and contain no information about the codeword.

The two conditions (7.39) and (7.43) can be unified: In an n-qubit QECC with code space C, any two errors described by Kraus operators E_a and E_b that can be rectified by the QECC must satisfy

$$PE_a^\dagger E_b P = \alpha_{ab} P, \qquad (7.44)$$

where $\alpha_{ab} \in \mathbb{C}$ and $\alpha_{ab} = \alpha_{ba}^*$, and P is the projector onto the code space C. For example, for three-qubit QECCs,

$$P = |000\rangle\langle 000| + |111\rangle\langle 111|. \qquad (7.45)$$

The proof of condition (7.44) is available in Nielsen and Chuang [13]. It turns out that verifying whether a set of errors of a QECC meet these conditions may be complicated. In the next section, you will see that stabilizer code simplifies such verifications.

A QECC is termed **nondegenerate** if $\alpha_{ab} = \delta_{ab}$; otherwise, it is **degenerate**. The three-qubit QECCs are examples of nondegenerate codes. The distinction between nondegenerate and degenerate codes is pivotal:

Nondegenerate codes:

- *Pros:* Simpler error diagnosis due to unique error syndromes for each correctable error, and often easier implementation.
- *Cons:* Less efficient in terms of qubit usage for error correction, potentially requiring more qubits to achieve a certain level of error correction compared to degenerate codes.

Degenerate codes:

- *Pros:* More efficient use of qubits, as they can correct more errors with fewer qubits, making them potentially better for large-scale quantum computers.
- *Cons:* More complex error diagnosis as multiple errors can map to the same syndrome, leading to more sophisticated and complex error correction mechanisms.

In summary, nondegenerate codes are simple and easy to implement but may require more qubits for equivalent error correction capabilities. Degenerate codes, conversely, provide greater efficiency in qubit usage but at the cost of increased complexity in error diagnosis and correction. The choice between these two types hinges on the specific needs and constraints of the quantum computing system.

Exercise 7.7 *Verify that Shor's nine-qubit QECC satisfies both conditions* (7.39) *and* (7.43), *or the unified condition* (7.44). *Determine whether Shor's QECC is nondegenerate.*

7.4.3 The Quantum Hamming Bound

While the three-qubit and nine-qubit QECCs we've explored encode only one logical qubit, it's generally possible to encode k logical qubits into n physical qubits, with $k < n$. A pertinent question arises: To encode k logical qubits, how many physical qubits are required, particularly when considering the correction of errors affecting up to t physical qubits? This leads us to the quantum Hamming bound, but first, we must discuss the notion of code distance and the standard notation for QECC.

Code Distance

Recall that the weight of an error is the number of Pauli matrices in the Kraus operator causing the error. The minimum weight of errors in $\mathcal{G}_n$ of an n-qubit QECC that can transform a codeword into an orthogonal codeword within the code space is known as the **code distance**.

Exercise 7.8 *An QECC capable of correcting errors up to weight t must have a code distance of at least $2t + 1$. Explain why.*

For example, the three-qubit bit-flip QECC has a code distance of 3, as demonstrated in Figure 7.5, indicating its ability to correct single-qubit bit-flip errors. Shor's nine-qubit QECC, though requiring nine bit flips to transform $|0_L\rangle = |+++\rangle$ to $|1_L\rangle = |---\rangle$, actually has a code distance of 3 due to its susceptibility to three phase flips. We typically denote a distance-d n-qubit QECC that encodes k logical qubits as

$$[[n, k, d]]. \tag{7.46}$$

Hence, $[[3, 1, 3]]$ and $[[9, 1, 3]]$ denote the three-qubit QECCs[4] and Shor's nine-qubit QECC, respectively.

Quantum Hamming Bounds for Nondegenerate QECCs

In our discussion, we focus on nondegenerate QECCs. The QHB for a nondegenerate QECC is the minimal number of physical qubits n required to encode k logical qubits, allowing for correctable errors that affect at most t physical qubits [209, 210]. To establish this bound, we first calculate the total number of errors $N_{\mathcal{E}}$ affecting up to t qubits, always including the case of no error. For any $0 \le j \le t$, there are

$$\binom{n}{j} \tag{7.47}$$

ways to choose the j faulty qubits, each with 3^j possible errors due to the three Pauli matrices. Thus,

$$N_{\mathcal{E}} = \sum_{j=0}^{t} \binom{n}{j} 3^j. \tag{7.48}$$

The logical space of k logical qubits is 2^k-dimensional. As each of the $N_{\mathcal{E}}$ errors is correctable and the QECC is nondegenerate, these errors map the 2^k-dimensional logical space to $N_{\mathcal{E}}$ orthogonal 2^k-dimensional subspaces within the 2^n-dimensional Hilbert space of n qubits (refer to Equation (7.14) for the three-qubit bit-flip QECC example). The aggregate dimensionality of these subspaces is constrained by 2^n, leading to a QHB of

$$2^k \sum_{j=0}^{t} \binom{n}{j} 3^j \le 2^n. \tag{7.49}$$

Example 7.2

Consider $t = 1$. The QHB simplifies to

$$2^k (3n + 1) \le 2^n. \tag{7.50}$$

For $k = 1$, it implies $n_{\min} = 5$, as $3n + 1 \le 2^{n-1}$. The three-qubit QECCs, which correct only one type of single-qubit error, violate this bound. For $k = 6$, we find $n_{\min} = 12$.

---

[4] This notation is generally reserved for QECCs that can correct both bit-flip and phase-flip errors, so the three-qubit QECCs do not qualify for using this notation. Here, I'm being somewhat sloppy.

In the example above, it is interesting to note that for $k = 6$, $n_{min}/k = 2$, which is less than $n_{min}/k = 5$ for $k = 1$. This observation is generally true. The QHB reveals that for a given t, the ratio n_{min}/k decreases with increasing k, suggesting more efficient but also more complex encoding for larger numbers of logical qubits.

> **Exercise 7.9** *Analyze Shor's nine-qubit QECC in the context of the quantum Hamming Bound. Determine if this QECC violates, satisfies, or saturates the bound. Provide reasoning to support your conclusion.*

7.5   Stabilizer Code

Having laid the groundwork with our general treatment of quantum error correction, we now turn to an important group-theoretic formalism within QECCs: the stabilizer code [203]. This formalism encompasses a variety of QECCs that are of theoretical interest or practical significance, providing a robust framework for understanding and constructing QECCs.

7.5.1   Shor's QECC as Stabilizer Code

To help you comprehend the stabilizer code formalism, we shall first reframe Shor's nine-qubit QECC within this context. Remember, the error syndromes in Shor's QECC are indicated by the measurement outcomes S_0, S_0' through S_3, S_3' from the eight syndrome qubits. For ease of reference, a part of the quantum circuit for error syndrome detection in Shor's QECC is depicted in Figure 7.16.

Let's determine the physical interpretation of these measurement outcomes, specifically considering S_2:

$$
S_2 = i_8 \oplus i_7 = \begin{cases} \underset{\substack{\downarrow \\ \text{no error}}}{0} & i_8 = i_7 \quad (\langle \sigma_z^{(8)} \rangle \langle \sigma_z^{(7)} \rangle = 1); \\[2em] \underset{\substack{\downarrow \\ \text{error}}}{1} & i_8 \neq i_7 \quad (\langle \sigma_z^{(8)} \rangle \langle \sigma_z^{(7)} \rangle = -1). \end{cases} \tag{7.51}
$$

Here, the superscripts label the physical qubits. Therefore,

$$
S_2 = \left\langle \sigma_z^{(8)} \otimes \sigma_z^{(7)} \otimes \mathbb{1}_2^{(6)} \otimes \cdots \mathbb{1}_2^{(0)} \right\rangle. \tag{7.52}
$$

The scenarios for S_1, S_1', S_0, and S_0' are analogous, yet the outcomes for S_3 and S_3' differ slightly. For instance,

$$
S_3 = \left\langle \sigma_x^{(8)} \otimes \sigma_x^{(7)} \otimes \cdots \otimes \sigma_x^{(3)} \otimes \mathbb{1}_2^{(2)} \otimes \mathbb{1}_2^{(1)} \otimes \mathbb{1}_2^{(0)} \right\rangle. \tag{7.53}
$$

Can you see why?

In essence, the eight syndrome qubits' outcomes correspond to measurements of the eight operators listed in Equation (7.54):

$$
M_1 = \sigma_z^{(8)} \otimes \sigma_z^{(7)} \otimes \mathbb{1}_2^{(6)} \otimes \mathbb{1}_2^{(5)} \otimes \mathbb{1}_2^{(4)} \otimes \mathbb{1}_2^{(3)} \otimes \mathbb{1}_2^{(2)} \otimes \mathbb{1}_2^{(1)} \otimes \mathbb{1}_2^{(0)},
$$

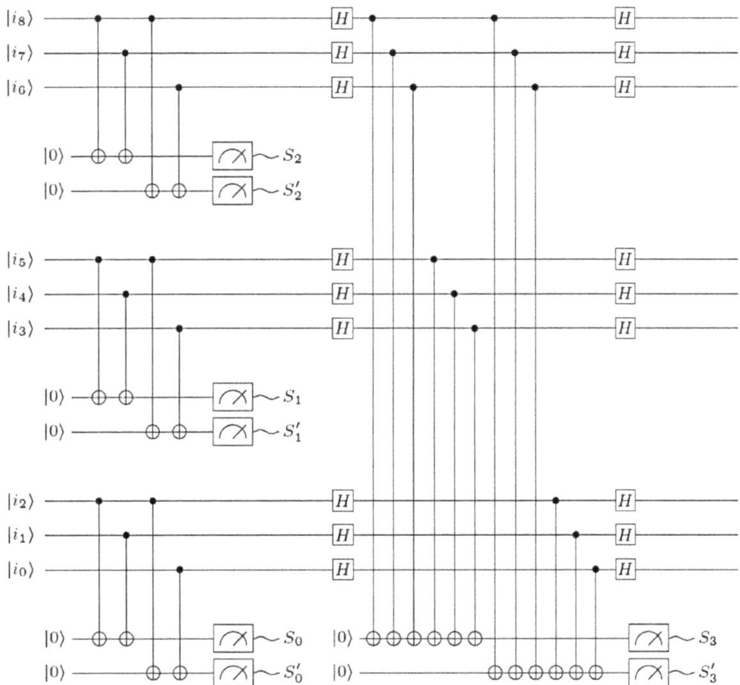

Figure 7.16 Quantum circuit of error syndrome detection in Shor's QECC, excluding the error correction stage.

$$M_2 = \sigma_z^{(8)} \otimes \mathbb{1}_2^{(7)} \otimes \sigma_z^{(6)} \otimes \mathbb{1}_2^{(5)} \otimes \mathbb{1}_2^{(4)} \otimes \mathbb{1}_2^{(3)} \otimes \mathbb{1}_2^{(2)} \otimes \mathbb{1}_2^{(1)} \otimes \mathbb{1}_2^{(0)},$$

$$M_3 = \mathbb{1}_2^{(8)} \otimes \mathbb{1}_2^{(7)} \otimes \mathbb{1}_2^{(6)} \otimes \sigma_z^{(5)} \otimes \sigma_z^{(4)} \otimes \mathbb{1}_2^{(3)} \otimes \mathbb{1}_2^{(2)} \otimes \mathbb{1}_2^{(1)} \otimes \mathbb{1}_2^{(0)},$$

$$M_4 = \mathbb{1}_2^{(8)} \otimes \mathbb{1}_2^{(7)} \otimes \mathbb{1}_2^{(6)} \otimes \sigma_z^{(5)} \otimes \mathbb{1}_2^{(4)} \otimes \sigma_z^{(3)} \otimes \mathbb{1}_2^{(2)} \otimes \mathbb{1}_2^{(1)} \otimes \mathbb{1}_2^{(0)},$$

$$M_5 = \mathbb{1}_2^{(8)} \otimes \mathbb{1}_2^{(7)} \otimes \mathbb{1}_2^{(6)} \otimes \mathbb{1}_2^{(5)} \otimes \mathbb{1}_2^{(4)} \otimes \mathbb{1}_2^{(3)} \otimes \sigma_z^{(2)} \otimes \sigma_z^{(1)} \otimes \mathbb{1}_2^{(0)},$$

$$M_6 = \mathbb{1}_2^{(8)} \otimes \mathbb{1}_2^{(7)} \otimes \mathbb{1}_2^{(6)} \otimes \mathbb{1}_2^{(5)} \otimes \mathbb{1}_2^{(4)} \otimes \mathbb{1}_2^{(3)} \otimes \sigma_z^{(2)} \otimes \mathbb{1}_2^{(1)} \otimes \sigma_z^{(0)},$$

$$M_7 = \sigma_x^{(8)} \otimes \sigma_x^{(7)} \otimes \sigma_x^{(6)} \otimes \sigma_x^{(5)} \otimes \sigma_x^{(4)} \otimes \sigma_x^{(3)} \otimes \mathbb{1}_2^{(2)} \otimes \mathbb{1}_2^{(1)} \otimes \mathbb{1}_2^{(0)},$$

$$M_8 = \sigma_x^{(8)} \otimes \sigma_x^{(7)} \otimes \sigma_x^{(6)} \otimes \mathbb{1}_2^{(5)} \otimes \mathbb{1}_2^{(4)} \otimes \mathbb{1}_2^{(3)} \otimes \sigma_x^{(2)} \otimes \sigma_x^{(1)} \otimes \sigma_x^{(0)}. \tag{7.54}$$

It is evident that these eight operators commute with each other:

$$[M_i, M_j] = 0 \quad \forall i, j = 1, 2, \dots, 8. \tag{7.55}$$

A correctly transmitted logical state, entangled with all syndrome qubit measurement outcomes being +1, is a simultaneous +1 eigenstate of all the eight operators in Equation (7.54). Thus,

$$M_i \,|0_L\rangle = |0_L\rangle, \quad M_i \,|1_L\rangle = |1_L\rangle \quad \forall i = 1, 2, \dots, 8, \tag{7.56}$$

where $|0_L\rangle$ and $|1_L\rangle$ are the logical basis states of Shor's QECC defined in Equation (7.26). The two-dimensional code space of Shor's QECC, being the +1 eigenspace of these operators, is invariant under their action. All operators on the nine physical qubits that have the code space as their common +1 eigenspace are products of these eight operators, forming a group. This group, denoted by $\mathcal{S}_{\text{Shor}}$, stabilizes the code space and is termed the **stabilizer group** of Shor's QECC. The elements of this group are the **stabilizers** of Shor's code, with the eight operators in Equation (7.54) being their generators. Notably, each stabilizer generator is not a product of any two other generators. The stabilizer group $\mathcal{S}_{\text{Shor}}$, being an Abelian[5] subgroup[6] of $\mathcal{G}_9$ – the group of all errors on Shor's code – contrasts with the non-Abelianness of $\mathcal{G}_9$. It's worth of emphasis that a stabilizer group preserves not only the code space but also every single codeword!

The difference between the number of physical qubits in Shor's code, which is 9, and the number of stabilizer generators, which is 8, equals 1, precisely the number of logical qubits. Each stabilizer generator imposes a constraint on a nine-qubit state to qualify as a codeword. With eight stabilizer generators providing eight constraints, we are left with one independent degree of freedom: the logical qubit. Hence, in general, a stabilizer code encoding k logical qubits into n physical qubits must have $n - k$ stabilizer generators.

Stabilizers simplify the determination of correctable errors in a QECC. By casting Shor's code as a stabilizer code, we identify operators in the Pauli group $\mathcal{G}_9$ that do not alter a correct logical state. This distinction arises because errors are generated qubit-wise by Pauli matrices, and some errors, when combined, negate each other. For instance, two phase flips on different physical qubits in a block of Shor's code nullify each other. Once we isolate the void errors from $\mathcal{G}_9$, the remaining set, $\mathcal{G}_9 \setminus \mathcal{S}_{\text{Shor}}$, consists of errors that displace a codeword from the code space.

Which errors in $\mathcal{G}_9 \setminus \mathcal{S}_{\text{Shor}}$ are correctable? An error in this set may still commute with all stabilizers. Consider the error that bit-flips all three physical qubits in the leftmost block of Shor's code, denoted by

$$\sigma_x^{(8)}\sigma_x^{(7)}\sigma_x^{(6)}, \tag{7.57}$$

where for convenience we have omitted the tensor product symbol $\otimes$ and the identity matrices on the remaining qubits. This error is not part of $\mathcal{S}_{\text{Shor}}$ (verify this yourself as an exercise). Yet, it commutes with all stabilizers. This error differs from the stabilizer group in that it doesn't preserve the codewords albeit preserving the codespace. This is a key point for Section 7.5.3. Measuring the stabilizers cannot detect this error because the measurement outcomes will be identically +1, the same result as for a correct codeword – let alone correcting the error.

Conversely, an error represented by an element of the Pauli group either commutes with all stabilizers or anticommutes with at least one. For example, the single-qubit bit-flip error

$$\sigma_x^{(8)} \tag{7.58}$$

---

[5] A group G is **Abelian** if for any $g, g' \in G$, $g \cdot g = g' \cdot g$, otherwise, it is **non-Abelian**.

[6] A **subgroup** of a group G is a subset of G that is itself a group. Note that any subgroup of G must contain the identity element of G.

commutes with most stabilizers, except M_1 and M_2: $\{\sigma_x^{(8)}, M_1\} = \{\sigma_x^{(7)}, M_2\} = 0$. Here, $\{,\}$ denotes the anticommutator.[7] If this error occurred, then arranging the outcomes of all stabilizer measurements in order as a vector, we would obtain a syndrome vector

$$\begin{pmatrix} -1 & -1 & 1 & 1 & 1 & 1 & 1 & 1 \end{pmatrix}, \tag{7.59}$$

which is a unique signature of this error. That is, the error can be detected.

Now, a detectable error is not necessarily correctable. For instance, a two-qubit bit-flip error

$$\sigma_x^{(7)} \sigma_x^{(6)} \tag{7.60}$$

is detectable but not correctable, as its correction is ambiguous: Should we bit-flip qubit 8 or bit-flip both qubits 7 and 6 to rectify the error? If this two-qubit error is permissible, its concurrence with the single-qubit error (7.58) leads to the undetectable three-qubit error (7.57).

Exercise 7.10 *Why is the error* (7.60) *detectable?*

The code distance of Shor's code is the least weight of the errors in $\mathcal{G}_9 \setminus \mathcal{S}_{\text{Shor}}$ that commute with the entire $\mathcal{S}_{\text{Shor}}$, which is 3. This analysis and the principles of stabilizer codes, drawn from Shor's code, are readily generalizable.

7.5.2 General Treatment

To lay the foundation for a detailed exploration of stabilizer codes, we first examine the properties of the n-qubit Pauli group $\mathcal{G}_n$.

The n-Qubit Pauli Group

We revisit the definition of $\mathcal{G}_n$, presented here with a slightly different notation:

$$\mathcal{G}_n = \{\pm 1, \pm i\} \times \{\mathbb{1}_2, \sigma_x, \sigma_y, \sigma_z\}^{\otimes n}, \tag{7.61}$$

where the superscript $\otimes n$ signifies the n-fold tensor product of the set. This Pauli group is $2n$-dimensional (see Remark 7.1), in the sense that it can be generated by the products of $2n$ generators:

$$\left\{\sigma_z^{(i)}, \sigma_x^{(j)} \,\middle|\, i, j = n - 1, n - 2, \ldots, 0\right\}, \tag{7.62}$$

where the shorthand notation as in Equation (7.57) is adopted.

Elements $M, M' \in \mathcal{G}_n$ exhibit the following properties:

- $M^{-1} = M^\dagger$.
- $M^2 = \pm \mathbb{1}_2^{\otimes n}$.
- $MM' = \pm M'M$, indicating that they either commute or anticommute.

---

[7] $\{A, B\} := AB + BA$.

> ### Remark 7.1 Symplectic Structure
>
> A Pauli group is a discrete group, which strictly speaking doesn't have the notion of dimension. I'm being sloppy here in terminologies. But essentially, the dimension of a Pauli group should be understood as the dimension of the **symplectic vector space** representing the Pauli group. Let me explain.
>
> **Symplectic structure** is fundamental to the study of stabilizer codes in quantum error correction. It provides a mathematical framework to represent Pauli operators and their commutation relations using vector spaces over the finite field $\mathbb{Z}_2$ (the field with two elements). Understanding this structure is crucial for systematic analysis of stabilizer codes and their properties.
>
> 1. **Pauli operators as binary vectors**
>   Each Pauli operator $M \in \mathcal{G}_n$ can be mapped to a binary vector in $\mathbb{Z}_2^{2n}$: $\mathbf{u} = (\mathbf{x} \mid \mathbf{z})$, where
>   - $\mathbf{x} = (x_1, x_2, \ldots, x_n)$ corresponds to the presence ($x_i = 1$) or absence ($x_i = 0$) of σ_x operators on qubit i.
>   - $\mathbf{z} = (z_1, z_2, \ldots, z_n)$ corresponds to the presence ($z_i = 1$) or absence ($z_i = 0$) of σ_z operators on qubit i.
>
>   Here is a single-qubit example:
> $$\begin{aligned}
>   \mathbb{1}_2 &\leftrightarrow (0 \mid 0), \\
>   \sigma_x &\leftrightarrow (1 \mid 0), \\
>   \sigma_z &\leftrightarrow (0 \mid 1), \\
>   \sigma_y &= i\sigma_x \sigma_z \leftrightarrow (1 \mid 1).
>   \end{aligned} \tag{7.63}$$
>
> 2. **Pauli group multiplication as vector addition and subtraction**
>   Two binary vectors $\mathbf{u} = (\mathbf{x} \mid \mathbf{z})$ and $\mathbf{u}' = (\mathbf{x}' \mid \mathbf{z}')$ can be added or subtracted modulo 2:
> $$\mathbf{u} \pm \mathbf{u}' = (\mathbf{x} \mid \mathbf{z}) \pm (\mathbf{x}' \mid \mathbf{z}') = (\mathbf{x} + \mathbf{x}' \mod 2 \mid \mathbf{z} + \mathbf{z}' \mod 2). \tag{7.64}$$
>
>   This rule of addition and subtraction reflects the multiplication of Pauli operators. For example,
> $$\begin{aligned}
>   (1 \mid 0) \pm (1 \mid 0) &= (0 \mid 0) \leftrightarrow \sigma_x^2 = \mathbb{1}_2, \\
>   (1 \mid 0) \pm (0 \mid 1) &= (1 \mid 1) \leftrightarrow i\sigma_x \sigma_z = \sigma_y.
>   \end{aligned} \tag{7.65}$$
>
>   Since subtraction is the same as addition in this construction, we can ignore subtraction. The vector space of such binary vectors $\mathbf{u}$ is defined over the **finite field** $\mathbb{Z}_2$, also known as the **Galois field** GF(2). That is, the linear combination coefficients of the binary vectors are either 0 or 1.
>
> 3. **The symplectic form**
>   The vector space $\{\mathbf{u}\}$ for the Pauli group $\mathcal{G}_n$ is endowed with an unusual inner product – a **symplectic inner product** ω, which is a bilinear form that encodes the commutation relations between Pauli operators: For two binary vectors $\mathbf{u} = (\mathbf{x} \mid \mathbf{z})$ and $\mathbf{u}' = (\mathbf{x}' \mid \mathbf{z}')$, the **symplectic form** is defined as
> $$\omega(\mathbf{u}, \mathbf{u}') = \mathbf{x} \cdot \mathbf{z}' + \mathbf{z} \cdot \mathbf{x}' \mod 2, \tag{7.66}$$

Remark 7.1, continued

where $\cdot$ denotes the usual dot product over $\mathbb{Z}_2$. Two Pauli operators M and M' corresponding to $\mathbf{u}$ and $\mathbf{u}'$ commute if $\omega(\mathbf{u}, \mathbf{u}') = 0$ and anticommute if $\omega(\mathbf{u}, \mathbf{u}') = 1$. For example,

$$\omega((1 \,|\, 0), (0 \,|\, 1)) = (1)(1) + (0)(0) = 1 \quad \mathrm{mod}\ 2, \tag{7.67}$$

corresponding to $\{\sigma_x, \sigma_z\} = 0$. The vector space of the binary vectors corresponding to the Pauli group $\mathcal{G}_n$ is referred to as the **symplectic vector space** $\mathcal{V}_{2n}$.

4. **The symplectic matrix** J

   The symplectic form can be represented using a $2n \times 2n$ matrix J, the **symplectic matrix**, which takes the form

   $$J = \begin{pmatrix} 0 & \mathbb{1}_n \\ \mathbb{1}_n & 0 \end{pmatrix}, \tag{7.68}$$

   where $\mathbb{1}_n$ is the $n \times n$ identity matrix, and 0 is the $n \times n$ zero matrix. The symplectic form can then be written as

   $$\omega(\mathbf{u}, \mathbf{v}) = \mathbf{u}^T J \mathbf{v}. \tag{7.69}$$

Let's check a two-qubit example to strengthen your understanding. Consider two Pauli operators $\sigma_x^{(1)} \sigma_z^{(2)}$ and $\sigma_z^{(1)} \sigma_x^{(2)}$. Their binary vectors are

$$\mathbf{u} = (1, 0 \,|\, 0, 1), \tag{7.70}$$
$$\mathbf{v} = (0, 1 \,|\, 1, 0). \tag{7.71}$$

The inner product between $\mathbf{u}$ and $\mathbf{v}$ reads

$$\omega(\mathbf{u}, \mathbf{v}) = (1, 0) \cdot (1, 0) + (0, 1) \cdot (0, 1) = 1 + 1 = 0 \quad \mathrm{mod}\ 2, \tag{7.72}$$

reflecting the commutativity: $\left[\sigma_x^{(1)} \sigma_z^{(2)}, \sigma_z^{(1)} \sigma_x^{(2)} \right] = 0$.

Therefore, the number of independent generators of the Pauli group $\mathcal{G}_n$ is indeed the dimension – the number of basis binary vectors – of the corresponding symplectic vector space $\mathcal{V}_{2n}$. Note that the phase factors ± 1 and $\pm i$ in $\mathcal{G}_n$ are irrelevant in $\mathcal{V}_{2n}$.

Exercise 7.11 *Represent the following Pauli operators acting on two qubits as binary vectors:* $\sigma_y^{(1)}, \sigma_x^{(1)} \sigma_y^{(2)},$ *and* $\sigma_y^{(1)} \sigma_y^{(2)}$.

Exercise 7.12 *For each pair of operators in parts (a)–(c), calculate the symplectic form ω and determine whether the operators commute or anticommute:*
(a) $\sigma_y^{(1)}$ *and* $\sigma_z^{(1)} \sigma_x^{(2)}$;
(b) $\sigma_x^{(1)} \sigma_y^{(2)}$ *and* $\sigma_y^{(1)} \sigma_x^{(2)}$;
(c) $\sigma_y^{(1)} \sigma_y^{(2)}$ *and* $\sigma_x^{(1)} \sigma_x^{(2)}$.

Stabilizer and Stabilizer Code

Insights from Shor's nine-qubit QECC, reinterpreted as a stabilizer code, guide us to define an n-qubit stabilizer code.

Definition 7.1. *Given an Abelian subgroup $S \subseteq \mathcal{G}_n$, the associated **stabilizer code** is the Hilbert subspace $\mathcal{H}_S \subset \mathcal{H}_{2^n}$, defined as*

$$\mathcal{H}_S = \{|\psi\rangle \in \mathcal{H}_{2^n} | M |\psi\rangle = |\psi\rangle \ \forall M \in S\}, \tag{7.73}$$

*with S being the **stabilizer** of the code $\mathcal{H}_S$. Note that $-\mathbb{1}_{2^n} \notin S$.*

An S with $n - k$ generators can encode k logical qubits, implying

$$\dim \mathcal{H}_S = 2^k. \tag{7.74}$$

To identify errors in a state $|\phi\rangle \in \mathcal{H}_{2^n}$, as opposed to codewords in $\mathcal{H}_S$, one measures the expectation values of all generators M_i, $i = 1, 2, \ldots, n - k$, in state $|\phi\rangle$. A negative expectation value, $\langle\phi|M_i|\phi\rangle = -1$, signals an error in $|\phi\rangle$.

A word of caution: In this book, we choose not to regard operators in a stabilizer S as error-inducing, considering only those in $\mathcal{G}_n \setminus S$ as potential sources of errors with respect to S. This approach, aimed at clarity and ease of understanding, particularly benefits newcomers to quantum error correction. It clearly delineates error operators in the context of a given QECC and simplifies the foundational aspects of stabilizer codes.

That said, this perspective might oversimplify more complex scenarios in quantum error correction, especially in the context of degenerate codes. More advanced materials often treat stabilizers as potential error operators, without a clear distinction. While our approach is effective for nondegenerate codes discussed in this book, such as surface codes and the five-qubit code to be explored, the distinction may not be as pronounced in more general or complex scenarios. We aim to provide a clear introduction to QECCs, and our presentation reflects this goal. Be prepared to adapt to more nuanced views as you progress in the field.

Key principles of stabilizer codes include:

- For the set of correctable errors $\mathcal{E}_c$ of S, two errors $E_a, E_b \in \mathcal{E}_c$ are correctable if and only if there exists at least one $M \in S$ such that

$$\{M, E_a^\dagger E_b\} = 0. \tag{7.75}$$

This condition mirrors Equation (7.44). Reflect on Shor's QECC as a stabilizer code for intuition.

> **Exercise 7.13** *Prove that condition (7.75) is equivalent to condition (7.44).*
>
> **Exercise 7.14** *In view of Equation (7.75), explain why Shor's QECC cannot correct two-qubit errors.*

- The **code distance** d of a stabilizer code $\mathcal{H}_S$ is the minimum weight of errors in $\mathcal{G}_n \setminus \mathcal{S}$ that commute with all elements of $\mathcal{S}$. A code of distance $2t+1$ can correct up to t-qubit errors, while a distance of $s+1$ allows for the detection of up to s-qubit errors or the correction of s-qubit errors on known qubits. Consider Shor's code for an example.

> **Exercise 7.15** *Prove that a stabilizer code of distance $s + 1$ can correct up to s-qubit errors on known qubits. (Not known states! Only the locations of qubits are known.)*

- For any correctable error $E_a \in \mathcal{E}_c$ and any $M \in \mathcal{S}$, we have

$$M_i E_a = (-1)^{s_{ia}} E_a M_i, \quad s_{ia} \in \{0, 1\}. \tag{7.76}$$

The binary string

$$s_{1a} s_{2a} \cdots s_{(n-k)a} \tag{7.77}$$

forms the error syndrome of E_a. Refer to the binary strings in the three-qubit QECCs and Shor's QECC for context. Note that a detectable (not necessarily correctable) error also anticommutes with at least one element in $\mathcal{S}$; however, the syndrome may coincide with that of a correctable error. For example, in the three-qubit bit-flip QECC, a single-qubit error, which is correctable, has the same error syndrome as a certain two-qubit error, which is detectable but uncorrectable. We didn't mention this for simplicity due to the much lower probability of two-qubit errors.

> **Exercise 7.16** *Find a single-qubit error and a two-qubit error in the three-qubit bit-flip QECC that share the same error syndrome.*

- A *nondegenerate* stabilizer code $\mathcal{H}_S$ is one that has distinct error syndromes for all correctable errors, enabling identification of errors solely by measuring the stabilizer generators.
- Different stabilizer codes can be equivalent. Two codes $\mathcal{H}_S$ and $\mathcal{H}_{S'}$ are equivalent if they can be transformed into each other through permutation of qubits and single-qubit basis transformations.

> **Exercise 7.17** *Find a stabilizer of Shor's nine-qubit QECC equivalent to that defined in Equation (7.54).*

7.5.3 Logical Operators

In our examination of QECCs, we have primarily focused on encoding, error detection, and correction, yet an essential aspect remains: understanding the operators that define and manipulate logical qubits. For a physical qubit, defined as the two-dimensional Hilbert space spanned by the eigenstates of σ_z, the Pauli matrix σ_x transforms the computational basis states $|0\rangle$ and $|1\rangle$ into each other. Analogously, we seek to identify the operators that define and perform similar transformations for the logical basis states $|0_L\rangle$ and $|1_L\rangle$ of a logical qubit. These operators, known as logical operators, are crucial for understanding the behavior of logical qubits, and the stabilizer formalism expedites the procedure for finding them.

Given an $[[n, k, d]]$ stabilizer code $\mathcal{H}_S$, we aim to determine the logical operators acting on the k logical qubits. First, a logical operator must be a member of $\mathcal{G}_n$, as it acts on the physical qubits individually to avoid unexpected entangling that may ruin the logical qubit. Second, it should not be an element of S, as stabilizers preserve every codeword. Third, a logical operator cannot be a detectable error, as such errors, represented by operators like E_a, anticommute with at least one element of S and displace codewords out of the codespace.

Therefore, a **logical operator** on a logical qubit is an undetectable and uncorrectable error operator, commuting with the entire stabilizer S. A familiar example is seen in Shor's code, where the basis states $|\pm\rangle$ of a block qubit are eigenstates of Z_B, and can be interconverted using X_B as defined in Equation (7.28). These are only prototypical logical operators because they cannot act on the entire logical qubit yet, but act instead on the block qubits comprising the logical qubit.

The set of operators in $\mathcal{G}_n$ that commute with the entire S forms a subgroup $Z(S) \subseteq \mathcal{G}_n$, known as the **centralizer** of S. Since $S \subseteq Z(S)$ and $\dim \mathcal{G}_n = 2n$, for an $[[n, k, d]]$ code with $\dim S = n - k$, we have

$$\dim Z(S) = 2n - (n - k) = n + k. \tag{7.78}$$

Thus, we can select $n - k$ generators of $Z(S)$ as the $n - k$ generators of S, leaving exactly $2k$ generators to act as logical operators on the k logical qubits.

Exercise 7.18 *The **normalizer** $N(S)$ is defined as all the elements $E \in \mathcal{G}_n$ meeting the condition $EME^\dagger \in S \; \forall M \in S$. Show that $N(S) \supseteq Z(S)$. Then, show that*

$$\dim N(S) = n + k.$$

Is there any difference between the $N(S)$ and $Z(S)$ of the three-qubit bit-flip code? Can a logical operator on the logical qubit be an element of $N(S)$ but not one of $Z(S)$?

Exercise 7.19 *A nondegenerate stabilizer code $\mathcal{H}_S$ is also one whose logical operators all have weights larger than or equal to its code distance d. Explain why this definition is equivalent to the definition of nondegenerate stabilizer code proposed after Exercise 7.16.*

Exercise 7.20 *Again, the dimension of $Z(S)$ should be understood as the dimension of the corresponding symplectic vector space, which also justifies the notation $S^\perp$ for $Z(S)$. Consider a stabilizer code with $n = 2$ qubits and $k = 0$ logical qubits (a fully stabilizing code). Answer the following questions:*

- *The stabilizer group $\mathcal{S}$ is generated by $M_1 = \sigma_x^{(1)} \sigma_x^{(2)}$ and $M_2 = \sigma_z^{(1)} \sigma_z^{(2)}$. Represent M_1 and M_2 as binary vectors $\mathbf{u}_1$ and $\mathbf{u}_2$.*
- *Verify that $\mathcal{S}$ is an **isotropic subspace** by showing that $\omega(\mathbf{u}_i, \mathbf{u}_j) = 0$ for all i, j.*
- *Determine the centralizer $\mathcal{Z}(\mathcal{S})$ in the Pauli group (modulo phases). List all its elements and represent them as binary vectors.*
- *Show that the **symplectic complement** $\mathcal{S}^\perp := \{\mathbf{v} \in \mathcal{V}_4 | \omega(\mathbf{v}, \mathbf{u}) = 0, \forall \mathbf{u} \in \mathcal{V}_\mathcal{S}\}$ of $\mathcal{S}$ is equal to $\mathcal{Z}(\mathcal{S})$.*
- *Explain why $\mathcal{S} \subseteq \mathcal{S}^\perp$ and discuss the implications for the dimensions of $\mathcal{S}$ and $\mathcal{S}^\perp$.*
- *For the stabilizer code in this exercise, confirm that $\dim \mathcal{S}^\perp = 2n - \dim \mathcal{S}$ and interpret this result in the context of symplectic geometry.*

These $2k$ generators can be denoted as single-logical-qubit operators

$$\{\Sigma_l^z, \Sigma_l^x | l = 1, 2, \ldots, k\}, \tag{7.79}$$

acting respectively on the k logical qubits. The basis states of the k logical qubits may be expressed as

$$\left| i_{L_{k-1}} i_{L_{k-2}} \cdots i_{L_l} \cdots i_{L_0} \right\rangle, \tag{7.80}$$

with $i_{L_l} \in \{0, 1\}$. These logical operators act on these basis states as follows:

$$\Sigma_l^z \left| i_{L_{k-1}} i_{L_{k-2}} \cdots i_{L_l} \cdots i_{L_0} \right\rangle = (-1)^{i_{L_l}} \left| i_{L_{k-1}} i_{L_{k-2}} \cdots i_{L_l} \cdots i_{L_0} \right\rangle, \tag{7.81}$$

$$\Sigma_l^x \left| i_{L_{k-1}} i_{L_{k-2}} \cdots i_{L_l} \cdots i_{L_0} \right\rangle = \left| i_{L_{k-1}} i_{L_{k-2}} \cdots \overline{i_{L_l}} \cdots i_{L_0} \right\rangle. \tag{7.82}$$

Let's consider an example.

Example 7.3

Take the three-qubit bit-flip QECC. Its stabilizer $\mathcal{S}_3$ has two generators:

$$M_1 = \sigma_z \otimes \sigma_z \otimes \mathbb{1}_2, \tag{7.83}$$

$$M_2 = \sigma_z \otimes \mathbb{1}_2 \otimes \sigma_z. \tag{7.84}$$

Thus, the two logical operators on the sole logical qubit are

$$\Sigma^z = \sigma_z \otimes \sigma_z \otimes \sigma_z, \tag{7.85}$$

$$\Sigma^x = \sigma_x \otimes \sigma_x \otimes \sigma_x, \tag{7.86}$$

acting as the three-qubit phase-flip and three-qubit bit-flip errors, respectively. They affect the logical basis states as follows:

$$\begin{aligned}
\Sigma^z |0_L\rangle &= \sigma_z \otimes \sigma_z \otimes \sigma_z |000\rangle = |000\rangle = |0_L\rangle, \\
\Sigma^z |1_L\rangle &= \sigma_z \otimes \sigma_z \otimes \sigma_z |111\rangle = -|111\rangle = -|1_L\rangle, \\
\Sigma^x |0_L\rangle &= \sigma_x \otimes \sigma_x \otimes \sigma_x |000\rangle = |111\rangle = |1_L\rangle, \\
\Sigma^x |1_L\rangle &= \sigma_x \otimes \sigma_x \otimes \sigma_x |111\rangle = |000\rangle = |0_L\rangle.
\end{aligned} \tag{7.87}$$

> **Exercise 7.21** *Identify the logical operators in Shor's nine-qubit QECC.*

7.5.4 A Physical Interpretation

Up to this point, our discussion of stabilizer codes has been predominantly mathematical. To gain a comprehensive understanding, it's crucial to consider the physical perspective as well. In this section, we explore stabilizer codes in terms of Hamiltonian and symmetry.

The Hamiltonian and Its Ground States

For an $[[n, k, d]]$ stabilizer code $\mathcal{H}_{\mathcal{S}}$, our aim is to construct an n-qubit Hamiltonian whose ground-state Hilbert space aligns with the code space of the stabilizer code. Recognizing that the code space is the $+1$ eigenspace of the stabilizer $\mathcal{S}$, generated by $n - k$ Pauli operators M_1 to M_{n-k}, we define the Hamiltonian as

$$H = -\sum_{i=1}^{n-k} M_i. \tag{7.88}$$

This Hamiltonian is *exactly solvable* due to the commutativity of its components. Its ground-state Hilbert space, $\mathcal{H}_0$, coincides with the simultaneous $+1$ eigenspace of M_1 to M_{n-k}. As a result, $\dim \mathcal{H}_0 = 2^k$ (contemplate the rationale behind this). Consequently, we establish that

$$\mathcal{H}_0 = \mathcal{H}_{\mathcal{S}}. \tag{7.89}$$

This physical interpretation not only bridges the conceptual gap between abstract quantum error correction and tangible quantum systems but also provides a deeper insight into the nature of stabilizer codes. The Hamiltonian perspective reinforces the understanding that stabilizer codes are fundamentally connected to the underlying quantum mechanics of the systems they are applied to.

The Symmetry

The codewords within $\mathcal{H}_{\mathcal{S}}$, being ground states of the Hamiltonian H, share identical energy eigenvalue. This raises a fundamental question: How can we distinguish these codewords? More broadly, what does energy degeneracy imply about a quantum system?

Recall the discussion on degeneracy and symmetry in Section 3.2.3. Energy degeneracy in a quantum system typically indicates the presence of a global symmetry of the system and, consequently, its Hamiltonian. Describing such a global symmetry involves identifying a set of symmetry generators. For our Hamiltonian (7.88), these generators must commute with the entire Hamiltonian and be elements of the n-qubit Pauli group $\mathcal{G}_n$. Therefore, the symmetry generators are encompassed by the centralizer $\mathcal{Z}(\mathcal{S})$ of the stabilizer $\mathcal{S}$.

The stabilizer generators M_1 to M_{n-k} are part of $\mathcal{Z}(\mathcal{S})$, representing the trivial aspect of the symmetry. The nontrivial symmetry, mathematically characterized by the quotient

$$\frac{\mathcal{Z}(\mathcal{S})}{\mathcal{S}}, \tag{7.90}$$

is $2k$-dimensional. Each element of $\mathcal{Z}(\mathcal{S})/\mathcal{S}$ forms an equivalence class, for which we can select a representative operator. Among these $2k$ representative operators, k are denoted as Σ_1^z to Σ_k^z, defining the basis states of the ground-state Hilbert space $\mathcal{H}_0$:

$$\left(\Sigma_1^z, \Sigma_2^z, \cdots, \Sigma_k^z\right) \left|i_{L_{k-1}} i_{L_{k-2}} \cdots i_{L_l} \cdots i_{L_0}\right\rangle$$
$$= \left[(-1)^{i_{L_{k-1}}}, (-1)^{i_{L_{k-2}}}, \cdots, (-1)^{i_{L_0}}\right] \left|i_{L_{k-1}} i_{L_{k-2}} \cdots i_{L_l} \cdots i_{L_0}\right\rangle. \tag{7.91}$$

The other k nontrivial symmetry generators, Σ_1^x to Σ_k^x, enable transitions between different basis ground states:

$$\Sigma_l^x \left|i_{L_{k-1}} i_{L_{k-2}} \cdots i_{L_l} \cdots i_{L_0}\right\rangle = \left|i_{L_{k-1}} i_{L_{k-2}} \cdots \overline{i_{L_l}} \cdots i_{L_0}\right\rangle, \tag{7.92}$$

where $i = 1, 2, \ldots, k$. Formally, Equations (7.91) and (7.92) correspond precisely with Equations (7.81) and (7.82). As such, we can equate the basis ground states of our Hamiltonian with the logical basis states of the stabilizer code $\mathcal{H}_\mathcal{S}$, and the nontrivial symmetry generators with the logical operators of the code.

The Basis Ground States

An important question arises in the context of stabilizer codes: Can we explicitly construct the basis ground states of the Hamiltonian? If achievable, these states would coincide with the logical basis states of the stabilizer code, a topic not systematically addressed in our mathematical treatment.

Quantum mechanics offers a standard answer to this query, rooted in linear algebra. The key lies in constructing projectors from the total Hilbert space to the ground-state Hilbert space.

Since the ground states are simultaneous $+1$ eigenstates of the stabilizer generators M_1 through M_{n-k}, the appropriate projector into $\mathcal{H}_0$ is given by

$$P_0 = \prod_{i=1}^{n-k} \frac{\mathbb{1}_{2^n} + M_i}{2}. \tag{7.93}$$

Exercise 7.22 *Verify that P_0 defined in Equation (7.93) is a projector, satisfying $P_0^2 = P_0$.*

Applying P_0 to any basis state in the total Hilbert space $\mathcal{H}_{2^n}$ yields a basis state of $\mathcal{H}_0$. We can define a logical basis state as follows:

$$|\underbrace{0_L 0_L \cdots 0_L}_{k}\rangle := P_0 |\underbrace{00 \cdots 0}_{n}\rangle. \tag{7.94}$$

This is merely a convenient choice; other n-qubit basis states can be used for defining the logical 0.

The nontrivial symmetry generators Σ_1^x through Σ_k^x commute with P_0 and can be employed to generate the remaining $2^k - 1$ basis ground states from Equation (7.94). These operators flip the logical qubits individually, and the resulting basis ground states form the logical basis states of the corresponding stabilizer code. The approach is clear, but the actual construction depends on the specific stabilizer $\mathcal{S}$. Example 7.4 illustrates this process.

Example 7.4

Consider the $[[5, 1, 3]]$ code. According to the quantum Hamming bound (7.49), a minimum of five physical qubits is required to protect a logical qubit from arbitrary single-qubit errors. The corresponding QECC is denoted $[[5, 1, 3]]$, indicating a code distance of 3. Let's analyze this code within the stabilizer formalism. Its stabilizer $\mathcal{S}_5$ has $5 - 1 = 4$ generators:

$$
\begin{aligned}
M_1 &= \sigma_x^{(4)} \sigma_z^{(3)} \sigma_z^{(2)} \sigma_x^{(1)}, \\
M_2 &= \sigma_x^{(3)} \sigma_z^{(2)} \sigma_z^{(1)} \sigma_x^{(0)}, \\
M_3 &= \sigma_x^{(4)} \sigma_x^{(2)} \sigma_z^{(1)} \sigma_z^{(0)}, \\
M_4 &= \sigma_z^{(4)} \sigma_x^{(3)} \sigma_x^{(1)} \sigma_z^{(0)},
\end{aligned}
\tag{7.95}
$$

employing the shorthand notation used in Equation (7.57).

Analyzing the generators reveals that any weight-1 or weight-2 errors in $\mathcal{G}_5 \setminus \mathcal{S}_5$ will anticommute with at least one generator, while certain weight-3 errors may commute with all generators. For example:

$$
\sigma_z^{(4)} \sigma_z^{(2)} \sigma_z^{(1)}, \quad \sigma_y^{(4)} \sigma_z^{(3)} \sigma_y^{(2)}.
\tag{7.96}
$$

Therefore, the code distance is confirmed to be 3.

> **Exercise 7.23** *This QECC is nondegenerate. Justify this conclusion.*

The logical operators can be chosen as follows:

$$
\Sigma^z = \sigma_z^{(4)} \sigma_z^{(3)} \sigma_z^{(2)} \sigma_z^{(1)} \sigma_z^{(0)}, \quad \Sigma^x = \sigma_x^{(4)} \sigma_x^{(3)} \sigma_x^{(2)} \sigma_x^{(1)} \sigma_x^{(0)}.
\tag{7.97}
$$

Note, however, that $|00000\rangle$, while a $+1$ eigenstate of Σ^z, is not an eigenstate of $\mathcal{S}_5$. The logical basis state $|0_L\rangle$ is derived using Equation (7.94):

$$
\begin{aligned}
|0_L\rangle :&= P_0 |00000\rangle \\
&= \prod_{i=1}^{4} \frac{\mathbb{1}_{2^5} + M_i}{2} |00000\rangle \\
&= \frac{1}{4} \big[|00000\rangle + |11000\rangle + |01100\rangle + |00110\rangle + |00011\rangle + |10001\rangle \\
&\quad - |10100\rangle - |01010\rangle - |00101\rangle - |10010\rangle - |01001\rangle \\
&\quad - |11110\rangle - |01111\rangle - |10111\rangle - |11011\rangle - |11101\rangle \big].
\end{aligned}
\tag{7.98}
$$

Also,

$$|1_L\rangle := \Sigma^x |0_L\rangle = \Sigma^x P_0 |00000\rangle = P_0 \Sigma^x |00000\rangle = P_0 |11111\rangle$$

$$= \frac{1}{16}\Big[\, |11111\rangle + |00111\rangle + |10011\rangle + |11001\rangle + |11100\rangle + |01110\rangle \qquad (7.99)$$

$$- |01011\rangle - |10101\rangle - |11010\rangle - |01101\rangle - |10110\rangle$$

$$- |00001\rangle - |10000\rangle - |01000\rangle - |00100\rangle - |00010\rangle\,\Big].$$

The encoding circuit for the five-qubit QECC is depicted in Figure 7.17.

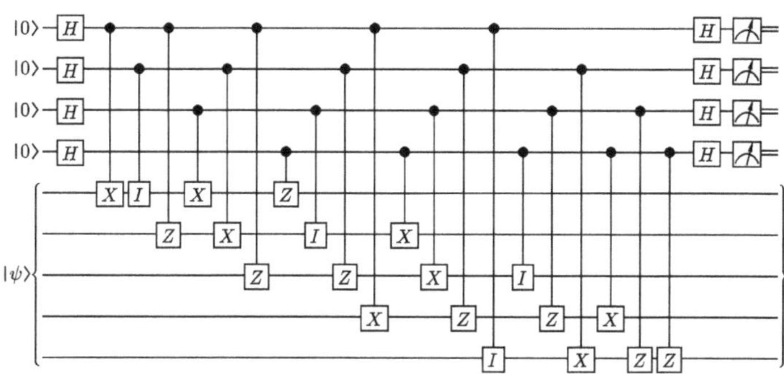

Figure 7.17 Encoding circuit for the five-qubit QECC.

> **Exercise 7.24** *Verify that the circuit in Figure 7.17 indeed works as it should. You may use a computer to assist you.*
>
> **Exercise 7.25** *Each of the two logical basis states constructed above is a superposition of 16 five-qubit basis states. Challenge yourself to create these logical basis states as superpositions of only 8 five-qubit basis states. Draw the corresponding quantum circuit for encoding. Hint: Explore the cyclic structure in the stabilizer $\mathcal{S}_5$.*

A more systematic and standard way of determining the logical operators of a stabilizer code is presented in Remark 7.2, which may be skipped upon first reading.

Remark 7.2 Standard Form for a Stabilizer Code*

A practical way of understanding a stabilizer code is casting the code in its *standard form*. Consider an $[n, k]$ stabilizer code, which utilizes n physical qubits to encode k logical qubits. This code is defined by $(n - k)$ generators, $M_1, M_2, \dots, M_{n-k}$. Each generator M_i, where $1 \le i \le (n - k)$, is a 2^n-dimensional matrix, constructed through the tensor products of n Pauli matrices $\mathbb{1}_i$, X_i, Y_i, and Z_i (related to the original Pauli matrices by in

---

Remark 7.2, continued

Figure 4.7), operating on the i-th physical qubit. The logical qubits are $+1$ eigenstates of all these generators M_i. For instance, the five-qubit QECC discussed above can be defined by four generators,

$$
\begin{aligned}
M_1 &:= X \otimes Z \otimes Z \otimes X \otimes \mathbb{1}, \\
M_2 &:= \mathbb{1} \otimes X \otimes Z \otimes Z \otimes X, \\
M_3 &:= X \otimes \mathbb{1} \otimes X \otimes Z \otimes Z, \\
M_4 &:= Z \otimes X \otimes \mathbb{1} \otimes X \otimes Z,
\end{aligned}
\tag{7.100}
$$

which determine one logical qubit.

Exercise 7.26 *Is this definition of the five-qubit stabilizer equivalent to that in Example 7.4?*

The central task in understanding a stabilizer code is finding its logical operators Σ_j^x and Σ_j^z, which acts on the j-th logical qubit, where $1 \leq j \leq k$. The logical Σ_j^z operators are those matrices that commute with and are independent of all M_i operators and are created only by the tensor products of Pauli matrices $\mathbb{1}_i$ and Z_i. The logical Σ_j^x operators are all other matrices that commute with all M_i operators and are independent of all M_i operators and Σ_j^z matrices. For the five-qubit QECC,

$$
\Sigma^z = Z \otimes Z \otimes Z \otimes Z \otimes Z,
\tag{7.101}
$$

$$
\Sigma^x = Z \otimes \mathbb{1} \otimes \mathbb{1} \otimes Z \otimes X.
\tag{7.102}
$$

Nevertheless, it is generally challenging to identify the logical Σ_j^x and Σ_j^z operators of a stabilizer code. To devise a general algorithm for calculating these logical operators, we express the generators M_i using a matrix. We arrange the generators $M_i, 1 \leq i \leq (n-k)$ into an $(n-k) \times 2n$ matrix, known as the *check matrix*:

$$
M = \left[M^X_{(n-k)\times n} \middle| M^Z_{(n-k)\times n} \right],
\tag{7.103}
$$

where M^X and M^Z are both $(n-k) \times n$ matrices filled with 0s and 1s. The j-th and $(n+j)$-th column elements represent the action on the j-th physical qubit as follows:

- If M_i has $\mathbb{1}$ on the j-th physical qubit, both the j-th and $(n+j)$-th column elements are 0.
- If M_i includes X on the j-th physical qubit, the j-th column is 1, and the $(n+j)$-th column is 0.
- If M_i includes Z on the j-th physical qubit, the j-th column is 0, and the $(n+j)$-th column is 1.
- If M_i contains Y on the j-th physical qubit, both the j-th and $(n+j)$-th columns are 1.

Remark 7.2, continued

For example, the check matrix of the five-qubit QECC is

$$\left[\begin{array}{ccccc|ccccc}
1 & 0 & 0 & 1 & 0 & 0 & 1 & 1 & 0 & 0 \\
0 & 1 & 0 & 0 & 1 & 0 & 0 & 1 & 1 & 0 \\
1 & 0 & 1 & 0 & 0 & 0 & 0 & 0 & 1 & 1 \\
0 & 1 & 0 & 1 & 0 & 1 & 0 & 0 & 0 & 1
\end{array}\right]. \tag{7.104}$$

Can you recognize each row in a check matrix is in fact a binary vector representing a Pauli operator, as introduced in Remark 7.1? Indeed, Remark 7.1 would help you comprehend the rest of the current remark. Write the k logical Σ_j^z operators and k logical Σ_j^x operators in the form of a check matrix

$$S = \left[S_{2k\times n}^X \middle| S_{2k\times n}^Z\right], \tag{7.105}$$

where S^X and S^Z are both $2k \times n$ matrices with elements that are either 0 or 1, determined by the same assignment rules described above. For instance, the two logical operators of the five-qubit code are encoded in this check matrix:

$$\begin{array}{c}
\Sigma^z \\
\Sigma^x
\end{array}
\left[\begin{array}{ccccc|ccccc}
0 & 0 & 0 & 0 & 0 & 1 & 1 & 1 & 1 & 1 \\
0 & 0 & 0 & 0 & 1 & 1 & 0 & 0 & 1 & 0
\end{array}\right]. \tag{7.106}$$

Finding the check matrix S is equivalent to solving the linear equation

$$M^X(S^Z)^T + M^Z(S^X)^T \equiv \mathbb{O} \ (\text{mod } 2). \tag{7.107}$$

The logical Σ_j^z operators can be read off from the row vectors, of which the elements in S^X are all 0. See for instance row 1 in the check matrix in Equation (7.106). The algorithm for solving Equation (7.107) involves applying Gaussian elimination to M. Swapping two rows corresponds to relabeling generators. Swapping the i_1-th and i_2-th columns, while simultaneously swapping the $(n+i_1)$-th and $(n+i_2)$-th columns, corresponds to relabeling physical qubits i_1 and i_2. Adding two rows (under the condition that $1 + 1 = 0$) corresponds to multiplying two generators. Importantly, we can replace a generator M_i with M_iM_j for $i \neq j$, as M_i can be reconstructed by $M_i = (M_iM_j)M_j$. This flexibility allows us to apply Gaussian elimination to M, bringing it to *row echelon form*:

$$\left[\begin{array}{ccc|ccc}
\mathbb{1}_r & A_{r\times(n-k-r)} & B_{r\times k} & C_r & \mathbb{O}_{r\times(n-k-r)} & D_{r\times k} \\
\mathbb{O}_{(n-k-r)\times r} & \mathbb{O}_{n-k-r} & \mathbb{O}_{(n-k-r)\times k} & E_{(n-k-r)\times r} & \mathbb{1}_{(n-k-r)} & F_{(n-k-r)\times k}
\end{array}\right], \tag{7.108}$$

where r is the rank of M^X, and the submatrices with a single subscript are square matrices. This row echelon form is called the *standard form* of a stabilizer code. It's noteworthy that this transformation process is not unique, but any stabilizer codes in this form are considered in standard form.

> **Remark 7.2, continued**
>
> For instance, the five-qubit QECC in its standard form is
>
> $$\left[\begin{array}{ccccc|ccccc} 1 & 0 & 0 & 0 & 1 & 1 & 1 & 0 & 1 & 1 \\ 0 & 1 & 0 & 0 & 1 & 0 & 0 & 1 & 1 & 0 \\ 0 & 0 & 1 & 0 & 1 & 1 & 1 & 0 & 0 & 0 \\ 0 & 0 & 0 & 1 & 1 & 1 & 0 & 1 & 1 & 1 \end{array}\right], \tag{7.109}$$
>
> which corresponds to the four generators in Equation (7.110):
>
> $$\begin{aligned} M_1' &= Y \otimes Z \otimes I \otimes Z \otimes Y, \\ M_2' &= I \otimes X \otimes Z \otimes Z \otimes X, \\ M_3' &= Z \otimes Z \otimes X \otimes I \otimes X, \\ M_4' &= I \otimes I \otimes Z \otimes Y \otimes Y. \end{aligned} \tag{7.110}$$
>
> Substituting the standard form of a stabilizer code into Equation (7.107) leads to k solutions of check matrix S. Each solution matrix S encodes a pair of logical operators according to the assignment rules of a check matrix: The first k rows would look like
>
> $$\left[\mathbb{0}_{k\times n} \,\middle|\, B_{k\times r}^T \quad \mathbb{0}_{k\times(n-k-r)} \quad \mathbb{1}_k\right], \tag{7.111}$$
>
> and yields a logical Σ_j^z operator, whereas the remaining k rows should read
>
> $$\left[\mathbb{0}_{k\times r} \quad F_{k\times(n-k-r)}^T \quad \mathbb{1}_k \,\middle|\, D_{k\times r}^T \quad \mathbb{0}_{k\times(n-k)}\right], \tag{7.112}$$
>
> producing a logical Σ_j^x operator. In summary, the structure and analysis of stabilizer codes, particularly their representation in standard form, provide a framework for understanding quantum error correction. This methodology enables the identification of logical qubits and their encoded operations within the quantum code, offering insights into the design and functionality of quantum error-correcting codes.
>
> **Exercise 7.27** *(1) Prove that Equations (7.111) and (7.112) are indeed the logical matrices of the stabilizer code. (2) Find the standard form of Shor's nine-qubit code, and find its logical operators.*

7.5.5 The Gottesman–Knill Theorem

In our exploration of stabilizer codes, we have exclusively dwelt on their role as QECCs. It's important to highlight, however, that stabilizer codes also represent a certain class of quantum dynamics or computation. While these codes are pivotal in safeguarding quantum information against errors, the quantum dynamics encompassed by stabilizer codes, particularly in the context of error correction, do not necessarily surpass classical computational capabilities. This is where the **Gottesman–Knill theorem** [211] offers particular insight.

Theorem 7.2 (Gottesman–Knill). *A quantum circuit that incorporates only the elements 1 through 4 can be efficiently simulated on a classical computer:*

1. *Initialization of qubits in the computational basis states*
2. *Quantum gates including the Pauli gates, CNOT, Hadamard, and phase gate S*
3. *Measurements in the computational basis*
4. *Classical control conditioned on the outcomes of these measurements*

This theorem suggests that certain quantum processes, despite their apparent complexity, do not fully utilize the computational power of quantum mechanics. Specifically, quantum circuits that adhere to the limitations set by the Gottesman–Knill theorem fall within the scope of classical simulatability.

The theorem's key lies in stabilizer states and **Clifford group operations**,[8] which include the Pauli gates, CNOT, Hadamard, and phase gate S. Stabilizer states are efficiently described by their stabilizers, providing a more compact representation than a full quantum state. Clifford group operations and Pauli measurements transform these stabilizers in a predictable manner. Specifically, the action of **Clifford gates** on stabilizer states results in new stabilizer states, and this transformation can be calculated with polynomial-time complexity on a classical computer. Consequently, the evolution of quantum states under such operations only requires polynomial computational resources.

Nonetheless, it is crucial to understand that the Gottesman–Knill theorem applies specifically to a subset of quantum operations. More general quantum computations, particularly those involving non-Clifford gates, retain the potential for exponential speedups compared to classical computation and remain outside the realm of efficient classical simulation. For example, Shor's algorithm utilizes the T gate, which is non-Clifford. Recall that the T gate is essential for universal quantum computation.[9]

> **Exercise 7.28** *Show that the T gate is non-Clifford.*

When considering the Hamiltonian formed as the sum of stabilizer generators, as previously discussed in the physical interpretation of stabilizer codes, the dynamics governed by such a Hamiltonian align with the classically simulatable scenarios outlined in the Gottesman–Knill theorem. This indicates that the quantum dynamics steered by this Hamiltonian, though quantum in nature, do not fully exploit the computational benefits of quantum mechanics and can be efficiently mimicked on a classical computer.

To conclude, while stabilizer codes and Clifford operations are fundamental to various quantum error correction techniques and provide deep insights into quantum mechanics, they represent a segment of quantum phenomena that can be

---

[8] A Clifford group element is an operator C that maps a Pauli operator P to a Pauli operator under conjugation. That is, $CPC^{\dagger}$ is also a Pauli operator.

[9] More generally, non-Clifford gates are essential for universal quantum computation.

simulated on classical computers. The Gottesman–Knill theorem aptly delineates the boundaries of classical and quantum computational capabilities, reminding us that the true quantum computational advantage resides in processes that exceed these simulatable limits.

Conclusion

In this chapter, we have introduced some basic concepts and examples of QECCs, which are essential for achieving reliable quantum computation using noisy physical qubits. We have seen how QECCs can protect logical qubits from various types of errors, such as bit-flip, phase-flip, and general Pauli errors, by encoding them into multiple physical qubits and performing syndrome measurements and recovery operations. We have also discussed some general properties and limitations of QECCs, such as the quantum no-cloning theorem and quantum Hamming bound.

That said, the QECCs we have presented in this chapter are not the only ones that exist. In fact, there is a rich and diverse landscape of QECCs that have been proposed and studied in the literature. One way to classify QECCs is based on whether they are linear or nonlinear. A **linear QECC** is a code that is a vector subspace of a finite field, which means that any linear combination of codewords is also a codeword. All the QECCs explored so far and the surface code to be discussed in the next chapter are linear codes. A **nonlinear QECC** is a code that does not satisfy this property, which means that some linear combinations of codewords are not valid codewords. Linear QECCs are easier to construct and analyze, and they have many advantages, such as being compatible with fault-tolerant quantum computation and having efficient decoding algorithms. Nonlinear QECCs are more difficult to construct and analyze, and they have some disadvantages, such as being incompatible with fault-tolerant quantum computation and having complex decoding algorithms. Yet, nonlinear QECCs also have some advantages, such as being able to correct more errors with fewer qubits, being more robust against certain types of noise, and being more suitable for certain quantum computing platforms. In this book, we focus on linear QECCs, as they are more widely used and understood in the field of quantum computation.

Note that the discussions and results in this chapter apply to linear codes only. For example, the precise form of the sufficient and necessary condition (7.44) on correctable qubit errors applies to linear codes only. Nonetheless, the key point of this condition is its focus on the ability of the QECC to distinguish between different error-affected states. This ability is crucial for the QECC to correctly identify and rectify errors without disturbing the encoded quantum information. Whether the QECC is linear or of a more general type, the central requirement for error correction is the distinguishability of error effects, which this condition addresses.

Another way to classify QECCs is based on the quantum computing platform that they are designed for. Different quantum computing platforms have different physical characteristics, such as the error model, the connectivity, the gate set, and the scalability, and these affect the performance and feasibility of QECCs. Therefore, some QECCs may work better on certain platforms than others, depending on these factors. For example, surface codes, which we will explore in the next chapter, are a type of topological QECC suitable for platforms with high error rates, low connectivity, and a limited gate set, such as superconducting qubits or trapped ions. **Low-density parity-check (LDPC) codes** [212] are a type of sparse QECC suitable for platforms with low error rates, high connectivity, and a universal gate set, such as photonic qubits or atomic ensembles. **Gottesman–Kitaev–Preskill (GKP) codes** [213], which are a type of nonlinear QECC, are suitable for platforms with bosonic modes, such as optical cavities or superconducting circuits.

In addition to active error correction strategies like those discussed in this chapter, it's important to consider **passive error correction** methods [214] such as **decoherence-free subspaces** [215, 216] **(DFS)** and dynamical decoupling. DFSs are subspaces of a quantum system's Hilbert space that are inherently immune to certain types of decoherence, often due to symmetries in the system-environment interaction. Nonetheless, DFS has limitations, such as requiring specific environmental conditions and offering protection against only certain types of errors. Implementing DFS can also be challenging, demanding precise control over system–environment interactions.

Dynamical decoupling is another strategy to combat decoherence [217]. This technique involves applying a sequence of fast, periodic control pulses to a quantum system, effectively averaging out its interaction with the environment over time. While dynamical decoupling can significantly reduce decoherence for certain error models, it requires precise timing and control and may not be universally applicable across all quantum computing platforms.

In conclusion, quantum error correction remains a vibrant and multifaceted field, encompassing a variety of strategies and codes to address the diverse challenges presented by different quantum computing platforms. As the field continues to evolve, we anticipate further developments in both active and passive error correction techniques, enhancing the feasibility and reliability of quantum computation.

8   Topological Orders and Surface Code

Building upon our comprehensive understanding of quantum error correction and the stabilizer formalism, which underpins a broad spectrum of QECCs, we now embark on exploring the surface code. This code represents one of the most promising strategies for implementing fault-tolerant quantum computation, particularly in prevalent quantum qubit systems like superconducting qubits and ion traps.

Remarkably, the surface code amalgamates stabilizer and topological codes. It inherits from stabilizer code the concept of a code space defined as the common eigenspace of a stabilizer group. Simultaneously, it exhibits a profound topological character. This is evidenced in its deployment on a two-dimensional lattice of physical qubits, where logical qubits manifest as topological defects within the qubit system. Logical operations in the surface code are conducted through string operators on the lattice, with logical gates enacted by maneuvering these defects in specific patterns. This interplay of physical and logical domains is anchored in the concept of topological phases of matter [218, 219] – states of matter characterized by their topological properties, rather than local, microscopic details.

In this chapter, our journey will begin with a concise overview of topological phases of matter, expounding on their fundamental attributes and significance. We will then examine these phases through the lens of their simplest lattice qubit model, which is intimately related to the physical interpretation of stabilizer codes in Section 7.5.4. This approach not only elucidates the core principles of topological phases but also seamlessly bridges to the domain of surface code, setting the stage for a deeper understanding of this pivotal concept in quantum computation. Moreover, an understanding of the characteristic properties of topological phases will facilitate our exploration in Chapter 9 of topological quantum computation, a realm drastically different from the quantum circuit model, which has mainly informed character of the book so far.

8.1   A Brief Introduction to Topological Phases of Matter

In this section, we shall first offer a brief historical account for our understanding of phases of matter from ancient times to the modern era. Following this history, we shall introduce succinctly what topological phases truly are, focusing on two spatial dimensions.

8.1.1 A Brief History

From ancient times, human understanding of the phases of matter and their transitions was based on everyday observations: solid, liquid, and gas. These were the primary classifications, understood through their distinct physical properties. This simplistic view persisted until the advent of modern physics, which brought a deeper comprehension of matter's behavior at molecular and atomic levels.

The Landau–Ginzburg paradigm [220], developed in the mid-twentieth century, revolutionized our understanding of phase transitions. Lev Landau and Vitaly Ginzburg explained phase transitions by means of spontaneous symmetry breaking. In this framework, phases of matter were understood through the lens of symmetry and its breaking. For instance, the transition from a liquid to a solid involves the breaking of continuous rotational and translational symmetry, as molecules in a liquid, which can move freely, become ordered in a solid. When a solid melts to be liquid again, the continuous symmetries will be restored.

This paradigm provided a powerful tool to describe many phase transitions, such as those between solid, liquid, and gas, or between magnetic and nonmagnetic states in materials. It was, however, predicated on the idea that different phases of matter were distinguishable by different symmetries describable by groups.

The discovery of the integer quantum Hall effect (IQHE) by Klaus von Klitzing in 1980 [221], and especially the subsequent discovery of the fractional quantum Hall effect (FQHE) by Daniel Tsui and Horst Störmer in 1982 [222], both experimentally observed in two-dimensional electron systems under strong magnetic fields, marked a turning point. These quantum Hall states exhibited unique quantized Hall conductance, a phenomenon that could not be explained by the Landau–Ginzburg paradigm. Intriguingly, these states exhibited the same $U(1)$ symmetry, suggesting that they represented a new form of matter phase that did not involve symmetry breaking.

This revelation led to the recognition of topological phases of matter, a groundbreaking concept that transcends the traditional symmetry-based understanding. Pioneers in this field, such as Xiao-Gang Wen [218, 219, 223, 224], Robert Laughlin [225], Yong-Shi Wu [226], Frank Wilczek [227], Anthony Zee [228], Duncan Haldane [229, 230], and others [231, 232] played crucial roles in developing the theoretical framework that describes these phases. They proposed that topological phases are characterized not by local order parameters and symmetry breaking, but by global properties and topological order.

These topological phases are exemplified by the quantum Hall states, where the phase is defined by the topological properties of the electron wavefunctions in the system. Unlike the Landau–Ginzburg paradigm, where phase transitions are marked by breaking or restoration of symmetries, transitions between different topological phases do not involve symmetry changes but rather changes in topological invariants.[1]

---

[1] Recent studies suggest that the conventional understanding of the Landau–Ginzburg paradigm for symmetry breaking may be incomplete. Phase transitions of topological orders could be described

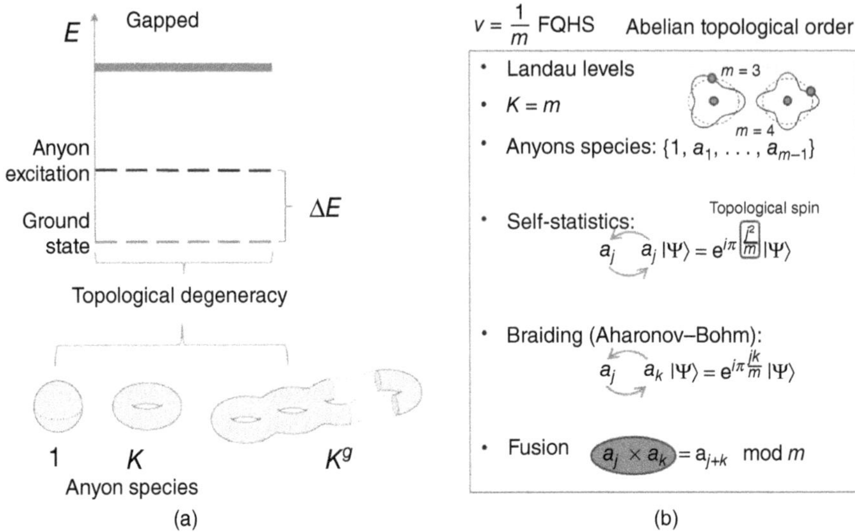

Figure 8.1 (a) Illustration of the defining and key properties of topological phase. (b) Filling-fraction $v = 1/m$ fractional quantum Hall state as an example.

This shift in understanding opened up a new realm in condensed matter physics, leading to the exploration of exotic states of matter such as topological insulators and superconductors, and paving the way for potential applications in quantum computing and other advanced technologies.

8.1.2 Topological Phases in Two Spatial Dimensions

This subsection may be bypassed if your primary interest is surface code.

Topological phases of matter, hereafter simply referred to as topological phases or topological orders,[2] exist in any spatial dimensions greater than 0; however, we are concerned about two spatial dimensions only, as the corresponding topological phases exhibit astonishing intrinsic properties that offer a fault-tolerant model of quantum computation drastically different from the quantum circuit model.

So, what exactly constitutes a topological phase? Figure 8.1(a) illustrates the defining properties of topological phases, which we will now explain.

- Primarily, a topological phase typically emerges in a strongly correlated electron system at nearly absolute zero temperature ($T = 0$ K), where all thermal fluctuations cease, leaving only quantum fluctuations. Thus, a topological phase is intrinsically a quantum phase.
- Second, a topological phase is a gapped phase: There is a system-size independent finite energy gap ΔE between the ground states and the lowest excited states of the system in the thermodynamic limit. Let's examine the ground states. A

within an expanded framework of the Landau–Ginzburg paradigm [233, 234, 235], wherein the symmetries involved are generalized to include categorical, algebraic [149, 151, 152, 236], and higher-form [150] symmetries. This broader interpretation is rapidly evolving.

2 Topological order is the order in a topological phase, but here we abuse the terminologies to use them interchangeably.

topological phase may have a topologically protected ground-state degeneracy (GSD). That is, given a system in a topological phase, its ground-state Hilbert space may be multidimensional, and this dimension is dictated by the topology rather than the microscopic detail of the system. If the system's topology is a sphere, its topological phase has a unique ground state. A toroidal system in a topological phase has a finite GSD $K > 1$. On higher-genus surfaces, the GSD grows as K^g, where g is the genus number; that is, the number of holes. Note, however, that this growth pattern K^g holds only for a simple type of topological phases – Abelian topological phases. General non-Abelian topological phases have more intricate growth patterns of their GSDs. We shall clarify these terminologies soon. The GSD of a topological phase cannot be altered by any local perturbations, which are exponentially suppressed by the system size, as long as the underlying system's topology remains [237, 238]. Such a GSD is thus topologically protected.

- We now discuss the elementary excitations in a topological phase. These excitations are collective particle-like manifestation of the microscopic degrees of freedom of the underlying system. They are categorized into a finite number of types, and this number – regardless of the underlying topology – equals the GSD K of the topological phase on the torus. This equality, a simple yet profound result of topological field theory, will be proven in Chapter 9. More remarkably, such excitations exhibit exotic statistical behavior: exchanging two of them may apply a phase factor to the many-body wavefunctions of the underlying system, and this phase factor is generally complex, neither 1 or -1. Therefore, such excitations are termed **anyons**, expanding the notions of bosons and fermions, exchanging two of which at most flips the sign of the wavefunctions.

  It's noteworthy that point-like anyons are exclusive to two spatial dimensions. In three dimensions (four spacetime dimensions), the **spin-statistics theorem** [239, 240] permits only bosons and fermions as elementary particles, while point-like quasiparticles cannot have nontrivial braiding either, excluding anyons. One-dimensional spaces lack the complexity to support anyons. Dimensions beyond three trivialize anyonic statistics.[3]

The explanation above is better corroborated by concrete examples:

Example 8.1

Figure 8.1(b) depicts the topological properties of the family of filling-fraction $v = 1/m$ fractional quantum hall (FQH) states, epitomizing topological phases.

- An FQH state is a (near) zero-temperature quantum phase and is gapped because of its Landau levels defined at the thermodynamical limit.
- Given an m, the FQH state has GSD $= m$ on the torus. This GSD has a rough, picturesque microscopic interpretation: The FQH state is a system of enormous number of strongly correlated electrons, which undergo cyclotron motions. Miraculously, in this hardly tractable soup of swirling electrons, any two of them, however

---

[3] In $d \geq 3$ spatial dimensions, physical entities of dimensions $d - 2$ may have nontrivial braiding statistics. But this is way beyond the scope of the book.

far apart, wind around each other in exactly m steps, which may be imagined as the wave-number of the standing wave of one electron around the other. This "dancing" pattern[4] of all electrons in the state is by no means a local phenomenon, implying the nonlocal nature of the GSD.

- For any m, the corresponding FQH state bear m types of anyons, labeled by 1 and a_1 through a_{m-1}, where 1 is the trivial anyon or vacuum, which is always included for mathematical completeness.

- Swapping two anyons a_j imparts a phase factor $\exp(i\pi j^2/m)$ to the many-body wavefunction of the state. The factor j^2/m in the exponent is dubbed the **topological spin** of a_j, generalizing of the notion of particle spins. Swapping two anyons of different types, say, a_j and a_k, a similar phase factor $\exp(i\pi jk/m)$ adheres to the wavefunction of the system. Such a phase factor can be understood as a Berry phase of the many-body wavefunction of which the coordinates of the two anyons are parameters.[5] Unlike the sign factors due to exchanging fermions, the phase factors due to exchanging two anyons can accumulate through multiple exchanges, manifesting as braiding of the anyon's world lines. Thus, the exchange statistics of anyons are aptly described as **braiding statistics**.

- Apart from exhibiting exotic braiding statistics, anyons can also interact in a way similar to particle interactions. Anyon interaction is termed fusion. For example, in a $v = 1/m$ FQHS two anyons a_j and a_k can fuse to be anyon $a_{j+k \bmod m}$, mathematically presented as the **fusion rule**:

$$a_j \otimes a_k = a_{j+k \bmod m}. \tag{8.1}$$

This fusion is analogous to the familiar spin coupling: $\frac{1}{2} \otimes \frac{1}{2} = 0 \oplus 1$. The difference is that in our current scenario, the fusion product is unique.

- This uniqueness causes a characteristic property of such anyons. Assume each such anyon has an internal Hilbert space and remember that a topological phase has only finite types of anyons. Since two such anyons always fuse to a unique such anyon, no matter how many of them are present in a topological phase, their total fusion product is still a unique such anyon. Consequently, the fusion rule of such anyons must consistently demand that the internal space of any such anyon is one-dimensional, such that the total Hilbert space of any number of such anyons must also be one-dimensional.

  Therefore, no matter how many of such anyons you braid or how you braid them, the effect is no more than an overall phase factor of the corresponding one-dimensional many-body wavefunction. The anyons in a $v = 1/m$ FQH state are thus christened **Abelian anyons**, and the state is said to be in an **Abelian topological phase**. Certainly, $v = 1/m$ FQH states are far from the only Abelian topological phases. In the next section, we shall construct a simple lattice model that describes an Abelian topological phase that is not any $v = 1/m$ FQH state.

---

4 This metaphor is attributed to Xiao-Gang Wen.
5 Equivalently, this phase factor is akin to an Aharonov–Bohm phase, in the context of the effective theory of the FQHS. The effective theory is blind to the many-body microscopic background and

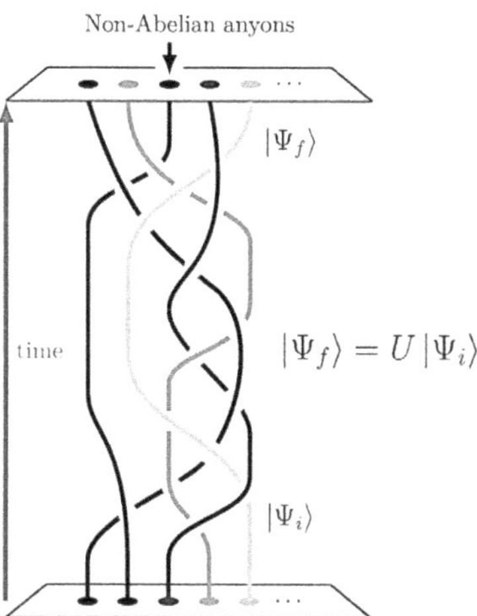

Figure 8.2 Sketch of topological quantum computation based on braiding Fibonacci anyons.

Surprisingly, the universe of anyons extends to non-Abelian anyons, which have internal spaces of dimensions greater than one. Hence, the fusion of two non-Abelian anyons can produce multiple anyons, though not all necessarily non-Abelian. A prominent example is the Fibonacci anyon, denoted by τ, governed by the fusion rule

$$\tau \otimes \tau = 1 \oplus \tau, \quad 1 \otimes \tau = \tau, \tag{8.2}$$

where 1 is the vacuum or trivial anyon. A topological phase that bears only the Fibonacci anyons is referred to as a Fibonacci topological phase, which is a non-Abelian topological phase. Fibonacci anyons, a central topic of Chapter 9, will not be dwelt on here. Instead, we offer a nutshell account of their importance as follows.

The fusion rule (8.2) implies that multiple Fibonacci anyons have a multidimensional Hilbert space. In contrast to braiding Abelian anyons, braiding multiple Fibonacci anyons does not merely cause an overall phase factor of the many-body wavefunctions but a unitary transformation on the Hilbert space. As to be detailed in the next chapter, a certain subspace of such a Hilbert space can be selected as a logical space that encodes numerous logical qubits, on which braiding the relevant Fibonacci anyons enacts a unitary transformation, as sketched in Figure 8.2. It can be shown that universal quantum computation can be accomplished by braiding Fibonacci anyons. This approach, known as **topological quantum computation** [241], represents a novel paradigm in quantum computing. The intricacies of this topic will be thoroughly explored in the subsequent chapter.

8.2 $\mathbb{Z}_2$ Toric Code Model

As discussed previously, topological phases typically manifest within strongly correlated electron systems where traditional perturbative methods fall short. This

perceives only the positions of the two anyons, which are the elementary excitations of the effective field. At the microscopic level, this phase is also understood as a Berry phase.

complexity necessitates alternative approaches for studying topological phases since conventional techniques – starting with a microscopic field or lattice theory followed by renormalization analysis – are often ineffective. Although the Kitaev honeycomb model [242], an exactly solvable microscopic lattice model, describes several topological phases, a general microscopic lattice model encompassing all two-dimensional topological phases remains elusive.[6] Furthermore, the Kitaev honeycomb model's lack of direct relevance to surface code precludes its detailed discussion in our context.

Nevertheless, if our interest lies solely in the topological characteristics of these phases, a microscopic model capturing the entirety of the underlying system's dynamics is unnecessary. The low-energy effective description of a topological phase provided by topological field theory (TFT) suffices to characterize the topological attributes of the phase fully. Three-dimensional spacetime hosts four types of TFTs: Reshetikhin–Turaev [243], Chern–Simons [244], Turaev–Viro [245, 246], and Dijkgraaf–Witten [247]. Collectively, these TFTs describe all conceivable two-dimensional topological phases.[7]

To seamlessly connect the stabilizer codes introduced in the preceding chapter with the imminent exploration of surface codes, in what follows we will introduce the lattice Hamiltonian model of the simplest Dijkgraaf–Witten TFT – the $\mathbb{Z}_2$ toric code model [17], also referred to as the $\mathbb{Z}_2$ Kitaev model. This model epitomizes an elementary yet seminal topological phase: the doubled $\mathbb{Z}_2$ phase, also known as the $\mathbb{Z}_2$ toric code phase. This phase is not merely of theoretical interest; it physically underpins surface codes.

The $\mathbb{Z}_2$ toric code model is an exactly solvable, lattice Hamiltonian model that swallows a bunch of input degrees of freedom and outputs the $\mathbb{Z}_2$ topological phase. We will construct this model in due steps and demonstrate the key properties of the output phase, including its GSD on the torus, anyon types, fusion rules, and braiding statistics.

8.2.1 Playground and Hilbert Space

The playground for our model is a two-dimensional square lattice.[8] We begin with an $n \times n$ square lattice configured with periodic boundary conditions in both dimensions, effectively mapping our system onto a torus (see Figure 8.3). Occupying each *edge* of this lattice is a qubit, all told $N = 2n^2$ qubits. These qubits represent the input degrees of freedom for our model. As such, the encompassing Hilbert space $\mathcal{H}$ of the model is the tensor product of these N two-dimensional qubit spaces, expressed as

---

[6] Topological phases have so far been observed primarily in strongly correlated electron systems, where the fundamental degrees of freedom are typically spin-1/2 particles, such as electrons. If, however, the fundamental degrees of freedom are not limited to electrons, the $\mathbb{Z}_2$ toric code model and its generalizations can also be considered as microscopic models of topological phases.

[7] There is an intersection among these TFT types. For instance, Dijkgraaf–Witten TFTs are a subset of Turaev–Viro TFTs, while Chern–Simons TFTs are a subset of Reshetikhin–Turaev TFTs.

[8] The model's topological invariance permits it to be defined on various lattice types, both regular and irregular. Nevertheless, square lattices are most convenient for our purposes.

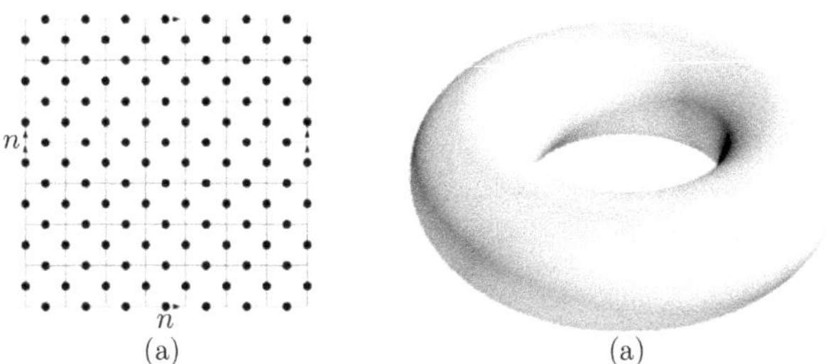

Figure 8.3 Illustration of the toric code model. A square lattice, depicted on a torus, consists of n rows and n columns of square faces. The lattice adheres to cyclic boundary conditions in both horizontal and vertical directions. Each edge of the lattice (marked by solid lines) hosts a qubit (indicated by solid dots), which can assume states of either 0 or 1. The lattice thus encompasses a total of $2n^2$ qubits.

$$\mathcal{H} = \operatorname{span}\{|j_1 j_2 \cdots j_r \cdots j_N\rangle \mid j_r = 0, 1 \ \forall r\}. \tag{8.3}$$

The order in which these N qubits are arranged within this basis is arbitrary; one simply selects an order and adheres to it consistently. The physics is independent of the chosen order.

Operators acting on $\mathcal{H}$ consist of Pauli matrices applied individually to each qubit. In fact, these operators form the Pauli group $\mathcal{G}_N$. For simplicity, we opt for the X, Y, and Z gates as defined in Figure 4.7, instead of the traditional Pauli matrices. Thus, the Pauli group $\mathcal{G}_N$ is generated by the following $2N$ Pauli operators:

$$\left.\begin{aligned}
X_r &:= \underbrace{\mathbb{1}_2 \otimes \mathbb{1}_2 \otimes \cdots \mathbb{1}_2}_{r-1} \otimes X \otimes \underbrace{\mathbb{1}_2 \otimes \cdots \otimes \mathbb{1}_2}_{N-r}, \\
Z_r &:= \underbrace{\mathbb{1}_2 \otimes \mathbb{1}_2 \otimes \cdots \mathbb{1}_2}_{r-1} \otimes Z \otimes \underbrace{\mathbb{1}_2 \otimes \cdots \otimes \mathbb{1}_2}_{N-r}
\end{aligned}\right\} \quad r = 1, 2, \cdots, N. \tag{8.4}$$

Their actions on the basis states are as follows:

$$X_r |j_1 \cdots j_r \cdots j_N\rangle = |j_1 \cdots \bar{j}_r \cdots j_N\rangle, \tag{8.5}$$

$$Z_r |j_1 \cdots j_r \cdots j_N\rangle = (-1)^{j_r} |j_1 \cdots j_r \cdots j_N\rangle, \tag{8.6}$$

where $\bar{j}_r := 1 - j_r$. In other words, X_r and Z_r exclusively influence the qubit on edge r. Notably, all basis states are eigenstates of the Z_r operator, each with eigenvalues of ± 1.

These Pauli operators exhibit the commutation and anti-commutation relations stated in Equations (8.7)–(8.9):

$$[X_r, X_{r'}] = [Z_r, Z_{r'}] = 0 \quad \forall r, r', \tag{8.7}$$

$$[X_r, Z_{r'}] = 0 \quad \forall r \neq r', \tag{8.8}$$

$$\{X_r, Z_r\} = 0. \tag{8.9}$$

8.2.2 Hamiltonian of the Model

The Hamiltonian is designed to couple qubits in a local manner; that is, it couples qubits that are adjacent as per the lattice spacing. On our square lattice, where qubits reside on the edges, there are two distinct types of local couplings: vertex type and plaquette type.

Vertex Operator

At each vertex s of the lattice, four incident edges, each hosting a qubit, can be locally coupled. Let us denote the set of edges attached to vertex s as $r \in s$. The local operator A_s, termed the **vertex operator**, facilitates this coupling:

$$A_s := \prod_{r \in s} X_r. \tag{8.10}$$

The operator A_s acts on the qubits located on the edges connected to vertex s, effectively flipping all four of them:

$$A_s \left| \begin{array}{c} j_1 \\ j_4 \quad s \quad j_2 \\ j_3 \end{array} \right\rangle = \left| \begin{array}{c} \bar{j}_1 \\ \bar{j}_4 \quad s \quad j_2 \\ \bar{j}_3 \end{array} \right\rangle. \tag{8.11}$$

Each ket state here is defined as a local-basis state, representing the tensor product of the four qubits at the edges meeting at s, while disregarding all other qubits irrelevant to A_s's action. Since A_s is a tensor product of four Pauli matrices, its eigenvalues are ± 1.

Plaquette Operator

A plaquette p is a square unit of the lattice, bounded by four edges, denoted as $r \in p$. The four qubits on the boundary of p can be locally coupled using the **plaquette operator** B_p:

$$B_p := \prod_{r \in p} Z_r. \tag{8.12}$$

Given that each basis state is an eigenstate of the Z_r operator, the action of B_p on a specific basis state $|j_1 j_2 \cdots j_{2n^2}\rangle$ yields a sign factor:

$$B_p \left| \begin{array}{c} j_1 \\ j_4 \quad p \quad j_2 \\ j_3 \end{array} \right\rangle = (-1)^{j_1+j_2+j_3+j_4} \left| \begin{array}{c} j_1 \\ j_4 \quad p \quad j_2 \\ j_3 \end{array} \right\rangle. \tag{8.13}$$

The eigenvalues of B_p are also ± 1.

Hamiltonian

Let Γ denote the lattice. It is evident that for all distinct vertices $s, s' \in \Gamma$ and plaquettes $p, p' \in \Gamma$, the commutation relations in (8.14) hold:

$$[A_s, A_{s'}] = [B_p, B_{p'}] = [A_s, B_p] = 0. \tag{8.14}$$

Thus, the Hamiltonian of the $\mathbb{Z}_2$ toric code model, which is exactly solvable, can be constructed as

$$H = -\sum_{s \in \Gamma} A_s - \sum_{p \in \Gamma} B_p. \tag{8.15}$$

The energy eigenstates of this Hamiltonian coincide with the common eigenstates of all A_s and B_p operators. In particular, the ground states are the simultaneous $+1$ eigenstates of all A_s and B_p. Does this Hamiltonian remind you of the Hamiltonian (7.88) of the stabilizer code? Indeed, as we will now demonstrate, this Hamiltonian defines a stabilizer code – specifically, the $\mathbb{Z}_2$ toric code, so named for its definition on the torus.

8.2.3 Ground States

In light of the Hamiltonian formulation of stabilizer code, and to demonstrate the topological properties of our model, we first examine the ground-state subspace $\mathcal{H}_0 \in \mathcal{H}$ of the Hamiltonian, which is the set of the common $+1$ eigenstates of all A_s and B_p:

$$\mathcal{H}_0 := \left\{ |\Phi\rangle \in \mathcal{H} \mid A_s |\Phi\rangle = B_p |\Phi\rangle = |\Phi\rangle, \forall s, p \in \Gamma \right\}. \tag{8.16}$$

Topological and Physical Picture of Ground States

As aforementioned, the GSD of a topological phase on the torus is a defining feature of the phase; however, before finding the GSD of our model on the torus, let's try to first understand a generic ground state from both topological and physical perspectives.

To find a generic ground state, we can construct the ground-state projector P_0 that projects the total Hilbert space $\mathcal{H}$ into the ground-state Hilbert space $\mathcal{H}_0$. Since $A_s^2 = B_p^2 = \mathbb{1}_{2N}$, we can define the projectors

$$P_s := \frac{I + A_s}{2}, \qquad P_p := \frac{I + B_p}{2} \tag{8.17}$$

with the properties

$$P_s^2 = P_s, \quad P_p^2 = P_p, \quad P_s P_p = P_p P_s, \quad A_s P_s = P_s, \quad B_p P_p = P_p. \tag{8.18}$$

Clearly, P_s projects any local-basis state defined at vertex s onto a $+1$ eigenstate of P_s. Likewise, P_p projects any local-basis state defined on a plaquette onto a $+1$ eigenstate of B_p. Therefore, the projector P_0 from $\mathcal{H}$ to $\mathcal{H}_0$ reads:

$$P_0 = \prod_{s \in \Gamma} P_s \prod_{p \in \Gamma} P_p. \tag{8.19}$$

Consider a simple state $|\mathbf{0}\rangle$ of all the N qubits on Γ:

$$|\mathbf{0}\rangle := |\underbrace{00\cdots0}_{N}\rangle = \Big| \cdots \Big\rangle, \qquad (8.20)$$

where on the RHS only part of the entire lattice Γ is shown. Obviouly, $|\mathbf{0}\rangle$ is not a ground state because $A_s |\mathbf{0}\rangle \neq |\mathbf{0}\rangle$, but rather a +1 eigenstate of B_p because $B_p |\mathbf{0}\rangle = |\mathbf{0}\rangle$, for all $s, p \in \Gamma$. But we can project $|\mathbf{0}\rangle$ onto a ground state using the projector P_0:

$$|\Psi_{\text{loop}}\rangle := \prod_{s \in \Gamma} \frac{\mathbb{1}_{2^N} + A_s}{2} \prod_{p \in \Gamma} \frac{\mathbb{1}_{2^N} + B_p}{2} |\mathbf{0}\rangle = \prod_{s \in \Gamma} \frac{\mathbb{1}_{2^N} + A_s}{2} |\mathbf{0}\rangle . \qquad (8.21)$$

This state $|\Psi_{\text{loop}}\rangle$ is truly a ground state of the Hamiltonian (8.15). Let's take a close look at the state and depict it topologically. Staring at the RHS of (8.21), one can see that $|\Psi_{\text{loop}}\rangle$ is a superposition of the products of any number (from 0 to n^2) of vertex operators at distinct vertices acting on the state $|\mathbf{0}\rangle$. We can get the topological picture by checking a few low-order terms in the superposition.

Recall that A_s by its definition (8.10) flips all four qubits on the four edges incident at s. Hence,

$$A_s |\mathbf{0}\rangle = A_s \Big| \cdots \Big\rangle = \Big| \cdots \Big\rangle, \qquad (8.22)$$

where in the resultant state, all qubits are in state $|0\rangle$ except the four qubits that are flipped by A_s. If we draw a line to virtually connect the four qubits in state $|1\rangle$, we obtain a loop centered at s, as the thick loop in Equation (8.23).

$$A_s |\mathbf{0}\rangle = A_s \Big| \cdots \Big\rangle = \Big| \cdots \Big\rangle, \qquad (8.23)$$

where we have neglected the labels of the qubit states. In other words, A_s creates a loop intersecting the four edges issuing at s. If A_s acts again, the loop disappears.

Next, consider what happens when two vertex operators defined at different vertices act on $|\mathbf{0}\rangle$. First consider two vertex operators A_{s_1} and A_{s_2} at two adjacent vertices s_1 and s_2. Because $X_r^2 = \mathbb{1}_2$, their combined action on $|\mathbf{0}\rangle$ effectively creates a virtual loop that surrounds both vertices s_1 and s_2:

$$A_{s_1} A_{s_2} |\mathbf{0}\rangle = \left| \;\cdots\; \right\rangle . \tag{8.24}$$

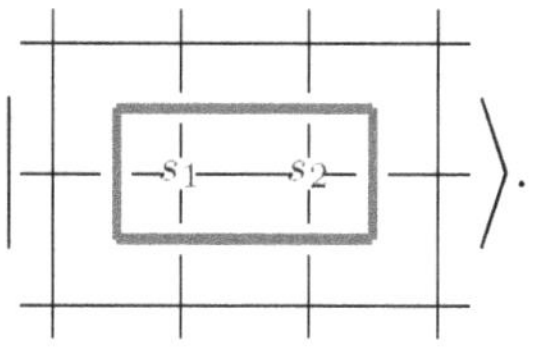

If s_1 and s_2 are not adjacent, the combined action results in two disjoint virtual loops respectively centered at s_1 and s_2, connecting the qubits in state $|0\rangle$. For instance, if s_1 and s_2 are two next-nearest vertices,

$$A_{s_1} A_{s_2} |\mathbf{0}\rangle = \left| \;\cdots\; \right\rangle . \tag{8.25}$$

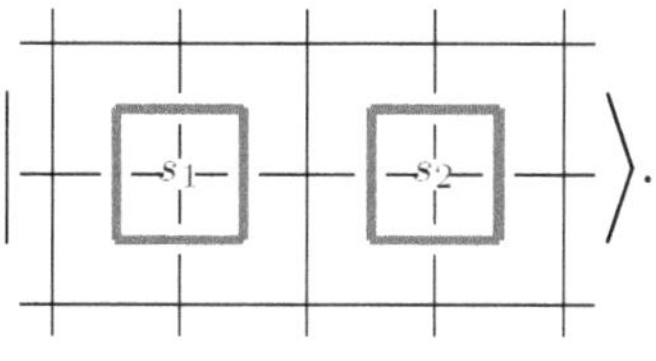

Therefore, the state $|\Psi_{\text{loop}}\rangle$ constructed in (8.21) is a superposition of the states respectively of all possible configurations of the vertex-encircling, edge-intersecting loops connecting the qubits in state $|1\rangle$ on the lattice, including the loop-free state:

$$
|\Psi_{\text{loop}}\rangle = \left| \;\cdots\; \right\rangle + \left| \;\cdots\; \right\rangle + \left| \;\cdots\; \right\rangle + \cdots
$$
$$
+ \left| \;\cdots\; \right\rangle + \cdots + \left| \;\cdots\; \right\rangle + \cdots
$$
$$
+ \left| \;\cdots\; \right\rangle + \cdots . \tag{8.26}
$$

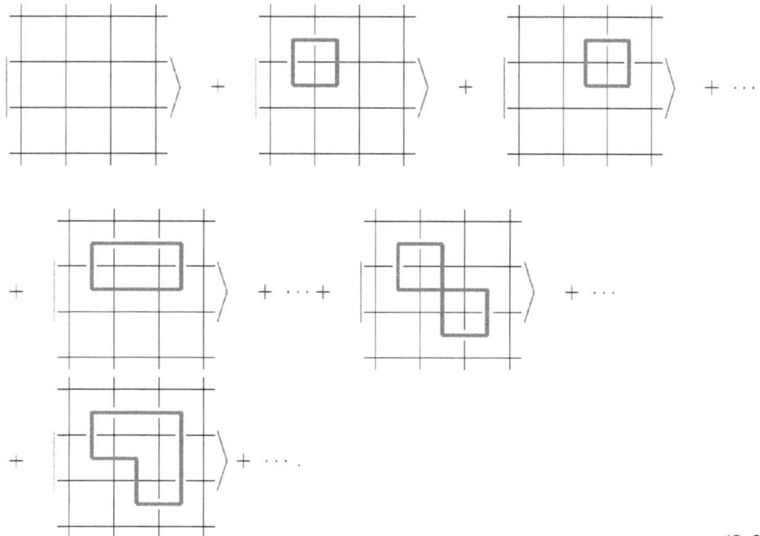

You may wonder, well, this $|\Psi_{\text{loop}}\rangle$ is only one particular ground state, would there be other possible ground states? The question is reasonable, and you can try to answer it by tackling Exercise 8.1.

Exercise 8.1 *Try to construct another ground state and check whether it is a superposition of all possible configurations of certain kinds of loops on the lattice. Hint: We obtained the state $|\Psi_{loop}\rangle$ by projecting the state $|\mathbf{0}\rangle$, which is already the common $+1$ eigenstate of all plaquette operators B_p, so you may try projecting a similar state. Is your resultant ground state orthogonal to $|\Psi_{loop}\rangle$?*

Upon working this exercise out, you will realize that any ground state of the model is a superposition of the states corresponding to all possible configurations of certain types of loops on the lattice, including the loop-free state. Since the Hamiltonian stabilizes the ground-state space $\mathcal{H}_0$, a ground state evolving under the Hamiltonian remains a ground state but with the loops fluctuating: splitting, merging, morphing, and so on. When you take a snap shot of the system in a ground state, you see a specific superposition of loop configurations. Physically, such a ground state is regarded as a *condensate of fluctuating loops*.

Astute readers may have noticed a caveat with the term "all possible loop configurations." If you do, hooray for you! But let's not worry about this caveat for now, as it will surface naturally in due time.

8.2.4 GSD and Ground-State Basis on the Torus

Having established the topological and physical characteristics of the generic ground states of the model, we now turn our attention to computing the model's GSD on the torus. We'll explore this through two distinct methods, each offering unique insights and a connection to stabilizer code.

Method 1: Counting Independent Degrees of Freedom

The most direct approach to ascertain the model's GSD on the torus is counting the number of independent degrees of freedom within the ground states. The system comprises $N = 2n^2$ quibts; however, in any ground state $|\Phi\rangle \in \mathcal{H}_0$, which is a simultaneous $+1$ eigenstate of all vertex operators A_s and plaquette operators B_p, these qubits are not entirely independent but constrained:

$$A_s |\Phi\rangle = |\Phi\rangle, \quad B_p |\Phi\rangle = |\Phi\rangle. \tag{8.27}$$

With n^2 vertices and n^2 plaquettes, supplying $2n^2$ constraints, a preliminary calculation might suggest that no independent degrees of freedom remain in the ground state since $2n^2 - n^2 - n^2 = 0$. Is this correct?

No, this initial estimate is flawed because it overlooks the interdependence of the constraints set by the vertex and plaquette operators. On the torus, we observe that

$$\prod_{s \in \Gamma} A_s = \mathbb{1}_{2N}, \quad \prod_p B_p = \mathbb{1}_{2N}, \tag{8.28}$$

as each edge r and the corresponding Pauli matrices X_r and Z_r appear twice in both $\prod_s A_s$ and $\prod_p B_p$. Therefore, the actual count of independent constraints is $2n^2 - 2$, leaving two independent degrees of freedom in the ground states. Two degrees of freedom, each being a qubit degree, implies that the GSD of the $\mathbb{Z}_2$ toric code model on a torus is $2^2 = 4$. Hence, the ground-state Hilbert space $\mathcal{H}_0$ is four-dimensional.

> **Exercise 8.2** *Demonstrate that if our model were defined on a sphere, the ground state would be unique. Also, show that on a genus-2 surface, the GSD would be 16.*

These two independent degrees of freedom, albeit qubit-like, are not associated with any specific two qubits out of the total N qubits. Instead, each represents a two-dimensional subspace of the entire Hilbert space, akin to encoding two logical qubits within multiple physical qubits. Consequently, it is apt to consider the N qubits on the toroidal lattice as encoding two logical qubits within their ground states. Therefore, the ground-state Hilbert space $\mathcal{H}_0$ forms a stabilizer code of the stabilizer group generated by all A_s and B_p operators, known as the **toric code** $\mathrm{TOR}(n)$, where n specifies the lattice size.

The toric code is foundational to the surface code, as will be discussed later. It has been argued [17] and subsequently proven [238] that such GSD is topologically protected. This protection is quantified as an exponential suppression of any local perturbation by the system size, described as $\sim \exp(-\alpha L)$, with α being a positive system-dependent parameter. Kitaev proposed leveraging the ground-state Hilbert space of the $\mathbb{Z}_2$ toric code phase for quantum memory applications [17, 248], recognizing its topological resilience.

This leads to two crucial questions: First, what constitutes the basis states of $\mathcal{H}_0$, namely the logical basis states of the toric code? Second, what are the logical operators that act on these two logical qubits? These intertwined questions will be elegantly addressed by the following alternative method of computing the GSD.

Method 2: Symmetry

As shown in Section 3.2.3 and exemplified in Section 7.5.4, systems with degenerate energy eigenstates, such as ground states, often indicate the presence of a global symmetry. The action of this symmetry commutes with the system's Hamiltonian and selects a basis for the degenerate energy eigenstates. This selection endows the states with distinct quantum numbers, distinguishing them despite their identical energy. Such global symmetries are generally described by groups, which are generated through multiplication by a set of elements known as symmetry generators.

While all symmetry generators commute with the system's Hamiltonian, they may not all commute among themselves. The mutually commuting symmetry generators yield the "good" quantum numbers and thus define the basis of the degenerate space of states. In contrast, the symmetry generators that do not commute with the mutually commuting set can transform these basis states into one another.

In the context of our $\mathbb{Z}_2$ toric code model on the torus, considering the degenerate Hilbert space $\mathcal{H}_0$, the mutually commuting symmetry generators correspond to logical Z operators. Conversely, those not commuting with the mutually commuting set assume the role of logical X operators. Our next step is to identify these symmetry generators. Although the GSD of the model is already known, we will proceed as if we were unaware of it and allow the symmetry generators to tell the story.

So, what are the symmetry generators of our model on the torus? Because our model is a qubit system, and the Pauli matrices X and Z on the qubits constitute the Hamiltonian, the symmetry generators should also be constructed by the Pauli matrices X and Z acting on certain qubits of the model. Any symmetry generator must commute with the Hamiltonian (8.15) consisting of the stabilizers A_s and B_p, each being a tensor product of four Pauli matrices acting on four neighboring qubits. In order to commute with this Hamiltonian, it suffices to consider two kinds of symmetry generators: the X-kind comprised of Pauli X matrices only, and the Z-kind comprised of Pauli Z matrices only. Mixed kinds can be obtained by multiplying the two kinds of generators.

An X-kind symmetry generator automatically commutes with all A_s; if it is to commute with all B_p, it ought to share with every B_p either zero or an even number of qubits being acted on. Upon close examination of the square lattice, it's evident that an X-kind symmetry generator must act on the qubits along certain edge-intersecting loops, such as any loop in (8.26). Similarly, a Z-kind symmetry generator would have to act on the qubits along certain loops formed by the edges of the lattice. (You should have seen this kind of loop in solving Exercise 8.1.)

Equation 8.29 defines a possible X-kind symmetry generator:

$$\mathcal{L}_C^x := \prod_{r \in C} X_r = \prod_{s \in C} A_s = \qquad\qquad , \qquad (8.29)$$

where $r \in C$ denotes the edges (thick edges) intersected by loop C, and $s \in C$ refers to the vertices (black dots) enclosed by the loop. It can be easily verified that these operators commute with the Hamiltonian:

$$[\mathcal{L}_C^x, H] = 0 \qquad (8.30)$$

because $\mathcal{L}_C^x$ is simply a product of several A_s operators. Thus, $\mathcal{L}_C^x$ operators preserve the ground-state subspace. In fact, this symmetry generator is trivial because it is actually a factor of the ground-state projector P_0 defined in (8.19) and hence cannot alter any ground state it acts on. It contributes to the identity element of the symmetry group.

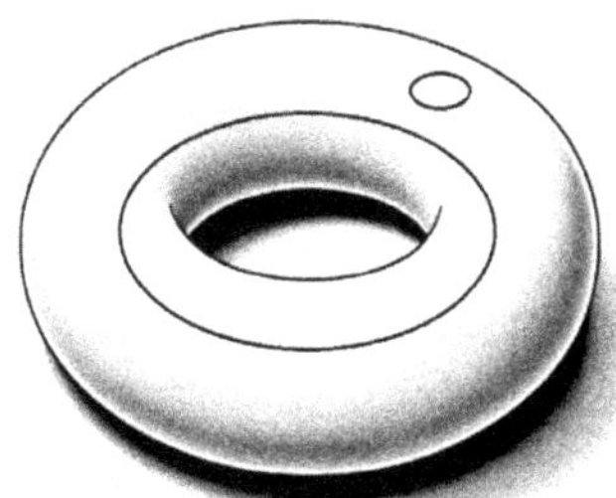

Figure 8.4 A torus with a noncontractible loop and a contractible one.

Similarly, an example of a Z-kind symmetry generator can be defined as

$$\mathcal{L}_{C^*}^z := \prod_{r \in C^*} Z_r = \prod_{p \in C^*} B_p = \quad , \tag{8.31}$$

which acts along the loop C^* lying on the edges (depicted by the thick lines). This loop, as encountered in Exercise 8.1, encircles the plaquettes $p \in C^*$. It is evident that this operator commutes with the Hamiltonian:

$$[\mathcal{L}_{C^*}^z, H] = 0. \tag{8.32}$$

Again, as with the X-kind symmetry generator, this Z-kind symmetry generator is also trivial, as it is likewise a factor of the ground-state projector P_0.

How can we then find the nontrivial symmetry generators? This brings us to the previously mentioned but overlooked caveat regarding "all possible loop configurations." The caveat lies in the nature of the loops present in the state $|\Psi_{\text{loop}}\rangle$ or the state you may have encountered in Exercise 8.1. In these configurations, the loops are contractible, meaning they can be completely eliminated by successive actions of A_s or B_p operators on the vertices or plaquettes enclosed by the loops. Figure 8.4 illustrates an example of such a contractible loop being shrunk.

The symmetry generators defined in Equations (8.29) and (8.31), and others like them, are also seen to be trivial because the loops they are defined along are topologically contractible, forming part of the continuously fluctuating contractible loops in the lattice.

Therefore, identifying nontrivial symmetry generators essentially comes down to finding noncontractible loops on the lattice. But do such loops exist on the torus? Indeed, they do, as the torus possesses two distinct holes, around which loops cannot be contracted. These noncontractible loops are those that wind around at least one of the torus's holes, in contrast to contractible loops, which do not encircle any hole. Figure 8.4 provides an example of both types of loops.

On the lattice, a noncontractible loop is characterized as either a horizontal straight line connecting the left and right identified boundaries or a vertical straight line stretching from the top to the bottom identified boundaries. Furthermore, any such straight line may either intersect the edges or lie completely along the edges, resulting in either an edge-intersecting loop or an edge-lying loop. Therefore, we can categorize four primary types of noncontractible loops[9] on the torus.

These noncontractible loops are inherently nonlocal, as they extend through the entire lattice. Unlike the contractible loops that can be created or eliminated by the local A_s and B_p operators, noncontractible loops cannot be formed or diminished in this manner. Consequently, such loops do not appear in the state $|\Psi_{\text{loop}}\rangle$, nor are they produced by the projector P_0. Their nonlocal nature makes them fundamentally different from the contractible loops and key to identifying the nontrivial symmetry generators of the $\mathbb{Z}_2$ toric code model on a torus.

On an edge-intersecting (edge-lying) loop, an X-type (Z-type) of nontrivial symmetry generators can be defined. Hence, four nontrivial symmetry generators can be defined:

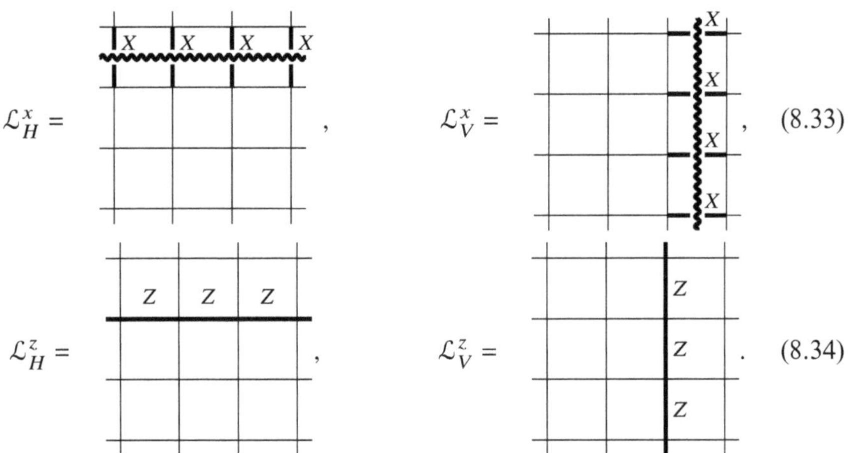

$$\mathcal{L}_H^x = \quad , \quad \mathcal{L}_V^x = \quad , \quad (8.33)$$

$$\mathcal{L}_H^z = \quad , \quad \mathcal{L}_V^z = \quad . \quad (8.34)$$

Note that the horizontal and vertical straight lines, which represent the noncontractible loops, extend from one boundary of the torus to the other. These loops cannot be expressed as products of A_s and B_p operators, and therefore do not belong to the stabilizer group of the toric code. Nevertheless, they obviously commute with the Hamiltonian. Consider, for instance, an operator $\mathcal{L}_H^x$ acting on a ground state; it creates a virtual noncontractible loop that links the qubits in state $|1\rangle$.

An astute reader might ponder the following question: If a horizontal straight line is shifted downward or upward by a grid, or even deformed, does it give rise to a different nontrivial symmetry generator? Fortunately, the answer is no. More rigorously, a horizontal straight line and its deformation lead to two equivalent nontrivial symmetry generators. They are equivalent in the sense that they can be transformed into each other by the action of A_s or B_p operators, which are stabilizer generators of the toric code. For instance, consider a deformed horizontal edge-intersecting loop and an undeformed one. They are equivalent due to an A_s action, as shown in Equation (8.35):

---

[9] While there are additional types of noncontractible loops, they can be constructed from these four elementary types.

Table 8.1 Four choices of the logical Z and logical X operators on $\mathcal{H}_0$. Two operators aligned vertically anticommute with each other.

	1	2	3	4
Logical Z	$\mathcal{L}_H^z,\ \mathcal{L}_V^z$	$\mathcal{L}_H^z,\ \mathcal{L}_H^x$	$\mathcal{L}_V^x,\ \mathcal{L}_V^z$	$\mathcal{L}_H^x,\ \mathcal{L}_V^x$
Logical X	$\mathcal{L}_V^x,\ \mathcal{L}_H^x$	$\mathcal{L}_V^x,\ \mathcal{L}_V^z$	$\mathcal{L}_H^z,\ \mathcal{L}_H^x$	$\mathcal{L}_V^z,\ \mathcal{L}_H^z$

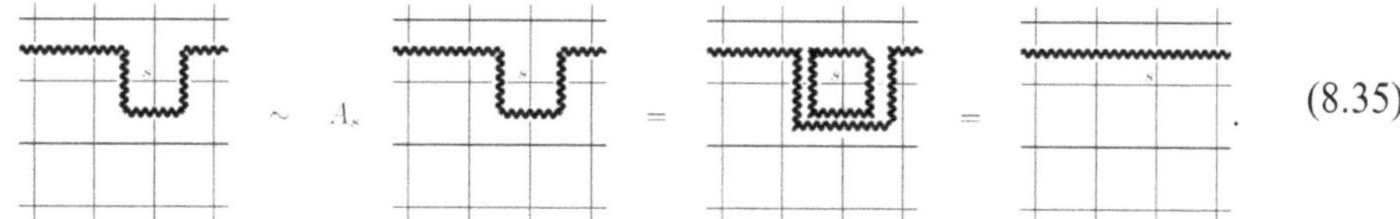

$$\tag{8.35}$$

The last equality is valid because $X^2 = \mathbb{1}_2$. Therefore, the four nontrivial symmetry generators defined by Equations (8.33) and (8.34) respectively represent four equivalence classes of nontrivial symmetry generators:[10]

$$\mathcal{L}_H^x, \qquad \mathcal{L}_V^x, \qquad \mathcal{L}_H^z, \qquad \mathcal{L}_V^z. \tag{8.36}$$

Now, as elucidated earlier, we need to identify the sets of mutually commuting nontrivial symmetry generators from these four candidates. There are four distinct sets, each containing two mutually commuting nontrivial symmetry generators. Selecting any one of these sets allows us to endow the degenerate ground states with additional quantum numbers. Since each nontrivial symmetry generator has eigenvalues ± 1, a set of two such generators provides two extra quantum numbers, each taking a value in $\{1, -1\}$. This confirms that $\dim \mathcal{H}_0 = 4$, which is consistent with the GSD obtained by directly counting the number of independent degrees of freedom in the ground states.

More importantly, within each set, the two mutually commuting nontrivial symmetry generators function as the two logical Z operators on the two-qubit four-dimensional space $\mathcal{H}_0$, namely the toric code space. The remaining two nontrivial generators, not part of the chosen set, each anticommute with one of the mutually commuting generators, and thus act as the two logical X operators. Table 8.1 lists these four choices of logical Z and X operators. It is worth noting that the two logical X operators in each choice also commute with each other, further reinforcing that the logical operators in each set define two independent logical qubits.

---

[10] Topologically, each such equivalence class is a homotopy class, as the equivalent loops are homotopic to one another.

To see what the logical basis states look like, let's take the first choice in Table 8.1 as a concrete example. We first define the +1 eigenstate, $\left|0_L^H\right\rangle \in \mathcal{H}_0$, of $\mathcal{L}_H^z$:

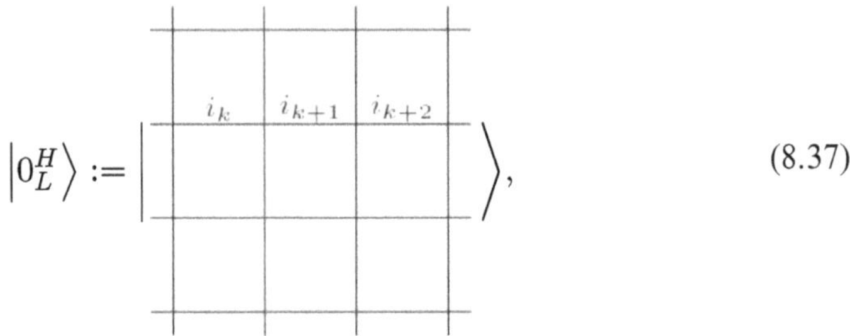

$$\left|0_L^H\right\rangle := \left| \right\rangle, \tag{8.37}$$

where we require that an even number of qubits among the qubits i_1 through i_n along a certain horizon line (a noncontractible loop, in fact) of lattice edges must be in state $|1\rangle$. We then enact the operator $\mathcal{L}_H^z$ defined along the noncontractible loop hosting the qubits i_1 through i_n on $\left|0_L^H\right\rangle$:

$$\mathcal{L}_H^z \left|0_L^H\right\rangle = \left| \right\rangle = (-1)^{i_1+i_2\cdots+i_n} \left| \right\rangle = \left|0_L^H\right\rangle, \tag{8.38}$$

where the last equality follows from the fact that an even number of the qubits i_1 through i_n are in state $|1\rangle$, the -1 eigenstate of Z. Note that the choice of the horizontal line along which $\mathcal{L}_H^z$ is defined is arbitrary because $\mathcal{L}_H^z$ represents an equivalence class of such operators. Similarly to what we did in Section 7.5.4 for stabilizer codes, we can apply the logical X operator $\mathcal{L}_V^x$ paired with the logical Z operator $\mathcal{L}_H^z$ to generate the -1 eigenstate $\left|1_L^H\right\rangle$ of $\mathcal{L}_H^z$, namely the logical $|1\rangle$ state in this basis:

$$\left|1_L^H\right\rangle = \mathcal{L}_V^x \left|0_L^H\right\rangle = \left| \right\rangle, \tag{8.39}$$

which is truly the -1 eigenstate of $\mathcal{L}_H^z$ because $\{\mathcal{L}_H^z, \mathcal{L}_V^x\} = 0$:

$$\mathcal{L}_H^z \left|1_L^H\right\rangle = \mathcal{L}_H^z \mathcal{L}_V^x \left|0_L^H\right\rangle = -\mathcal{L}_V^x \mathcal{L}_H^z \left|0_L^H\right\rangle = -\mathcal{L}_V^x \left|0_L^H\right\rangle = -\left|1_L^H\right\rangle. \tag{8.40}$$

This can be alternatively understood: The noncontractible loop operator $\mathcal{L}_V^x$ flips all the qubits, in particular qubit i_{k+2} on the edges cut through by the wiggly vertical line (red, a noncontractible loop) in Equation (8.39), so the resultant state $\left|1_L^H\right\rangle$ has an odd number of qubits among the qubits i_1 through i_n in state $|1\rangle$ and is a -1 eigenstate of $\mathcal{L}_H^z$.

Defining the ± 1 eigenstates, or the logical basis states $|0_L^V\rangle$ and $|1_L^V\rangle$, of the logical Z operator $\mathcal{L}_V^z$ works likewise. Nevertheless, since there are two logical qubits, we should find the basis states of the two-qubit logical space. If the state $|0_L^H\rangle$ has also an even number qubits among the qubits j_1 through j_n in state $|1\rangle$, it is also a $+1$ eigenstate of $\mathcal{L}_V^z$. We denote this simultaneous $+1$ eigenstate of both $\mathcal{L}_H^z$ and $\mathcal{L}_V^z$ by

$$\left|0_L^H 0_L^V\right\rangle := \left| \vcenter{\hbox{[lattice diagram with qubits i_k, i_{k+1}, i_{k+2}, j_l, j_{l+1}, j_{l+2}]}} \right\rangle, \tag{8.41}$$

where qubits i_1 through i_n have an even number of state $|1\rangle$, so do qubits j_1 through j_n. By Equation (8.39), we can define

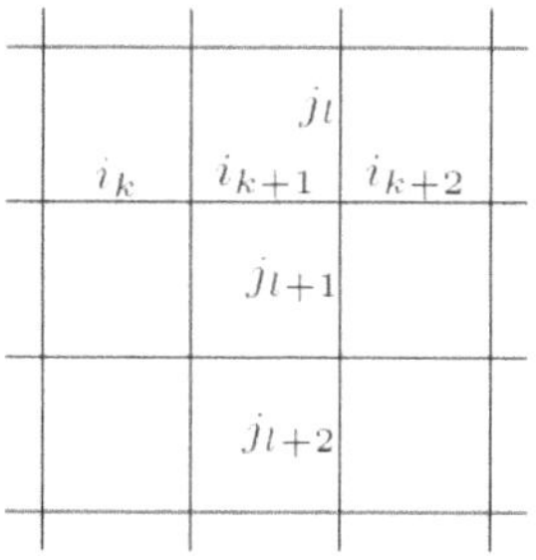

$$\left|1_L^H 0_L^V\right\rangle := \mathcal{L}_V^x \left|0_L^H 0_L^V\right\rangle = \left| \vcenter{\hbox{[lattice diagram with qubits i_k, i_{k+1}, i_{k+2}, j_l, j_{l+1}, j_{l+2} and a vertical X loop]}} \right\rangle. \tag{8.42}$$

Hence, by an $\mathcal{L}_H^x$ operator defined along any horizontal noncontractible loop, we have

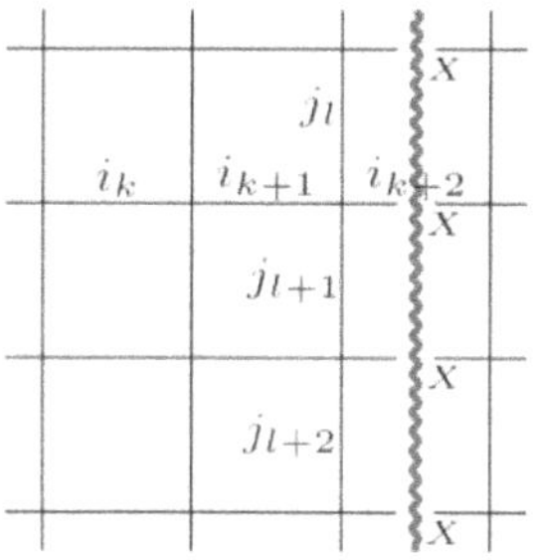

$$\left|0_L^H 1_L^V\right\rangle := \mathcal{L}_H^x \left|0_L^H 0_L^V\right\rangle = \left| \vcenter{\hbox{[lattice diagram with qubits i_k, i_{k+1}, i_{k+2}, j_l, j_{l+1}, j_{l+2} and a horizontal X loop]}} \right\rangle, \tag{8.43}$$

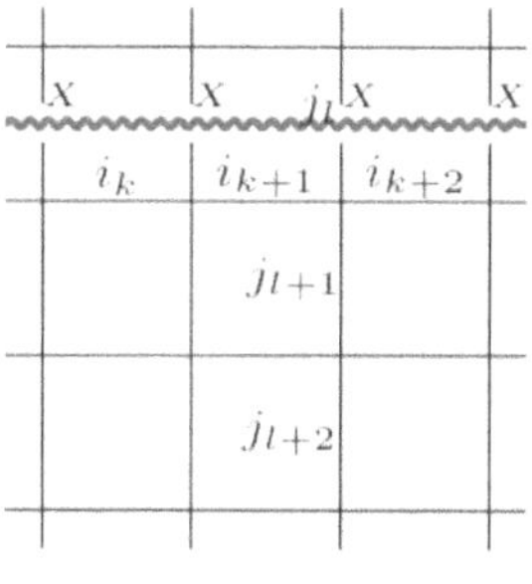

which is correct because $\{\mathcal{L}_V^z, \mathcal{L}_H^x\} = 0$. That is,

$$\mathcal{L}_V^z \left|0_L^H 1_L^V\right\rangle = \mathcal{L}_V^z \mathcal{L}_H^x \left|0_L^H 0_L^V\right\rangle = -\mathcal{L}_H^x \left|0_L^H 0_L^V\right\rangle = -\left|0_L^H 1_L^V\right\rangle. \tag{8.44}$$

Finally, since $[\mathcal{L}_V^x, \mathcal{L}_H^x] = 0$, we can define

$$\left|1_L^H 1_L^V\right\rangle := \mathcal{L}_V^x \mathcal{L}_H^x \left|0_L^H 0_L^V\right\rangle = \left| \quad \right\rangle. \tag{8.45}$$

We can collectively denote the four logical basis states as $\left|z_L^H z_L^V\right\rangle$. As such, we have

$$\mathcal{L}_H^z \left|z_L^H z_L^V\right\rangle = (-1)^{z_L^H} \left|z_L^H z_L^V\right\rangle, \quad \mathcal{L}_V^x \left|z_L^H z_L^V\right\rangle = \left|\overline{z_L^H} z_L^V\right\rangle,$$
$$\mathcal{L}_V^z \left|z_L^H z_L^V\right\rangle = (-1)^{z_L^V} \left|z_L^H z_L^V\right\rangle, \quad \mathcal{L}_H^x \left|z_L^H z_L^V\right\rangle = \left|z_L^H \overline{z_L^V}\right\rangle. \tag{8.46}$$

The procedure from Equations (8.37) through (8.46) suggests that the four basis ground states, or logical basis states, are generated by the corresponding two logical X operators $\mathcal{L}_H^x$ and $\mathcal{L}_V^x$ defined along two noncontractible loops on the torus. As such, the basis $\{\left|z_L^H z_L^V\right\rangle\}$ can be depicted in Figure 8.5(a), where the two wiggly noncontractible loops indicate the corresponding logical X operators.

The basis ground states corresponding to the other three choices of logical operators in Table 8.1 can be obtained analogously. (As an exercise, I encourage you to find these states on your own.) These states can thus be depicted as comprising two noncontractible loops, as shown in Figure 8.5(b) to (d). For all the four sets of basis states, each noncontractible loop also represents a logical qubit i_L, where $i = z, x$. If i_L is in the state $|0_L\rangle$, you can consider the corresponding noncontractible loop to be absent.

A committed reader may puzzle over why only four bases are constructed for our four-dimensional $\mathcal{H}_0$, when a finite-dimensional Hilbert space should have an infinite number of orthonormal bases, transformable into one another by unitary transformations. This puzzle is valid. Indeed, there are infinitely many bases for $\mathcal{H}_0$. The four bases illustrated in Figure 8.5 are simply convenient examples. Section 8.2.7 will explore the unitary transformations that can transform these four bases into one another and into infinitely many other bases. To keep you further engaged, these unitary transformations are more than mere transformations; they profoundly characterize the $\mathbb{Z}_2$ toric code phase in a unique way. For a deeper understanding of these transformations, two more important aspects of the $\mathbb{Z}_2$ toric code phase and topological phases at large should be addressed in the next two subsections.

8.2.5 Topological Entanglement Entropy

Recall the state $|\Psi_{\text{loop}}\rangle$, which represents a condensation of fluctuating contractible loops. Similarly, any ground state of our model on the torus can be seen as a condensation of fluctuating loops, both contractible and noncontractible, linking the

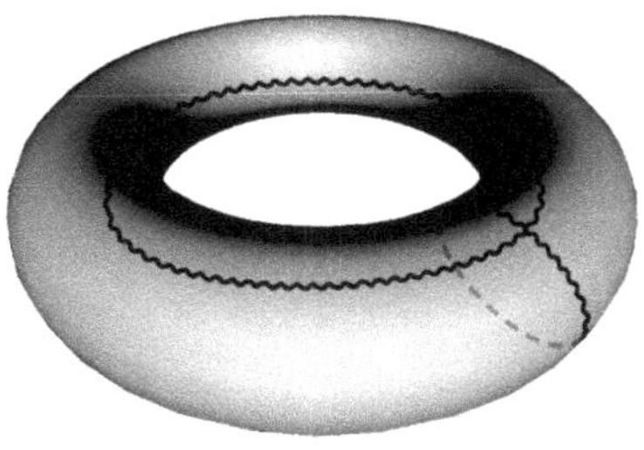

(a) Logical basis states $\left| z_L^H z_L^V \right\rangle$, with logical Z operators $\{\mathcal{L}_H^z \mathcal{L}_H^z\}$ and logical X operators $\{\mathcal{L}_V^x, \mathcal{L}_H^x\}$.

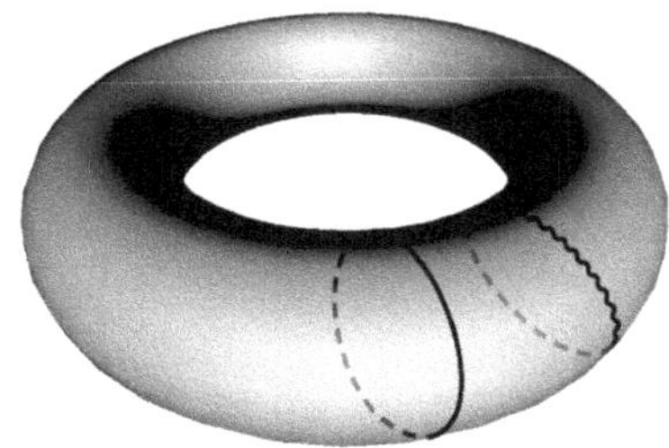

(b) Logical basis states $\left| z_L^H x_L^H \right\rangle$, with logical Z operators $\{\mathcal{L}_H^z \mathcal{L}_H^x\}$ and logical X operators $\{\mathcal{L}_V^x, \mathcal{L}_V^z\}$.

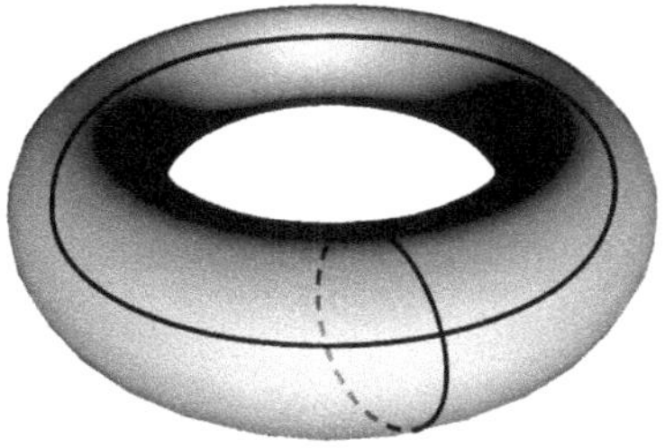

(c) Logical basis states $\left| x_L^H x_L^V \right\rangle$, with logical Z operators $\{\mathcal{L}_H^x \mathcal{L}_V^x\}$ and logical X operators $\{\mathcal{L}_V^z, \mathcal{L}_H^z\}$.

(d) Logical basis states $\left| z_L^V x_L^V \right\rangle$, with logical Z operators $\{\mathcal{L}_H^z \mathcal{L}_H^x\}$ and logical X operators $\{\mathcal{L}_H^x, \mathcal{L}_H^z\}$.

Figure 8.5 The four sets of two basis ground states, or logical basis states, corresponding to the four choices of logical operators in Table 8.1.

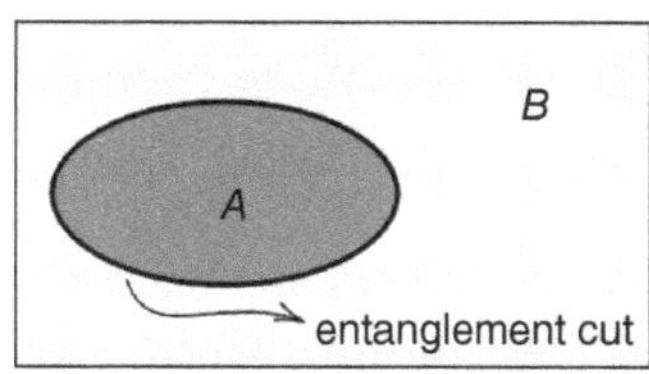

Figure 8.6 A system divided into subsystems A and B by an entanglement cut.

qubits in identical states. This imagery naturally raises questions about the entanglement properties of such states, where the qubits on the lattice exhibit complex correlations.

Let us briefly revisit the concept of entanglement, particularly bipartite entanglement, which is easier to grasp compared to multipartite entanglement. Figure 8.6 illustrates a system divided into two subsystems, A and B, separated by a boundary known as the entanglement cut or entanglement boundary. Given the density matrix ρ_{AB} of the entire system, the entanglement entropy between A and B is defined as in Equation (3.82), replicated here for convenience:

$$S_A = -\mathrm{Tr}[\rho_A \ln \rho_A], \tag{8.47}$$

where $\rho_A = \mathrm{Tr}_B(\rho_{AB})$ is the reduced density matrix of subsystem A. Typically, this entanglement entropy is a local or short-range attribute, measuring the intensity of entanglement of degrees of freedom near and across the entanglement cut. Consequently, S_A generally adheres to an area law:

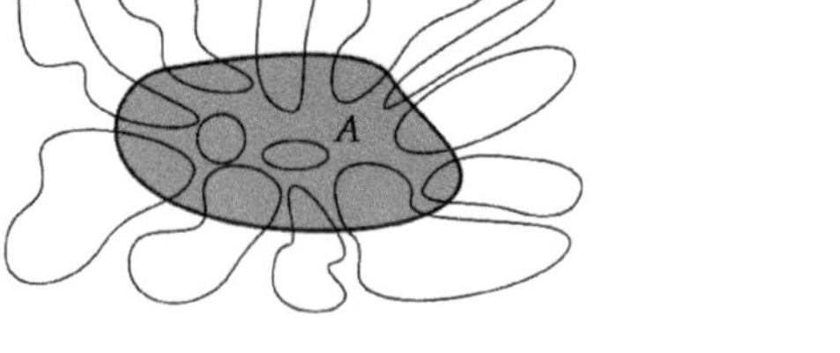

Figure 8.7 Topologically ordered system divided into subsystems A and B by an entanglement cut.

$$S_A \propto L, \tag{8.48}$$

where L represents the circumference of the entanglement cut, effectively a one-dimensional measure of area.

To compute the entanglement entropy of a system in a topological phase, one divides it into two subsystems, as exemplified in Figure 8.7. A key characteristic of entanglement in topological phases, such as the $\mathbb{Z}_2$ toric code phase, is that a state (not necessarily a ground state[11]) is a superposition of various loop configurations, a notion applicable in any topological phase. Consequently, no matter how carefully a topologically ordered state is divided, the entanglement cut inevitably intersects certain loops. Since each loop connects degrees of freedom in the same state, the entanglement entropy between the two subsystems encapsulates this "linkage." Notably, these loops can span the system size, signifying nonlocal or long-range entanglement, in contrast to the local or short-range entanglement in nontopological systems.

In the thermodynamic limit, which defines a topological phase, this nonlocality is expected to manifest as a system-size independent component in the entanglement entropy. For the $\mathbb{Z}_2$ toric code phase, the entanglement entropy is given by[12]

$$S_A = \alpha L - \ln 2, \tag{8.49}$$

where the leading term αL follows an area law, with $\alpha \geq 1$ depending on the anyonic excitations involved. The subleading term, $\ln 2$, is a constant and system-size independent, and is known as the **topological entanglement entropy** (TEE) [237]. This term quantifies the long-range entanglement between the two subsystems. For a general topological phase, the TEE is proportional to

$$\sim \ln D,$$

where D is the **total quantum dimension**, a concept to be introduced in Chapter 9. Although the TEE $\ln D$ is characteristic of a topological phase, it does not uniquely define them: Two topological phases with distinct total quantum dimensions are different, but two phases with the same total quantum dimension may still be distinct. This limitation arises because entanglement entropy, being a numerical value, cannot fully capture the complexity of quantum entanglement, which might be represented in various entanglement patterns.

---

[11] Anyonic excitations are localized, so fluctuating loops predominantly occupy the system.

[12] The specific form of the entanglement entropy depends on the topology. Equation (8.49) assumes a spherical topology.

Indeed, as proposed by Xiao-Gang Wen [249], topological orders themselves can be viewed as measures of patterns of long-range entanglement. Wen posits that a matter phase exhibits long-range entanglement if and only if it is topologically ordered.[13] In other words,

$$\text{long-range entangled} \iff \text{topologically ordered}.$$

Long-range entanglement presents a fascinating avenue of research, not only within the realm of topological phases but also in the wider field of (topological) quantum (gauge) field theories. Readers interested in delving deeper into this topic are encouraged to consult the relevant literature.

> **Exercise 8.3** *Can the entanglement entropy defined in Equation (8.49) be negative? Explain why.*

8.2.6 Anyonic Excitations

After a detailed exploration of the ground states of the model, our next step is to investigate the first excitations and ascertain whether they manifest as anyons. Recall that the ground states are the simultaneous +1 eigenstates of all A_s and B_p operators constituting the Hamiltonian (8.15). To generate a first excitation, one might intuitively consider constructing a creation operator that anticommutes with either an A_s or a B_p, leading to a state with an energy unit higher than that of the ground state. While this approach seems straightforward, the actual creation operators in this model are somewhat more complex.

Given that the qubits reside on the edges of the lattice and our tools are limited to Pauli matrices, it is not feasible to construct a creation operator that anticommutes with exactly one A_s or B_p operator. Instead, such a creation operator will necessarily anticommute with a pair of vertex operators A_s and $A_{s'}$, or a pair of plaquette operators B_p and $B_{p'}$. This nuance in the construction of the creation operators is pivotal to the understanding of excitations in the $\mathbb{Z}_2$ toric code model.

Charge

Consider, for instance, the operator $Z_1 Z_2$ acting on the qubits on the two edges (thick lines) in Figure 8.8(a). This operator commutes with all vertex and plaquette operators except for A_s and $A_{s'}$ shown in the figure, with which it anticommutes due to sharing exactly one edge with each. Therefore, for any ground state $|\Psi_0\rangle$ in

---

[13] This perspective is sometimes referred to as the "East-Coast" definition of topological order, while Alexei Kitaev's "West-Coast" definition states that a matter phase is topologically ordered if and only if it supports anyon excitations. These definitions generally align except in cases like the integer quantum Hall states, which exhibit long-range entanglement but lack anyons.

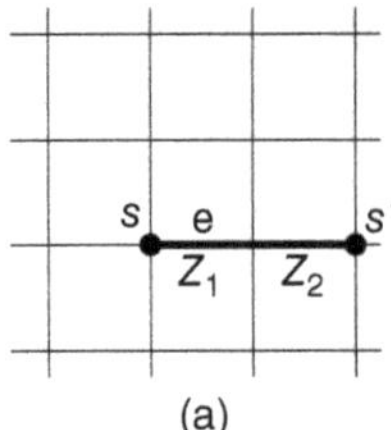
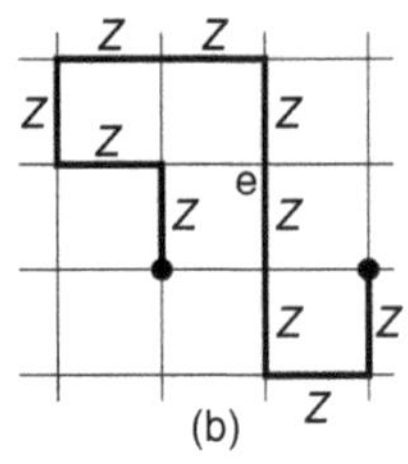

(a)   (b)

Figure 8.8 (a) Operator $Z_1 Z_2$ creates two charge excitations at vertices s and s'. (b) An e-string operator defined along a path $\mathbf{e}$ creates two charge excitations at the two ends of the path.

$\mathcal{H}_0$, we find

$$A_s(Z_1 Z_2 |\Psi_0\rangle) = -(Z_1 Z_2)A_s |\Psi_0\rangle = -Z_1 Z_2 |\Psi_0\rangle,$$
$$A_{s'}(Z_1 Z_2) |\Psi_0\rangle = -(Z_1 Z_2)A_{s'} |\Psi_0\rangle = -Z_1 Z_2 |\Psi_0\rangle. \tag{8.50}$$

Consequently, $Z_1 Z_2 |\Psi_0\rangle$ is not in $\mathcal{H}_0$; it represents an excited state. Letting $E_0 = -N = -2n^2$ be the ground-state energy, and using Equations (8.15) and (8.50), we get

$$H(Z_1 Z_2 |\Psi_0\rangle) = (E_0 + 4)(Z_1 Z_2 |\Psi_0\rangle). \tag{8.51}$$

This implies that the energy gap between $Z_1 Z_2 |\Psi_0\rangle$ and $|\Psi_0\rangle$ is $\Delta E = 4$, which is independent of the system size. Hence, the operator $Z_1 Z_2$ indeed creates an excited state, with $\Delta E = 4$ being the smallest energy gap achievable in this model, making $Z_1 Z_2 |\Psi_0\rangle$ a first excited state. What, then, are the elementary excitations in this excited state?

The violation of the ground-state constraints at vertices s and s' suggests the creation of two point-like excitations at these vertices by the operator $Z_1 Z_2$, each carrying 2 units of energy. We shall refer to such point-like excitations at vertices as **charges**, which are elementary excitations of the $\mathbb{Z}_2$ toric code phase. The justification for naming them charges goes beyond the scope of quantum mechanics; however, for completeness and to satisfy the curiosity of the reader, a brief explanation is provided in Remark 8.1.

Remark 8.1   Understanding Charge Excitations*

To comprehend why point-like excitations at vertices are regarded as charges in the $\mathbb{Z}_2$ toric code model, we dive into the physics behind the violation of ground-state constraints at vertices s and s'. Specifically, we consider the cases where $A_s(Z_1 Z_2 |\Psi_0\rangle) \neq Z_1 Z_2 |\Psi_0\rangle$ and $A_{s'}(Z_1 Z_2 |\Psi_0\rangle) \neq Z_1 Z_2 |\Psi_0\rangle$. The $\mathbb{Z}_2$ toric code model serves as a lattice Hamiltonian model of the effective topological field theory for the $\mathbb{Z}_2$ toric code phase, essentially a $\mathbb{Z}_2$ topological gauge theory. In this framework, the qubits on the lattice edges are analogous to a $\mathbb{Z}_2$ electric field, akin to the conventional $U(1)$ electric field.[a] In this analogy, a $\mathbb{Z}_2$ electric field takes values in the group $\mathbb{Z}_2 = \{0, 1\}$, indicating that

> **Remark 8.1, continued**
>
> each lattice edge contains a group element from $\mathbb{Z}_2$. The operation of A_s, comprising four Pauli X matrices, flips the qubits on the four edges incident at s. This action, in the context of our $\mathbb{Z}_2$ field analogy, equates to multiplying the group element "1" to each existing group element on these edges. Therefore, A_s functions as a gauge transformation of the $\mathbb{Z}_2$ field at the vertex s. As a result, the ground-state constraint $A_s |\Psi_0\rangle = |\Psi_0\rangle$ ensures the preservation of local gauge invariance of the $\mathbb{Z}_2$ field in its ground states.
>
> In standard $U(1)$ gauge theories, local gauge invariance is upheld everywhere except in the presence of an electric charge, or excitation. Analogously, in our lattice model, the disturbance of local gauge invariance at vertices s and s' due to the operator $Z_1 Z_2$ indicates the emergence of two $\mathbb{Z}_2$ "gauge/electric charges" at these vertices. It is essential to understand, however, that these "charges" in the $\mathbb{Z}_2$ toric code model are not akin to matter charges encountered in conventional gauge field theories. Rather, they emerge from the topological characteristics of the lattice model. These gauge charges signify the breakdown of local gauge invariance resulting from specific configurations of the lattice gauge field, as opposed to representing intrinsic properties of fundamental matter particles. Therefore, while they mirror electric charges in disrupting local gauge invariance, they are fundamentally distinct, stemming from the topological and geometric features of the gauge field configuration rather than from the inherent qualities of matter particles.
>
> ---
> [a] If the term "topological field theory" seems daunting, focus instead on this analogy.

Given that the operator $Z_1 Z_2$ is defined along a string (thick edges in the figure) and results in the creation of charges, it is an example of a charge- or e-**string operator**. An e-string operator $W_{\mathbf{e}}$ can be defined along any open path $\mathbf{e}$ comprising certain edges of the lattice, as illustrated in Figure 8.8(b). The definition of such an operator is

$$W_{\mathbf{e}} = \prod_{r \in \mathbf{e}} Z_r . \tag{8.52}$$

This e-string operator creates a pair of charge excitations at the two endpoints of the path $\mathbf{e}$. All charges generated in this manner are of the same type, as they are created by identical types of creation operators, each violating the same ground-state constraint and possessing the same energy. Consequently, we shall denote a charge excitation simply by e.

Flux

Similar to charge excitations, we can also construct another type of string operator. Consider $X_1 X_2$ as shown in Figure 8.9(a). This operator commutes with all vertex and plaquette operators, except for B_p and $B_{p'}$, with which it anticommutes. For any ground state $|\Psi_0\rangle$ in $\mathcal{H}_0$, the state $X_1 X_2 |\Psi_0\rangle$ is not part of $\mathcal{H}_0$, and we have

$$B_p (X_1 X_2 |\Psi_0\rangle) = -X_1 X_2 |\Psi_0\rangle, \tag{8.53}$$

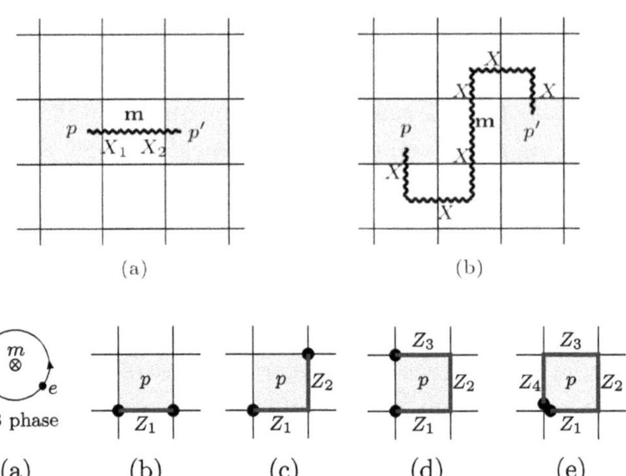

Figure 8.9 (a) Operator $X_1 X_2$ creates two flux excitations at plaquettes p and p'. (b) An m-string operator defined along a path **m** creates two flux excitations at the two ends of the path.

Figure 8.10 (a) Aharonov–Bohm phase. (b) Create a pair of charges by Z_1. (c)–(e) Move one charge around the plaquette with a flux excitation until it's annihilated by the other charge to measure the AB phase.

$$B_{p'}(X_1 X_2 |\Psi_0\rangle) = -X_1 X_2 |\Psi_0\rangle. \tag{8.54}$$

Therefore, $X_1 X_2 |\Psi_0\rangle$ is an excited state with an energy gap $\Delta E = 4$ above the ground states. The ground-state constraints being violated in plaquettes p and p' implies the presence of point-like excitations at these locations. To understand the nature of these excitations, we interpret the action of B_p on $X_1 X_2 |\Psi_0\rangle$ as sequential actions of four e-string operators along the boundary of p (see Figure 8.10(b) to (e)). This process ultimately results in a state $-X_1 X_2 |\Psi_0\rangle$, acquiring a minus sign compared to the original state $X_1 X_2 |\Psi_0\rangle$. This minus sign is an Aharonov–Bohm phase (Figure 8.10(a)), leading us to interpret the excitation in plaquette p as a **flux excitation**, or simply **flux**.

The creation operator $X_1 X_2$, therefore, is termed a flux- or m-string operator. An m-string operator $W_\mathbf{m}$ can be defined along any path **m** cutting through certain edges of the lattice (see Figure 8.9(b)):

$$W_\mathbf{m} = \prod_{r \in \mathbf{m}} X_r, \tag{8.55}$$

composed of Pauli X matrices flipping the qubits on the edges intersected by **m**. An m-string operator $W_\mathbf{m}$ invariably creates a pair of fluxes at the plaquettes on either end of **m**, as it anticommutes with the corresponding plaquette operators. The process depicted in Figure 8.10 illustrates that moving a charge around one of the end-plaquettes in the state $W_\mathbf{m} |\Psi_0\rangle$ imparts a minus sign to the state, resulting in $-W_\mathbf{m} |\Psi_0\rangle$ (refer to Figure 8.11). In fact, the minus sign emerges whenever a charge crosses the m-string operator, due to the anticommutation relation $\{W_\mathbf{e}, W_\mathbf{m}\}$.

Given that there is only one type of charge (e), and that the phase factor induced by moving a charge around a flux is consistently -1, it follows that there is also only one type of flux, denoted as m.

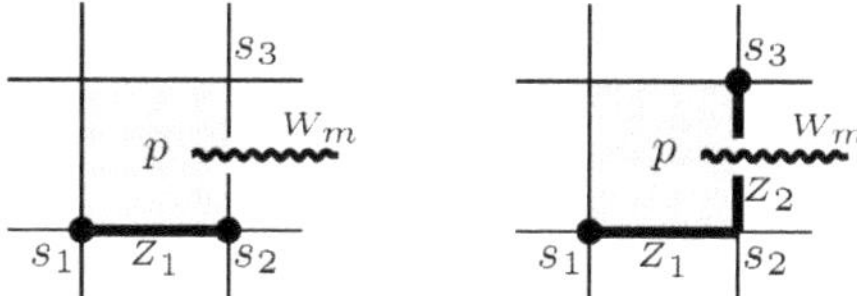

Figure 8.11 Creating a pair of charges by Z_1 on the two bottom vertices s_1 and s_2 of plaquette p at the end of the m-string operator W_m; moving the charge from vertex s_2 to s_3 with Z_2.

Dyons

Having identified three types of anyons – the charges e, the fluxes m, and the trivial vacuum 1 – in the $\mathbb{Z}_2$ toric code model, we are now poised to introduce the fourth type. Given that the ground-state degeneracy (GSD) on the torus is 4, this fourth type is expected to be a composite or a bound state of e and m. This composite anyon is termed a **dyon** and denoted as ϵ or em. Unlike the charge and flux excitations created by string operators, a dyon is created by a more complex operator, aptly named a ribbon operator, which combines the characteristics of both e-string and m-string operators. The construction of such an ϵ-ribbon operator is an insightful exercise:

> **Exercise 8.4** *Devise an ϵ-ribbon operator capable of creating a pair of dyons in two distinct plaquettes.*

For those acquainted with standard gauge field theories, such as quantum electrodynamics, which lack elementary flux excitations, the existence of flux excitations and consequently dyons in the $\mathbb{Z}_2$ toric code model may seem perplexing. This model, serving as an effective topological gauge field theory, diverges from conventional gauge theories in this aspect. Remark 8.2 aims to clarify this distinction but may be skipped by readers primarily focused on quantum computation.

> **Remark 8.2   Understanding the Flux and Dyon Excitations**
>
> In standard gauge field theories, such as quantum electrodynamics (QED) and quantum chromodynamics (QCD) that form the foundation of the Standard Model of particle physics, the notion of flux is associated with the integral of a gauge field over a surface, rather than localized, quantized excitations. These theories operate within a continuous spacetime framework and are anchored in the principle of local gauge invariance, demanding that physical observables remain unchanged under local transformations of gauge fields. Here, gauge fields themselves are not direct physical observables; instead, it is the field

Remark 8.2, continued

strengths, indicating curvature or variations in these fields, that are physically meaningful.

Unlike topological quantum field theories, where fluxes are associated with the topological characteristics of field configurations and can be manifested as quantized excitations such as anyons in two-dimensional systems, conventional gauge theories do not exhibit a similar quantization of flux. For example, in QCD, which describes the strong interaction, the gauge fields are confining, meaning that they do not exist in free space as isolated fluxes but are instead always bound within composite particles such as protons and neutrons. There are no elementary particles in these theories that carry an isolated gauge flux; particles either carry charges, like electrons, or mediate forces, like photons for the electromagnetic interaction, but none serve as isolated carriers of gauge flux. Therefore, in the realm of conventional gauge field theories, elementary flux excitations do not arise due to the continuous nature of these theories, the absence of topological quantization, and the confining properties of the gauge fields.

In the $\mathbb{Z}_2$ toric code model, however, the scenario is different. As a Hamiltonian lattice model embodying the effective topological field theory of the $\mathbb{Z}_2$ toric code phase, it allows for point-like flux quanta to appear as localized excitations in the first excited states, where the ground state's local constraints are disrupted. These broken constraints, defined on plaquettes, can be detected by the $\mathbb{Z}_2$ gauge field (current, or a moving charge) around the plaquette via an Aharonov–Bohm phase factor (in this case, a minus sign). This phase factor corresponds to a gauge flux piercing through the plaquette in a conventional gauge theory, where the gauge field becomes undefined in the region traversed by the flux. The key distinction lies in the nature of the flux; in standard gauge theories, it is a classical flux and not an elementary excitation, whereas in the topological context of our model, an elementary excitation carrying a flux quantum is feasible. This unique aspect of the model leads to its designation as the doubled $\mathbb{Z}_2$ phase, highlighting the coexistence of gauge charges and fluxes, in contrast to standard gauge theories where only gauge charges exist as elementary excitations. The presence of both charges and fluxes naturally gives rise to dyons, bound states of these two entities. An additional $\mathbb{Z}_2$ gauge invariance, whose breach leads to flux excitations, is seen as emerging from the primary $\mathbb{Z}_2$ invariance associated with the gauge field on lattice edges, thereby earning the effective topological field theory the name of doubled $\mathbb{Z}_2$ gauge field theory.

For your information, in a standard gauge field theory, a region through which a gauge flux is piercing is said to be nonflat, as the corresponding fiber bundle above the region is curved. As such, the constraint $B_p |\Psi_0\rangle = |\Psi_0\rangle$ is also termed a **flatness condition**.

> **Remark 8.2, continued**
>
> In gauge theories, the flatness condition refers to a state where the gauge field is locally flat or trivial, meaning the field strength (or curvature) is zero everywhere. This is described using the language of fiber bundles, where the base manifold typically represents spacetime, and the fibers at each point represent the internal gauge degrees of freedom. The field strength tensor, an essential element in gauge theories, is interpreted as the curvature tensor of the principal bundle. A nonzero curvature implies a deviation from the flatness condition and indicates the presence of gauge field strength.
>
> Flux, in this framework, is understood as the integral of the gauge field strength over a surface or, equivalently, as the nontrivial holonomy around a loop in the base manifold. In the case of $U(1)$ gauge field theory, a nontrivial holonomy is merely a phase factor – the Aharonov–Bohm phase. The presence of flux signifies a region where the gauge field is not flat, as indicated by the nonzero curvature or field strength in that region. This situation can be visualized as a "curvature" of the fibers over the base manifold, representing the gauge field's action in spacetime.
>
> In physical terms, nonzero curvature or field strength is associated with forces or interactions, such as electromagnetic fields in QED or gluon fields in QCD. The breaking of the flatness condition due to flux has significant implications in both continuum and lattice gauge theories. In topological gauge theories, this phenomenon also carries topological significance, reflecting nontrivial global properties of the gauge field configuration.
>
> Overall, the interplay between flux and the flatness condition in gauge theories beautifully illustrates how the geometric and topological properties of gauge fields are intricately tied to their physical manifestations, providing a deep understanding of the forces and interactions in the quantum realm.

Fusion Interaction

In summary, we have found the set of four types of point-like elementary excitations of the $\mathbb{Z}_2$ toric code model, namely

$$\{1, e, m, \epsilon\}. \tag{8.56}$$

But we haven't justified that they are anyons yet. To do so, we have to nail down their fusion rules and statistical behaviors.

Because $X^2 = Z^2 = \mathbb{1}_2$, the fusion rules of these anyon types are as simple as

$$e \otimes e = m \otimes m = \epsilon \otimes \epsilon = 1, \quad e \otimes m = \epsilon, \quad e \otimes \epsilon = m, \quad m \otimes \epsilon = e. \tag{8.57}$$

The last three fusion rules follow from the fact that ϵ is a bound state of e and m.

Self and Braiding Statistics

Figure 8.11 also illustrates the braiding statistics, which is -1, of an e and an m, so the mutual statistics between e and m are termed **semionic**, signifying that they are

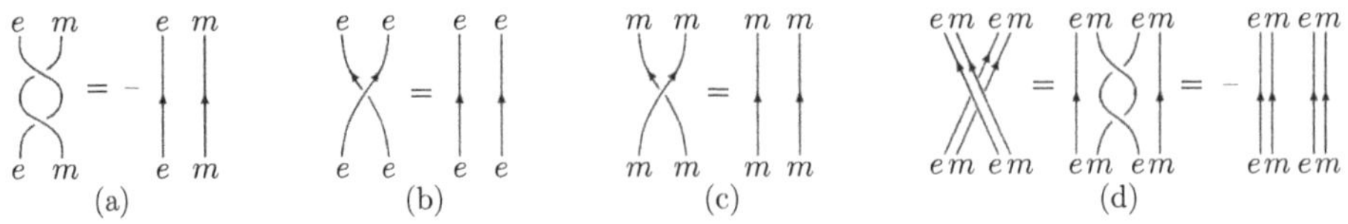

Figure 8.12 Self and braiding statistics of the nontrivial anyons e, m, and ϵ.

half fermionic.[14] The question then arises: What happens during the exchange of two identical anyons, such as two charges or two fluxes?

Given that charges are generated by e-string operators, which mutually commute, and fluxes by m-string operators, which also commute, exchanging two identical anyons of either type should leave the corresponding state unaltered. Thus, both e and m exhibit **self-boson** statistics. This behavior, where a charge e and a flux m each display bosonic statistics individually but semionic statistics in mutual interaction, is a characteristic absent in conventional bosons and fermions, highlighting their anyonic nature. Nevertheless, it is important to note that e and m do not exhibit exotic statistical phases, placing them among the simpler types of anyons.

Determining the braiding statistics of two dyons involves a more intricate analysis. One can ascertain these statistics by employing a procedure akin to that depicted in Figures 8.10 and 8.11. This analysis serves as an insightful exercise for readers seeking to deepen their understanding of anyonic behavior. The results of such an investigation, encompassing both the self and braiding statistics of e and m, are concisely presented in Figure 8.12.

> **Exercise 8.5** *Prove the self statistics of ϵ in Figure 8.12(d). This result shows that ϵ is **self-fermionic**.*
>
> **Exercise 8.6** *Find the mutual statistics between e and ϵ and those between m and ϵ. You may simply provide a convincing reasoning without going through the procedure in Figures 8.10 and 8.11.*

We have confirmed that the point-like excitations e, m, and ϵ of the $\mathbb{Z}_2$ toric code phase are anyons. Since the fusion product of any two anyons in the $\mathbb{Z}_2$ toric code phase is exactly one anyon and unique, the anyons e, m, and ϵ are all Abelian, and the $\mathbb{Z}_2$ toric code is an Abelian topological phase.

Anyon Basis

Our journey through the ground states of the $\mathbb{Z}_2$ toric code model has yielded four distinct bases for the ground-state Hilbert space $\mathcal{H}_0$. Each serves as logical basis states for the four-dimensional stabilizer code space, congruent with $\mathcal{H}_0$. In each

---

[14] In the context of fermions, exchanging two particles results in a minus sign. Here, a similar minus sign emerges when e and m undergo a double exchange, justifying their characterization as semionic with respect to each other. Note that e and m here are not semions per se.

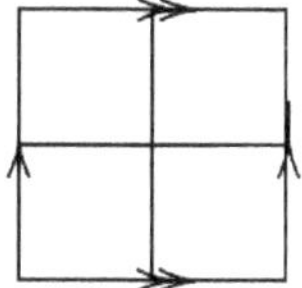

Figure 8.13 A 2×2 periodic square lattice.

basis, except for the trivial state without any noncontractible loop, the basis states are crafted by applying the corresponding logical X operators to a selected ground state. As a result, each basis state, barring the trivial one, is represented by one or two noncontractible loops, along which the logical X operators are defined. A closer look reveals that if we "open up" these logical X operators, they effectively transform into either e-string or m-string operators. Thus, we can associate a basis state with a specific anyon type: 1 (no non-contractible loop), e, or m. Yet, in the four bases we've constructed so far, none offers a basis state that aligns with the dyon ϵ. This observation naturally leads us to wonder: Is it possible to construct a ground-state basis where each basis state corresponds to one of the four anyon types 1, e, m, and ϵ?

Indeed, such a basis, known as an **anyon basis**, can be constructed. If you've tackled Exercise 8.4 and successfully created an ϵ-ribbon operator, you're well on your way. You can extend this ribbon to traverse both sides of the square lattice, allowing it to close into a loop under periodic boundary conditions. Given the topological invariance of the ground-state Hilbert space, let's explore this basis in a specific scenario.

Consider our model on a 2×2 periodic square lattice, with $n = 2$, comprising $N = 2 \times 2^2 = 8$ qubits on the eight edges (Figure 8.13). While the total Hilbert space of this system is 2^8-dimensional, the ground-state Hilbert space $\mathcal{H}_0$ is consistently four-dimensional. Thus, each ground state represents a superposition of certain eight-qubit basis states. Your challenge is to transform an appropriate ground state with noncontractible loop and/or ribbon operators, whose "open" forms generate pairs of anyons of a specific type. The states you obtain will constitute the anyon basis. Here's an exercise to guide you through this process:

Exercise 8.7 *Formulate an explicit anyon basis for $\mathcal{H}_0$ on the lattice shown in Figure 8.13. Your task is to express each basis state $|a\rangle$, where $a \in \{1, e, m, \epsilon\}$, as a superposition of select eight-qubit basis states.*

8.2.7 The Modular S and T Matrices

Having already constructed four distinct bases for the four-dimensional ground-state Hilbert space $\mathcal{H}_0$ of the $\mathbb{Z}_2$ toric code model, each depicted by two noncontractible loops in Figure 8.5, we now turn our attention to the unitary transformations that interconnect these bases and enable the generation of all other (infinitely many) possible bases for $\mathcal{H}_0$. The key to understanding these transformations lies in the modular S and T matrices.

Each basis in Figure 8.5 is formed through the action of a specific set of two logical X operators. The task at hand is to explore the full spectrum of possible logical X operators that can generate the entire range of bases for $\mathcal{H}_0$. While direct construction of these sets is conceivable, it proves to be a complex undertaking. Instead, we will approach this challenge from a different angle, leveraging lattice transformations.

A pivotal observation is that rotating the square lattice by $\pi/2$ transforms a horizontal noncontractible loop into a vertical one. This suggests that lattice transformations can effectively convert one basis of $\mathcal{H}_0$ into another. The complete set of transformations for a two-dimensional lattice constitutes the group $SL(2,\mathbb{Z})$, defined by the presentation:

$$SL(2,\mathbb{Z}) := \left(S, T \middle| S^4, (ST)^3 S^{-2} \right), \tag{8.58}$$

where S and T are the fundamental generators of the group. The relations defining this group are given as

$$S^4 = 1, \quad (ST)^3 = S^2. \tag{8.59}$$

All elements of $SL(2,\mathbb{Z})$ are generated through various combinations of S, T, and their inverses, in accordance with the relations expressed in Equation (8.59). The defining representation of these generators is

$$S = \begin{pmatrix} 0 & -1 \\ 1 & 0 \end{pmatrix}, \quad T = \begin{pmatrix} 1 & 1 \\ 0 & 1 \end{pmatrix}, \tag{8.60}$$

These are 2×2 matrices, apt for transforming two-dimensional lattices. The matrix entries are integers ($\mathbb{Z}$), as they map one lattice unit cell to another. Figure 8.14 illustrates how the S and T transformations affect a 2×2 square lattice unit cell. These visual representations of the transformations can be deduced by assigning coordinates (x, y) from $\mathbb{Z} \times \mathbb{Z}$ to each vertex of the unit cell and applying the matrices in Equation (8.60) to these coordinates. The S transformation effectively rotates a square unit cell counterclockwise by $\pi/2$, transforming a horizontal noncontractible loop into a vertical one. On the other hand, the T transformation shears the unit

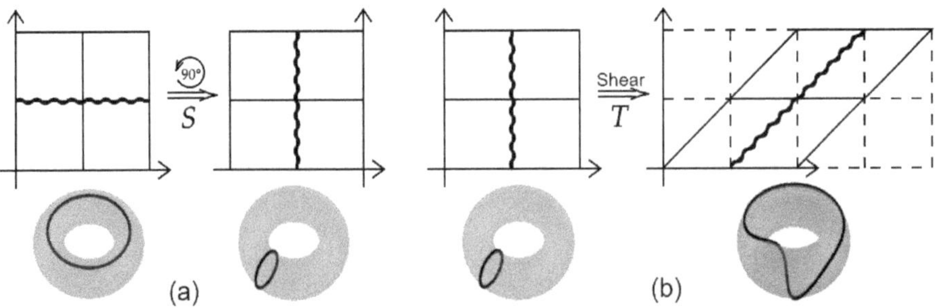

Figure 8.14 (a) S and (b) T transformations on a 2×2 unit cell of a periodic square lattice.

cell, as shown in Figure 8.14(b), causing a vertical noncontractible loop to wrap around both holes of the torus, akin to a twist. Consequently, the T transformation is also referred to as a **Dehn twist** [250], named after the German mathematician Max Dehn.

Combining the S and T transformations in various ways generates all possible transformations of the noncontractible loops on the torus, and consequently, all the bases of the four-dimensional ground-state Hilbert space $\mathcal{H}_0$. Nevertheless, to facilitate these transformations on the bases, we must represent the S and T operations as 4×4 unitary matrices acting on the four-dimensional $\mathcal{H}_0$. To achieve this, we select a specific basis of $\mathcal{H}_0$ to calculate the entries of these matrices. An anyon basis, potentially obtained in Exercise 8.7, serves as a convenient choice for this purpose. In an anyon basis, the S and T transformations are represented by the matrices

$$
S = \begin{array}{c} \begin{matrix} 1 & e & m & \epsilon \end{matrix} \\ \begin{pmatrix} 1 & 1 & 1 & 1 \\ 1 & 1 & -1 & -1 \\ 1 & -1 & 1 & -1 \\ 1 & -1 & -1 & 1 \end{pmatrix} \begin{matrix} 1 \\ e \\ m \\ \epsilon \end{matrix} \end{array}, \qquad
T = \begin{array}{c} \begin{matrix} 1 & e & m & \epsilon \end{matrix} \\ \begin{pmatrix} 1 & 0 & 0 & 0 \\ 0 & 1 & 0 & 0 \\ 0 & 0 & 1 & 0 \\ 0 & 0 & 0 & -1 \end{pmatrix} \begin{matrix} 1 \\ e \\ m \\ \epsilon \end{matrix} \end{array}. \tag{8.61}
$$

Exercise 8.8* *Attempt to calculate the S and T matrices within an anyon basis on a 2×2 periodic square lattice. The expected outcome should align with the matrices presented in Equation (8.61). To guide your process, apply the S and T transformations, as defined in Equation (8.60) and illustrated in Figure 8.14, on each of the four states in the anyon basis. Then, determine the inner product between the resulting state and each anyon-basis state.*

The matrices provided in Equation (8.61), known as the S and T matrices of the $\mathbb{Z}_2$ toric code phase, hold a dual significance. First, as we have come to understand, they underpin all conceivable unitary basis transformations within $\mathcal{H}_0$. More crucially, they encapsulate the self and braiding statistics of the four types of anyons characteristic of the $\mathbb{Z}_2$ toric code phase. The question then arises: Why and how do these matrices convey information about the anyons' statistics?

An S matrix entry S_{ab} is the probability amplitude of converting the anyon-basis state $|b\rangle$ to the state $|a\rangle$. Since S is also a 90° rotation of the lattice, S_{ab} can be interpreted as the phase factor induced by the process shown in Figure 8.15(a).

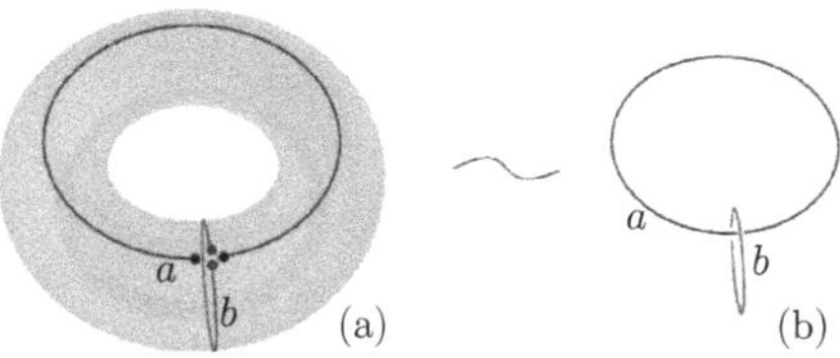

Figure 8.15 Physical interpretation of S_{ab}.

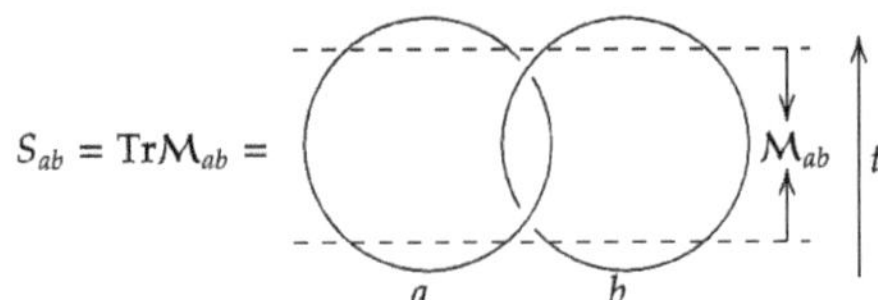

Figure 8.16 The entry S_{ab} of the S matrix defined as the trace of the monodromy (braiding) operator $\mathcal{M}_{ab}$ over the state space of two anyons a and b.

$$S_{ab} = \mathrm{Tr}\mathcal{M}_{ab} =$$

The process consists of several steps: 1) A pair of anyons a and a pair of anyons b are created somewhere on the torus. 2) One anyon a moves around one of the two holes and comes back to annihilate with the other a, while one anyon b travels around the other hole and also comes back to annihilate with the other b. 3) Supposing one of the two travelling anyons returns earlier than the other, the process is equivalent to the picture – **Hopf link** – in Figure 8.15(b). Slightly rearranging the two linked loops in Figure 8.15(b), we can obtain the equivalent definition of S_{ab} in Figure 8.16, where the process is appropriately presented in a spacetime diagram: A pair of anyons a and a pair of anyons b are created somewhere in spacetime, then one a and one b move around each other and return to their original positions to annihilate with the nonmoving a and b, such that their world lines are a Hopf link.

In this spacetime diagram, the part after the two pairs of anyons are created but before the two pairs annihilate can be regarded as a quantum operator that braids an anyon a with an anyon b in the state with two anyons a and two anyons b. This operator is denoted by $\mathcal{M}_{ab}$ and termed the **monodromy** of a and b.

Bearing in mind that all anyon types in the $\mathbb{Z}_2$ toric code phase are Abelian, the Hilbert space of any number of these anyons is one-dimensional. Consequently, the monodromy $\mathcal{M}_{ab}$, which braids anyons a and b, results in at most a phase factor. This phase factor is precisely the matrix element S_{ab} of the S matrix:

$$S_{ab} = \langle a \otimes b | \mathcal{M}_{ab} | a \otimes b \rangle, \tag{8.62}$$

where $|a \otimes b\rangle$ denotes the state of the two anyons a and b acted on by the monodromy.

In the case of non-Abelian anyons a and b, however, the state space is multidimensional, say, k-dimensional. Here, the monodromy $\mathcal{M}_{ab}$ is represented on this space as a $k \times k$ matrix. The representation matrix depends on the chosen basis, so $\mathcal{M}_{ab}$ itself does not have an invariant meaning for characterizing the topological feature of braiding anyons a and b. Nonetheless, the trace of $\mathcal{M}_{ab}$ is invariant under basis transformations and thus serves as a characteristic of the braiding. This trace is precisely the corresponding S matrix element:

$$S_{ab} := \mathrm{Tr}\mathcal{M}_{ab}, \tag{8.63}$$

where the trace is taken over any set of basis states in the state space with two anyons a and b. The diagram in Figure 8.16 is a standard graphical expression of this trace. Equation (8.63) naturally reduces to Equation (8.62) for Abelian anyons.

To recap, the monodromy matrix $\mathcal{M}_{ab}$ captures the quantum amplitude of anyons a and b braiding around each other. In an Abelian system, this matrix reduces to a phase factor, representing the quantum amplitude for the interaction of anyons. The trace of this matrix, a crucial concept in quantum mechanics, sums up the probabilities of returning to each possible initial state after braiding. It condenses the entire impact of anyon braiding into one value – the corresponding S matrix element S_{ab} – essential in non-Abelian systems where interactions yield a multidimensional matrix.

In contrast to the S matrix, the T matrix characterizes the self statistics of the anyons, as evident in its diagonal elements in an anyon basis. It is noteworthy that $T_{aa}^2 = S_{aa}$, since exchanging two identical anyons is equivalent to moving one of them around the other in a full circle.

The S and T matrices in Equation (8.61) uniquely characterize the $\mathbb{Z}_2$ toric code phase. Indeed, S and T matrices serve as fingerprints of topological phases, as two topological phases are distinct if and only if they have inequivalent S and T matrices.[15]

8.2.8 Concluding Remarks

Topological phases of matter represent an enchanting research area, standing at the forefront of condensed matter physics and intersecting with other domains like quantum field theory, high energy physics, and quantum information. The sections above have only scratched the surface, presenting rudimentary concepts and a basic model of topological phases. For those intrigued by this field, there is much more to explore and discover.

The application of topological phases in quantum computation is a vibrant area of research. The simplicity of the $\mathbb{Z}_2$ toric code model belies its profound impact, having inspired ideas in topological quantum memory, potential avenues for fault-tolerant quantum computation, and the development of one of the most promising quantum error correction codes – surface code. In the upcoming section, we will expound on the core principles of surface code. A thorough understanding of the $\mathbb{Z}_2$ toric code, particularly its topological characteristics such as ground state degeneracy, anyonic excitations, and their braiding statistics, lays a solid foundation for grasping non-Abelian anyons. This includes Fibonacci anyons, which are pivotal to topological quantum computation, a subject we will explore in detail in Chapter 9.

8.3   Surface Code Quantum Computation

As suggested by Kitaev, a $\mathbb{Z}_2$ toric code system on the torus, in its ground states, may serve as a robust quantum memory. For it to function as a memory, however, it is essential to have mechanisms to write and read information. This leads us to ponder: Can we possibly perform quantum computation on such a system?

Regrettably, quantum computation on the ground states is unfeasible, as the logical operators on the two logical qubits encoded in these states are limited to single-logical-qubit operations. To enable quantum computation on the $\mathbb{Z}_2$ toric code system, exploration of either the excited states with anyons or alternative methods becomes necessary.

---

[15] Rare exceptions to this rule may exist. [251, 252]

Given that the $\mathbb{Z}_2$ toric code anyons are Abelian, with their braiding only inducing an overall phase factor to the system's state, constructing nontrivial quantum gates within the Hilbert space of these anyons proves unattainable. Thus, an alternative strategy is required. Furthermore, considering the physical qubits' inherent susceptibility to faults, it is desirable for this strategy to leverage the model's stabilizer nature for error detection and correction. Ideally, we need an approach that encodes logical qubits into the $\mathbb{Z}_2$ toric code system, facilitating the fault-tolerant implementation of nontrivial logical gates, particularly the two-logical-qubit entangling gates, such as the CNOT. This approach, known as surface code quantum computation, is both viable and well-established [58, 253, 254]. This section aims to guide the reader through the emergence of surface code from the $\mathbb{Z}_2$ toric code, the construction of logical qubits and their operators, and the implementation of the logical CNOT gate.

It must be acknowledged that this section is by no means an exhaustive introduction to surface code quantum computation, being constrained by the scope of the book. Numerous technical details, particularly those concerning error correction, will be omitted. Nevertheless, the key concepts and aspects, especially the inherent topological features of surface code quantum computation, outlined in this section are expected to captivate readers and inspire further exploration of the subject.

8.3.1 From $\mathbb{Z}_2$ Toric Code to Surface Code

The journey from the $\mathbb{Z}_2$ toric code to surface code requires a few intermediate stages for full appreciation, which we now explore.

From Torus to Plane

In practice, realizing the torus topology with a large number of qubits is challenging. Therefore, let's redefine our $\mathbb{Z}_2$ toric code model on the plane. Consider the lattice in Figure 8.17. This lattice is nonperiodic and has an open boundary, giving it a disk topology. It contains $n \times n$ plaquettes, with the four corner plaquettes each having three edges. This boundary is termed a **smooth boundary**, as it consists of smoothly connected edges. On this boundary, vertex operators A_s are still defined, but each boundary A_s involves a tensor product of three Pauli X matrices, since

Figure 8.17 $\mathbb{Z}_2$ toric code model on a lattice with disk topology, featuring a smooth, or X-boundary.

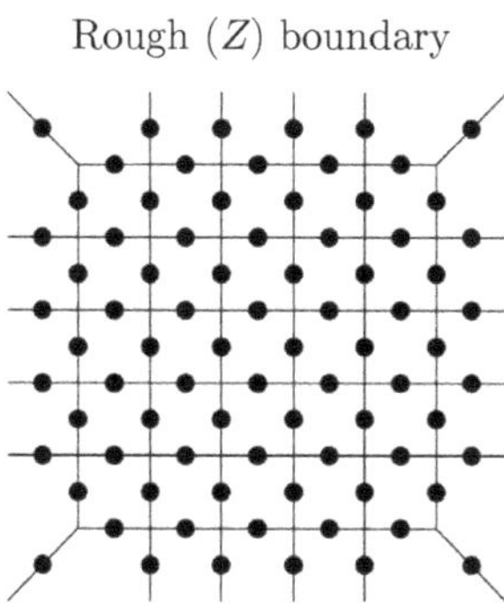

Figure 8.18 $\mathbb{Z}_2$ toric code model on a lattice with disk topology, featuring a rough, or Z-boundary.

each boundary vertex has only three incident edges. Thus, a smooth boundary is also known as an X-**boundary**.

Unfortunately, on the disk, no independent degrees of freedom remain in the ground states. The total number of qubits is $2n(n+1) - 4 = 2(n-1)(n+2)$, with $(n+1)^2 - 4$ vertex constraints and n^2 plaquette constraints. While $\prod_s A_s = \mathbb{1}, \prod_p B_p \neq \mathbb{1}$. Thus, the number of independent degrees of freedom is

$$2(n-1)(n+2) - (n+1)^2 + 4 - n^2 + 1 = 0, \tag{8.64}$$

leading to

$$\mathrm{GSD}_{\mathrm{disk}}^{\mathrm{smooth}} = 1 \implies \text{no logical qubit.} \tag{8.65}$$

Is the smooth boundary the only option for our square lattice with disk topology? Certainly not. Another boundary type, as shown in Figure 8.18, is the **rough boundary** or Z-**boundary**. It comprises open plaquettes, each bounded by three edges, with each boundary plaquette operator B_p containing three Pauli Z matrices. As you can confirm in Exercise 8.9, we also have

$$\mathrm{GSD}_{\mathrm{disk}}^{\mathrm{rough}} = 1 \implies \text{no logical qubit.} \tag{8.66}$$

Exercise 8.9 *Count the number of independent degrees of freedom in the ground states of the $\mathbb{Z}_2$ toric code model on the lattice with the rough boundary in Figure 8.18.*

Thus, to achieve a GSD on the disk greater than 1, further efforts are needed.

From Single Boundary Condition to Alternating Boundary Conditions

An intriguing approach involves setting alternating boundary conditions on the disk's perimeter. Figure 8.19 illustrates an instance with four alternating smooth and rough boundaries. We now calculate the GSD for this configuration. With $2n^2$ qubits in total and $2n^2 - 1$ vertex and plaquette constraints, which do not multiply to be the identity operator, we ascertain that there is precisely one independent

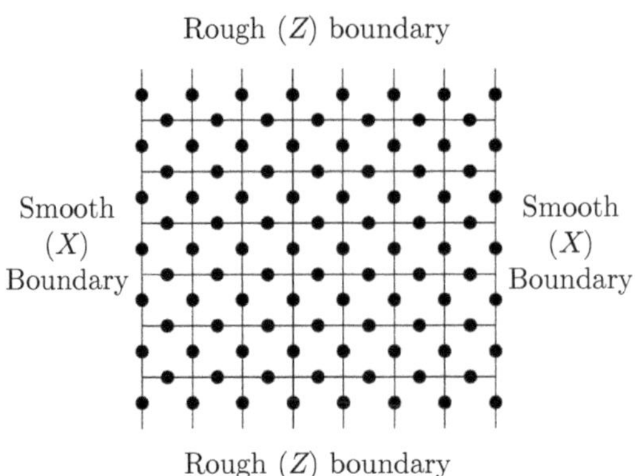

Figure 8.19 $\mathbb{Z}_2$ toric code model on a lattice with disk topology, featuring four alternating smooth and rough boundaries.

degree of freedom in the ground states. This translates to a GSD of 2, signifying the presence of one logical qubit.

Though this is an improvement, it still falls short of the ideal. A discerning reader might recognize that increasing the number of alternating boundary conditions – smooth, rough, smooth, rough, and so forth – could elevate the GSD on the disk.

Exercise 8.10 *Determine the GSD of the $\mathbb{Z}_2$ toric code model on a disk with $2n$ alternating smooth and rough boundaries.*

This strategy allows encoding an increasing number of logical qubits into the model's ground states, provided the lattice is sufficiently large. These logical qubits are challenging to operate because they are not localized but rather are superpositions of all physical qubits on the lattice. The logical operators, drawing inspiration from those on the torus, would have to be string operators of the lattice size, which are both inconvenient and resource-intensive.

The pressing question, then, is how we might use boundaries to create logical qubits that are more readily manipulable. Addressing this challenge requires additional groundwork.

Remark 8.3   Anyon Condensation at the Boundary

The smooth boundary or X-boundary and the rough boundary or Z-boundary bear other interesting names that are physically profound. Let's look closely at Figure 8.20. In the figure, the area outside the lattice is physically regarded as the vacuum in the context of topological phase. Note that the vacuum shouldn't be confused with a ground state of the model; it is understood as a trivial phase, in which all excited states are infinitely gapped above the vacuum.

Suppose you wish to excite a pair of e anyons by an e-string operator with one end in the bulk of the lattice but the other end touching a rough boundary, as showcased by the vertical thick string in Figure 8.20. This e-string operator can indeed create an e anyon in the bulk; however, there won't be any e anyon

> **Remark 8.3, continued**
>
> at the other end touching the rough boundary because there is no A_s operator at the boundary to anticommute with the e-string operator. In this scenario, we say that a charge can disappear at the rough boundary, or can disappear into the vacuum. In more physical terms, charges can condense at a rough boundary and become part of the vacuum. In contrast, you can create a flux m at the rough boundary, as indicated by the vertical wiggly m-string operator in the figure. That is, fluxes cannot disappear or condense at a rough boundary but can remain as boundary anyons there. Therefore, a rough boundary is also referred to as a **flux boundary**, or an m**-boundary**.
>
> Similarly, as showcased by the horizontal e-string and m-string operators in Figure 8.20, anyons m can disappear or condense at a smooth boundary and become part of the vacuum, whereas anyons e cannot, and can be smooth boundary anyons. Hence, a smooth boundary is also christened a **charge boundary**, or an e**-boundary**.
>
> 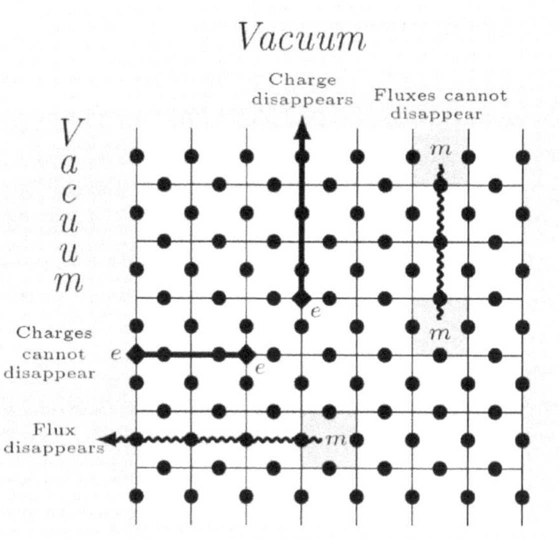
>
>
> **Figure 8.20** Respective anyon condensation at the smooth and rough boundaries.

8.3.2   Basic Idea of Surface Code

The $\mathbb{Z}_2$ toric code model can be viewed as a Hamiltonian system whose ground-state Hilbert space aligns with the code space of a certain stabilizer code. In this model, physical qubits are ideally arranged in a square lattice, mirroring the structure of the $\mathbb{Z}_2$ toric code model. This arrangement naturally lends itself to identifying the stabilizer generators of the surface code with the vertex and plaquette operators (A_s and B_p) from the $\mathbb{Z}_2$ toric code model.

A key distinction lies in the interaction dynamics. In the $\mathbb{Z}_2$ toric code model, the vertex and plaquette operators facilitate four-body interactions, where four qubits interact simultaneously. Nevertheless, the Hamiltonian governing the surface code qubit system diverges from the $\mathbb{Z}_2$ Hamiltonian. Realistically achievable Hamiltonians in surface code systems typically involve at most two-body interactions.

For each vertex operator $A_s = X_a X_b X_c X_d$ in the $\mathbb{Z}_2$ toric code model, there corresponds an X-stabilizer in the surface code, represented as $X_a X_b X_c X_d$. Similarly, each $B_p = Z_a Z_b Z_c Z_d$ translates to a Z-stabilizer, $Z_a Z_b Z_c Z_d$, in the surface code. Measuring these stabilizers in the state of the associated qubits (a, b, c, d) does not necessitate the four-body interactions characteristic of the $\mathbb{Z}_2$ toric code model. Instead, these measurements are compatible with simpler interaction schemes.

The code space of the surface code is comprised of the simultaneous $+1$ eigenstates of all its stabilizers. Alternatively, depending on specific requirements, it might consist of the simultaneous -1 eigenstates. This space forms the foundational structure upon which surface code quantum computation is built, leveraging the inherent stabilizer properties for effective quantum computation and error correction.

Data Qubits and Measurement Qubits

Now, let's ponder a critical question: How can we tell if the lattice of physical qubits is actually in a codeword state? Your first thought might be to measure all the X-stabilizers and Z-stabilizers, right? But, here's the catch – doing this directly would mess up the state we're trying to read. To dodge this hurdle, remember our trusty friends, the ancilla qubits? We're going to use them cleverly here. For each local vertex state of four qubits, we'll entangle it with an ancilla qubit in such a way that measuring this ancilla qubit gives us the same result as measuring the corresponding X-stabilizer. And we do the same thing with the plaquette states for the Z-stabilizers.

So, from now on, we'll refer to the original physical qubits on the lattice as *data qubits*, and these helpful ancilla qubits will be known as *measurement qubits*. Where do we put them? Well, each measurement qubit for an X-stabilizer gets cozy right at the corresponding vertex. As for the Z-stabilizers, their measurement qubits chill at the center of the respective plaquettes. Check out Figure 8.21 for a visual on this. You'll see a couple of example measurement qubits (black dots) added to our square lattice with alternating Z- and X-boundaries. Data qubits are the gray dots.

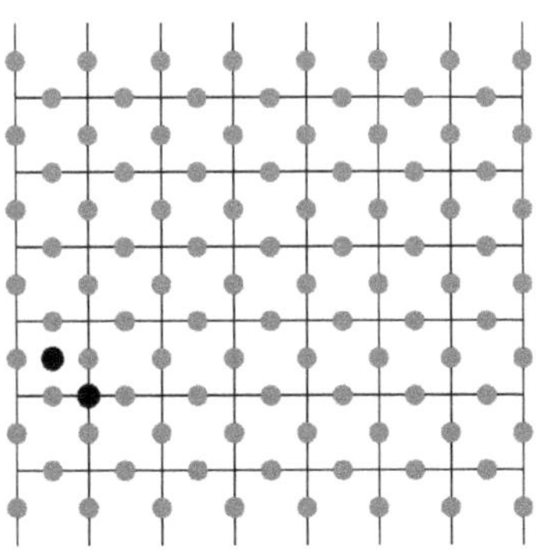

Figure 8.21 This figure shows a couple of example measurement qubits (black dots) on a square lattice that has alternating Z-boundaries and X-boundaries. Data qubits are the gray dots.

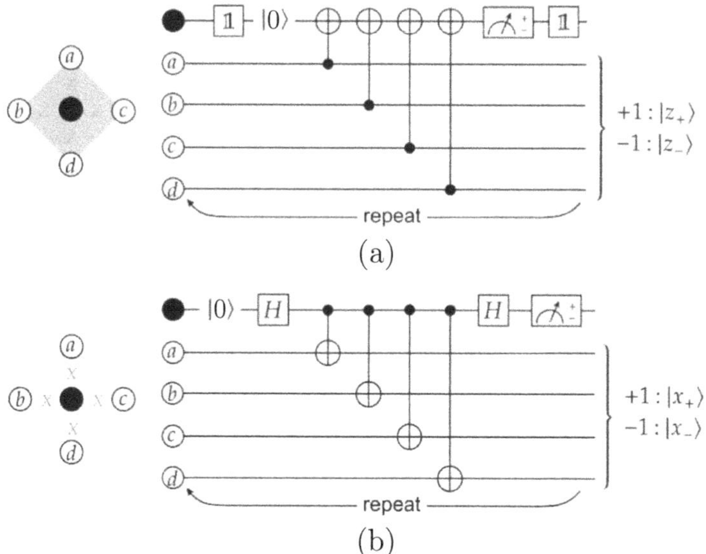

Figure 8.22 (a) Quantum circuit for measuring a Z-stabilizer. (b) Quantum circuit for measuring an X-stabilizer.

Let's explore how measurement qubits are entangled with their corresponding stabilizers. Figure 8.22 illustrates the representation of a Z-stabilizer and an X-stabilizer, along with the quantum circuits entangling them with their respective measurement qubits:

- A Z-stabilizer is depicted as a deep gray lozenge, encasing four Pauli Z matrices around its central measurement qubit (shown as a black dot). This measurement qubit, responsible for gauging the Z-stabilizer, is also termed a Z-syndrome qubit. The circuit manifests the Z-syndrome qubit and four data qubits, with the following procedural steps:
  - Step 1: The Z-syndrome qubit is initialized with an identity operator $\mathbb{1}_2$. This step synchronizes the Z-stabilizer measurement with the X-stabilizer measurement, necessitating an additional Hadamard gate for the latter.
  - Step 2: The Z-syndrome qubit is set to its Hamiltonian's ground state, $|0\rangle$. Each measurement qubit in the system possesses its own Hamiltonian, typically proportional to σ_z.
  - Step 3: Each of the four data qubits controls the Z-syndrome qubit through four CNOT gates. Assuming each data qubit is in a generic state, the combined five-qubit system is represented as

$$|0\rangle \otimes (a\,|0\rangle + a'\,|1\rangle) \otimes (b\,|0\rangle + b'\,|1\rangle) \otimes (c\,|0\rangle + c'\,|1\rangle) \otimes (d\,|0\rangle + d'\,|1\rangle), \quad (8.67)$$

with each single-qubit state normalized and the coefficients being complex numbers. Post-CNOT gates, the resultant state is

$$|0\rangle \otimes \sum \propto |\text{even number of 1's}\rangle + |1\rangle \otimes \sum \propto |\text{odd number of 1's}\rangle, \quad (8.68)$$

where the coefficients in each superposition do not affect the outcome.

- Step 4: Measuring the Z-syndrome qubit yields two possible outcomes, collapsing the state of the four data qubits into either of the two superpositions corresponding to the $+1$ and -1 eigenstates of the Z-stabilizer:

$$Z_a Z_b Z_c Z_d \sum \propto |\text{even number of 1's}\rangle = \sum \propto |\text{even number of 1's}\rangle, \quad (8.69)$$

$$Z_a Z_b Z_c Z_d \sum \propto |\text{odd number of 1's}\rangle = -\sum \propto |\text{odd number of 1's}\rangle. \quad (8.70)$$

These eigenstates are respectively labeled as $|z_+\rangle$ and $|z_-\rangle$.

A subsequent $\mathbb{1}_2$ operator ensures synchronization with the X-stabilizer's circuit.

- An X-stabilizer is drawn as a lozenge in light gray, encapsulating four Pauli X's surrounding the measurement qubit (a black dot) at the center. Such a measurement qubit is also referred to as an X-syndrome qubit. The circuit is examined likewise:
  - Step 1: Prepare the X-syndrome qubit in state $|0\rangle$.
  - Step 2: Apply a Hadamard gate to turn the X-syndrome qubit into $(|0\rangle + |1\rangle)/\sqrt{2}$. This step is the reason that in the circuit for measuring a Z-stabilizer, there is a wait time, such that the CNOT gates in the circuit of each Z-stabilizer are roughly synchronized with those in the circuit of each X-stabilizer. We shall explain this rough synchronization shortly.
  - Step 3: Enact the CNOT gates and another Hadamard gate.

> **Exercise 8.11** *Show that the resultant total state is an entangled state, where the state $|0\rangle$ ($|1\rangle$) of the X-syndrome qubit is paired with the $+1$ (-1) eigenstate of $X_a X_b X_c X_d$.*

Measure the X-syndrome qubit.

- The precise design of the above circuits is critical for ensuring accurate outcomes of stabilizer measurements. Any modification could potentially impair the desired results.
- The procedures we've outlined apply to every Z-stabilizer and X-stabilizer on the lattice. It is essential to measure all Z-stabilizers and X-stabilizers across the lattice in lockstep and repeatedly. The completion of measurements for all the Z-stabilizers and X-stabilizers signifies the end of a **surface-code cycle**, after which a new cycle commences.
- During each surface-code cycle, careful timing of interactions is key to preventing measurement conflicts. Specifically, a CNOT gate involving a Z-syndrome qubit and a data qubit should not concur with a CNOT gate involving an X-syndrome qubit or another Z-syndrome qubit connected to the same data qubit. This strategic timing prevents interference between the interactions of syndrome qubits and data qubits. Although these CNOT gates are executed with slight temporal differences to avoid concurrent interactions, the overall process of measuring all Z- and X-stabilizers across the lattice must be completed in a coordinated and roughly synchronized manner before starting the next surface-code cycle. This method

Figure 8.23 An illustrative surface code lattice with four alternating X- and Z-boundaries.

ensures efficient and error-resistant operation of the surface code, safeguarding the integrity and accuracy of the quantum error correction process.

- With a firm grasp of the surface code concept, including its fundamental units such as X- and Z-stabilizers and their measurement circuits, we are now ready to transition from the $\mathbb{Z}_2$ toric code model to the surface code paradigm. As an initial step, let's reinterpret the square lattice from Figure 8.21 as a standard surface code lattice, as depicted in Figure 8.23.

  In this visual representation, we disregard all lattice edges. Data and measurement qubits are represented by unfilled and filled circles, respectively. Each X-syndrome (or Z-syndrome) qubit, surrounded by four Pauli X (or Z) matrices, forms an X-stabilizer (or Z-stabilizer) that acts on the four data qubits located at the corners of a lozenge shape colored in deep (or light) gray. Note that boundary X-stabilizers (or Z-stabilizers) comprise only three Pauli X (or Z) matrices, hence their triangular, rather than lozenge, shape. Behind each lozenge, a network of quantum circuits (and possibly other types of circuits) connects the four corner data qubits to the central measurement qubit, coordinating the measurements of all X- and Z-stabilizers.

- To initialize the surface code, one sets all data qubits to specific initial states and then allows them to pass through the quantum circuits that measure all the X- and Z-stabilizers, thus completing a full surface-code cycle. At the conclusion of this initial cycle, the overall state of all data qubits randomly aligns as a simultaneous $(+1$ or $-1)$ eigenstate $|\Psi\rangle$ of all stabilizers. This state $|\Psi\rangle$ represents a complex entangled state of all data qubits. Given that all stabilizers commute, subsequent surface-code cycles will maintain this state $|\Psi\rangle$, provided no errors occur. In surface-code terminology, such a state $|\Psi\rangle$ is known as a **quiescent state**. Any logical qubits are encoded within this quiescent state.

We shall next turn our attention to discussing the logical qubits and gates.

Inconvenient Logical Qubits and Operators

As proven earlier, the square lattice with four alternating Z- and X-boundaries is able to supply the surface code exactly one logical qubit. The question is: How do we identify the logical basis states $|0_L\rangle$ and $|1_L\rangle$? To be specific, let's consider the lattice in Figure 8.24, where there are 41 data qubits all told.

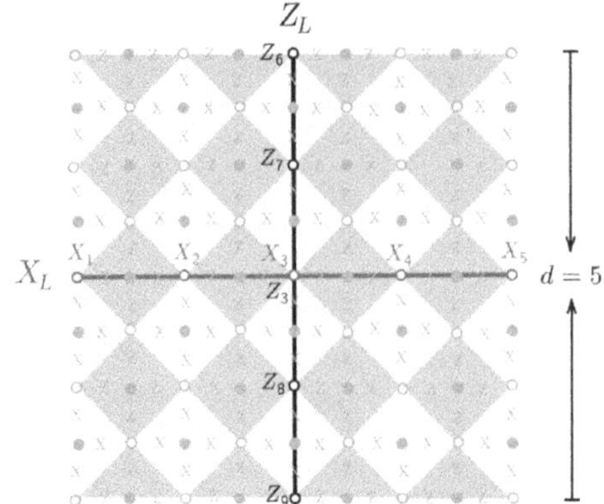

Figure 8.24 A sample surface code lattice, 4×5 plaquettes (including the top and bottom boundary plaquettes, 41 data qubits), with four alternating X- and Z-boundaries.

A logical qubit state is not localized anywhere on the lattice but is encoded in a quiescent state $|\Psi\rangle$. No local operator can manipulate such a logical qubit state, according to the lesson drawn from the logical operators of the two logical qubits encoded in the ground states of the $\mathbb{Z}_2$ toric code model on the torus. The lesson also tells that the logical operators on the single logical qubit in the current case should be long string operators that commute with all the stabilizers and cannot be products of the stabilizers. The lattice in Figure 8.24 lacks noncontractible loops; however, we have paths stretching from one boundary of the lattice to another boundary. By the same topological invariance as that in the $\mathbb{Z}_2$ toric code mode, we can construct two possible equivalence classes of logical operators on the single logical qubit in our current surface code. Two representatives can be respectively chosen from these two equivalence classes: as shown in Figure 8.24, a long Z-string operator Z_L connecting the top and bottom Z-boundaries and a long X-string operator X_L connecting the left and right X-boundaries.[16]

It's evident that these two operators X_L and Z_L commute with all X- and Z-stabilizers. Besides, $\{X_L, Z_L\} = 0$. So, Z_L can be taken as the logical Z operator, and X_L as the logical X operator. The code distance is evidently $d = 5$, as dictated by the weight of the logical X operator X_L, which in this configuration is as long as the lattice size. In order to be able to detect and correct more concurrent qubit errors, one would have to enlarge the lattice and hence increase the weight of X_L. One can also increase the number of alternating boundary conditions to add more logical qubits, whose logical operators are likewise lattice-size string operators. But the price is ever-increasing lattice size, qubit resources, and complexity in manipulating the logical qubits.

But the idea of boundaries does inspire a way out of this conundrum: We may forget about the outer boundaries of the lattice but create inner boundaries enclosing holes – defects – on the lattice, as we are going to explore now.

---

[16] In the context of the $\mathbb{Z}_2$ toric code, these two string operators are also regarded as defined on noncontractible loops because the two ends of each path are secretly "connected" through the corresponding vacuum outside the lattice.

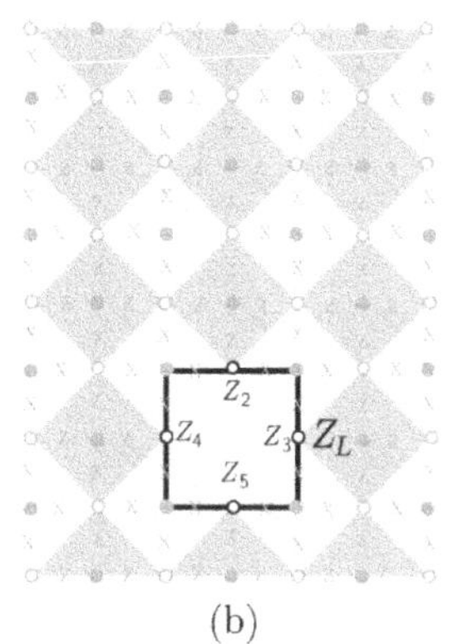

Figure 8.25 (a) A Z-defect qubit. (b) Logical Z operator Z_L on the defect qubit.

(a)   (b)

8.3.3 Defect Logical Qubits and Operators

What does it mean to create a defect on the surface code lattice? Consider the lattice in Figure 8.25(a). Suppose after a certain surface-code cycle, we stop measuring the Z-stabilizer $Z_2Z_3Z_4Z_5$, depicted as a dashed square in the figure, we say we create a **Z-defect**. This Z-defect is not a new concept but merely a hole on the lattice, whose boundary – an inner boundary of the lattice – is a smooth or X-boundary. In the literature, a Z-defect may also be dubbed a **smooth** or **dual defect**. Note that:

stop measuring a Z-stabilizer = remove the stabilizer from the group of stabilizers

$$\neq \text{ violate the } Z\text{-stabilizer.}$$

That is, we have one fewer constraint on the quiescent state, hence one more independent degree of freedom, yielding one more logical qubit. Although this extra logical qubit is also a collective state of all the data qubits on the lattice, it can be associated with the Z-defect and thus termed a **Z-defect qubit**, which can be treated as a "localized" logical qubit.

When a Z-defect is created, a surface-code cycle must be executed to ensure a new quiescent state $|\Psi\rangle$ that encodes the Z-defect qubit. We'll get back to this shortly. What about the logical operators on this Z-defect qubit?

Logical Z operator

The logical Z operator on a Z-defect qubit is straightforward: The stopped Z-stabilizer, such as the $Z_2Z_3Z_4Z_5$ in Figure 8.25(a), doesn't belong to the stabilizer group of the code any longer, yet it commutes with the entire stabilizer group. Hence, it is precisely the logical Z operator, Z_L, on the Z-defect qubit. In our example, $Z_L = Z_2Z_3Z_4Z_5$, and

$$Z_L |\Psi\rangle = \pm |\Psi\rangle. \tag{8.71}$$

Logical X Operator and Need for Another Defect

Constructing the logical X operator is a bit more subtle. To flip a Z-defect qubit, one can choose to bit-flip any of the four data qubits on the boundary of the defect. To

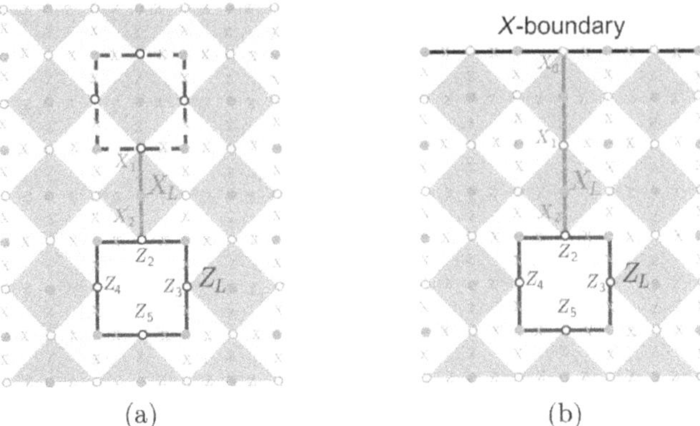

Figure 8.26 (a) A tentative logical X operator X_L on the Z-defect qubit; it inevitably anticommutes with the Z-stabilizer at the other end of the string, causing a bit-flip error there. (b) Another tentative X_L, which ends at an X-boundary; it causes no error elsewhere but attaches the defect to the boundary.

Figure 8.27 (a) Logical X operator $X_L = X_1 X_2$ and logical Z operator $Z_L = Z_2 Z_3 Z_4 Z_5$ on a Z-defect qubit. The Z-defect in the dashed square is ignored. (b) Logical Z operator $Z_L = Z_1 Z_2$ and logical X operator $X_L = X_2 X_3 X_4 X_5$ on an X-defect qubit. The X-defect in the dashed square is ignored.

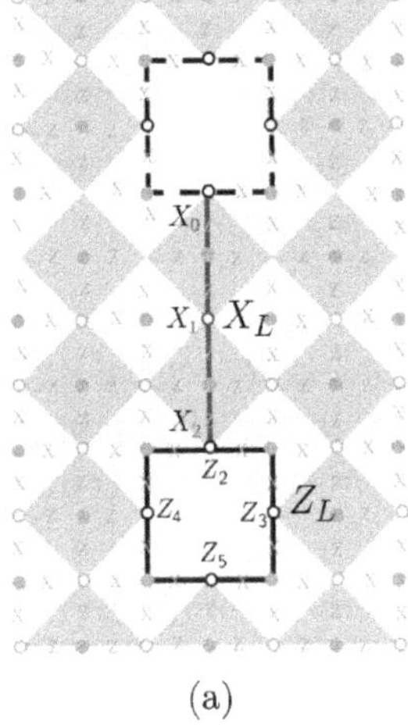

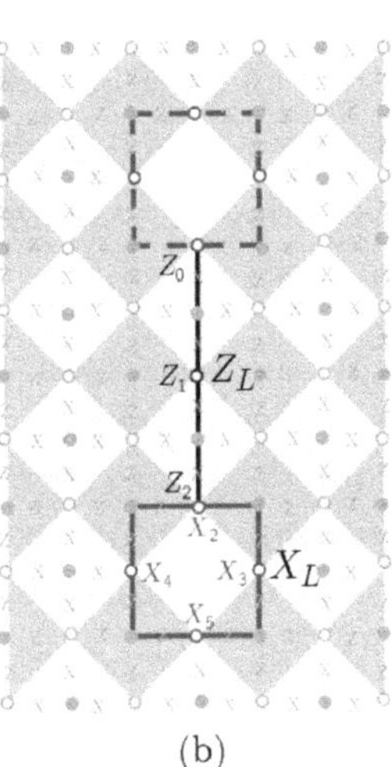

be specific, consider the Z-defect in Figure 8.26(a). The X-string operator $X_L = X_1 X_2$ can certainly flip the data qubit 2 on the boundary of the Z-defect and is thus a logical X operator; however, it simultaneously causes a bit-flip error on the qubit (qubit 1) at the other end of the string.

Another tentative logical X operator is the X_L in Figure 8.26(b). This string operator also flips qubit 2 on the boundary of the Z-defect, but it ends on a data qubit (qubit 0) at the outer X-boundary and causes a bit-flip error on that data qubit.

So, how to construct a logical X operator on a Z-defect qubit? Similarly to forgetting the outer boundaries of the surface code lattice, we can create a pair of Z-defects, which add two logical qubits, but use only one of them while ignoring the other one. See Figure 8.27(a) for an example: Two Z-defects are created but only the lower one is taken as the Z-defect qubit, whereas the upper one is forgotten. Then one can take any X-string operator X_L, with one end on a boundary data

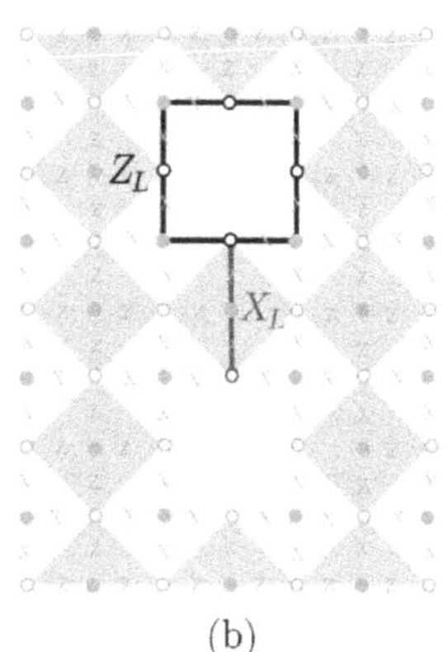

(a)   (b)

Figure 8.28 (a) Initialization and (b) measurement of a Z-defect qubit.

qubit of the Z-defect qubit and the other end on a boundary data qubit of the forgotten Z-defect; for instance, $X_L = X_1 X_2$. This operator X_L bit-flips the data qubit 2 and thus the Z-defect qubit. It also bit-flips the data qubit 1 on the boundary of the forgotten Z-defect, but since the corresponding Z-defect qubit is ignored and the Z-stabilizer is stopped, no error would occur. Therefore, such an X-string operator is a well-defined and well-behaved logical X operator.

In the example shown in Figure 8.27(a), the logical X operator X_L has weight 3; hence, the code distance of this surface code with one Z-defect qubit (the other one is forgotten) is $d = 3$.

Likewise, one can create a pair of X-**defects** by stopping measuring two X-stabilizers. This would add two X-**defect qubits**, but one of them would be forgotten as in the case of Z-defect qubit. Figure 8.27(b) depicts an example. The logical X operator X_L of an X-defect qubit is the stopped X-stabilizer, for instance, $X_L = X_2 X_3 X_4 X_5$ in the figure.[17] A logical Z operator Z_L is a Z-string operator connecting two boundary data qubits respectively of the two X-defects, such as $Z_L = Z_1 Z_2$ in the figure. In the literature, an X-defect may also be called a **rough** or **primal defect**. This example also has code distance $d = 3$.

Defect Qubit Initialization and Measurement

Contrary to what you might think, initializing and measuring defect qubits is not complicated. Depicted in Figure 8.28(a), for initializing a Z- or X-defect qubit, it's essential to measure the Z- or X-stabilizer during the surface-code cycle prior to ceasing the stabilizer measurement that leads to defect creation. This preparatory measurement determines the initial state of the Z- or X-defect qubit, integrating it into the quiescent state $|\Psi\rangle$.

Measuring a Z- or X-defect qubit is straightforward, as in Figure 8.28(b). It involves measuring their respective logical Z operators, which are the halted Z- or X-stabilizers. The measurement outcome, either ± 1, reveals whether the defect qubit is in the state $|0_L\rangle$ or $|1_L\rangle$.

---

[17] This definition of the logical Z and X operators of an X-defect qubit is commonly adopted in the literature.

8.3.4 Logical CNOT on Defect Qubits

The remarkable aspect of defect qubits lies in their ability to implement the CNOT gate through the process of braiding two defects. This process is akin to introducing defects in the $\mathbb{Z}_2$ toric code model by removing certain vertex and/or plaquette operators. These defects expand the ground-state Hilbert space, adding extra dimensions and thus creating defect logical qubits. Interestingly, braiding these defects in the $\mathbb{Z}_2$ toric code model can also facilitate the CNOT gate on the corresponding logical qubits. In this scenario, defects take on the properties similar to non-Abelian anyons, despite the model originally hosting only Abelian anyons. Specifically, these defects in the $\mathbb{Z}_2$ toric code model are recognized as Ising anyons. Nevertheless, our focus here will not be on this broader context but specifically on the surface code.

To fully grasp the mechanism behind braiding defect qubits to execute the CNOT gate, some preparatory understanding is essential.

CNOT in the Heisenberg Picture

Revisiting the CNOT gate is necessary to understand its operation in the context of defect qubits in a surface code lattice. Although the gate's action on qubit states is straightforward, assessing its effect on the complex, encoded quiescent state of defect qubits is less direct.

In quantum mechanics, the Schrödinger and Heisenberg pictures offer equivalent formulations. The former emphasizes state evolution over time, while the latter focuses on the time evolution of operators. Given the clarity of logical operators for defect qubits, analyzing the effect of braiding on these operators is a feasible strategy.

Let's recall how the CNOT gate transforms the two-qubit Pauli operators, the generators of the Pauli group $\mathcal{G}_2$. These are $\mathbb{1} \otimes X$, $X \otimes \mathbb{1}$, $\mathbb{1} \otimes Z$, and $Z \otimes \mathbb{1}$. With the CNOT gate defined as $\mathrm{CNOT} = |0\rangle \langle 0| \otimes \mathbb{1} + |1\rangle \langle 1| \otimes X$, we derive the transformations

$$
\begin{aligned}
\mathrm{CNOT}(\mathbb{1} \otimes X)\,\mathrm{CNOT} &= \mathbb{1} \otimes X, \\
\mathrm{CNOT}(X \otimes \mathbb{1})\,\mathrm{CNOT} &= X \otimes X, \\
\mathrm{CNOT}(\mathbb{1} \otimes Z)\,\mathrm{CNOT} &= Z \otimes Z, \\
\mathrm{CNOT}(Z \otimes \mathbb{1})\,\mathrm{CNOT} &= Z \otimes \mathbb{1}.
\end{aligned}
\tag{8.72}
$$

Equipped with these transformations, we can examine the effect of braiding two defect qubits on their logical operators. But hold on, what does it mean by braiding two defect qubits on the lattice? To braid two defect qubits, we must be able to move them on the lattice. So, let's discuss moving before braiding.

Moving a Z-Defect Qubit

To demonstrate the process of moving a Z-defect on a surface code lattice, let's consider a specific example. We aim to move the Z-defect qubit, initially in the square bounded by the solid black edges in Figure 8.29(a), downward by one plaquette. This movement involves a series of steps, each of which must be executed with precision to ensure the defect qubit's integrity throughout the process.

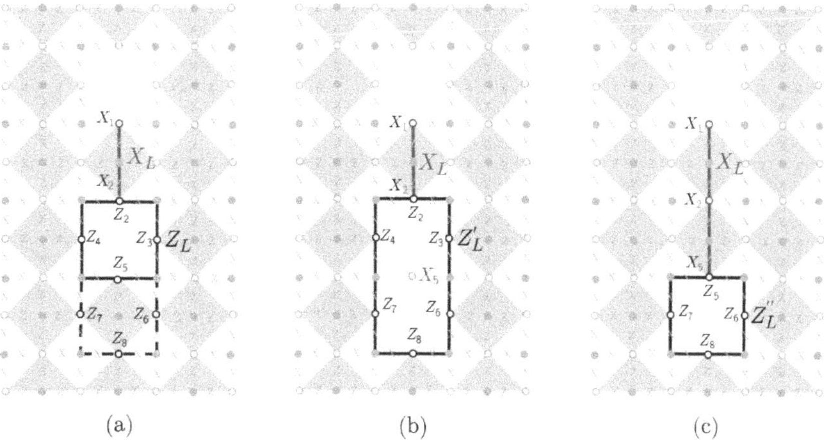

$$(a) \qquad\qquad (b) \qquad\qquad (c)$$

Figure 8.29 Moving a Z-defect downward by one plaquette.

While this explanation focuses on moving a defect by a single plaquette, the underlying principles and actions can be extended to movements over multiple plaquettes. The steps for such extended movements maintain a similar structure but involve repeating certain actions to achieve the desired displacement.

Here are the detailed steps:

1. Stop measuring the Z-stabilizer $Z_5 Z_6 Z_7 Z_8$ at the plaquette directly below the current Z-defect (see Figure 8.29(a)). This action enlarges the defect to include both plaquettes. The enlarged Z-defect is shown in Figure 8.29(b), and the corresponding logical Z operator of this enlarged Z-defect qubit becomes $Z'_L = Z_L Z_5 Z_6 Z_7 Z_8 = Z_2 Z_3 Z_4 Z_6 Z_7 Z_8$.

2. Concurrently, modify the X-stabilizer measurements around the boundary of the enlarged defect. Now, two three-qubit X-stabilizers are present on the left and right sides of the defect, and an isolated data qubit in the center leads to a single-qubit X-stabilizer X_5.

3. Perform a surface-code cycle with the modified stabilizer configuration. This cycle includes the preparation, entanglement, and measurement of the modified stabilizers.

4. Measure the logical Z operator Z'_L at the end of the modified surface-code cycle and record the outcome.

5. Restore the Z-stabilizer measurement for the original defect location (adding back the Z-stabilizer $Z_2 Z_3 Z_4 Z_5$) while continuing not to measure the Z-stabilizer $Z_5 Z_6 Z_7 Z_8$. This step moves the Z-defect qubit downward by one plaquette.

6. Reset the modified X-stabilizers to their original form and perform an additional surface-code cycle to stabilize the lattice in this new configuration.

7. To ensure proper error correction, especially on a $d \times d$ lattice, an additional $d - 1$ surface-code cycles are necessary, totaling $d + 1$ cycles for moving and stabilizing the state after the move. The new logical Z and X operators become $Z''_L = Z'_L Z_2 Z_3 Z_4 Z_5 = Z_5 Z_6 Z_7 Z_8$ and $X''_L = X_1 X_2 X_3 X_5$.

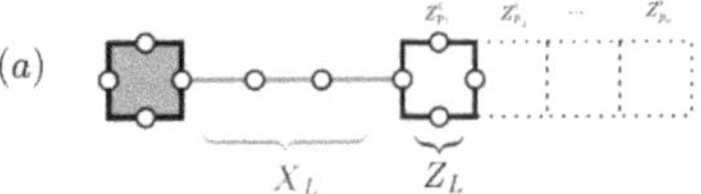

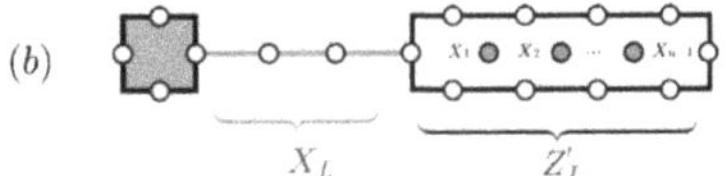

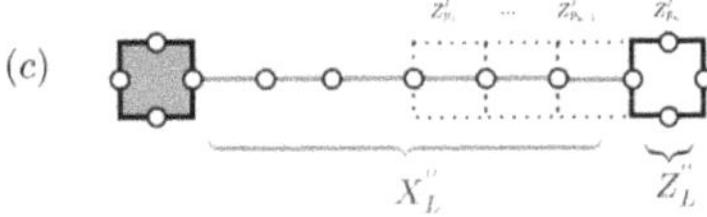

Figure 8.30 Moving a Z-defect qubit to its right by $n-1$ plaquettes. (a) The original Z-defect at plaquette p_1, with logical operators $Z_L = Z^i_{p_1}$ and X_L. (b) Stop measuring all the $n-1$ Z-stabilizers to the right of the original Z-defect, resulting in an enlarged Z-defect, with intermediate logical operators $Z'_L = Z_L Z^i_{p_2} Z^i_{p_3} \cdots Z^i_{p_n}$ and X_L. (c) Restore the Z-stabilizers on plaquettes p_1 through p_{n-1}, resulting in the final moved Z-defect qubit with logical operators Z''_L and X''_L. The superscripts i and f mean initial and final.

8. Be aware that Z''_L and X''_L might differ from the original Z_L and X_L by a sign, based on the measurement outcomes of the Z-stabilizer $Z_2 Z_3 Z_4 Z_5$, Z'_L, and X_5. These signs should be tracked by the software and corrected when determining the state of the moved Z-defect qubit.

Moving a Z-defect by multiple plaquettes follows similar steps to moving it by a single plaquette. For example, consider moving a Z-defect to its right by $n-1$ plaquettes, as showcased in Figure 8.30. After completing the surface-code cycle prior to the move, one can stop measuring the $n-1$ Z-stabilizers to the right of the original Z-defect, all at once. Thus, only one surface-code cycle is necessary for this step. In total, $d+1$ surface-code cycles are sufficient to move a Z-defect by $n-1$ plaquettes, the same as in the case of moving by one plaquette. After completing the move, the logical operators become

$$Z''_L = Z'_L \prod_{j=1}^{n-1} Z^f_{p_j} = Z^f_{p_n}, \tag{8.73}$$

$$X''_L = X_L \prod_{j=1}^{n-1} X_j, \tag{8.74}$$

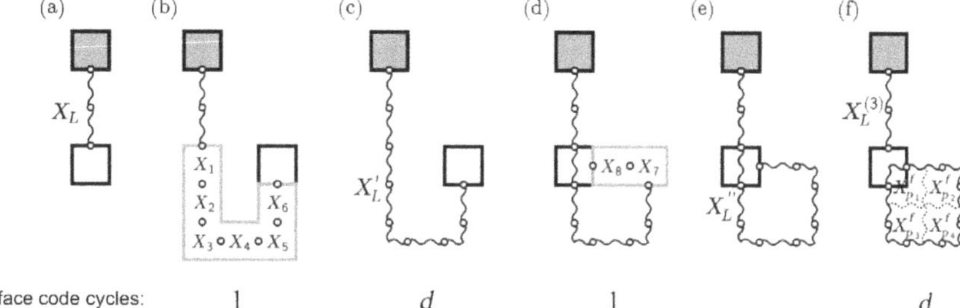

Figure 8.31 Braid a Z-defect (unfilled square) with itself by moving it along a closed loop path. (a) The original Z-defect and the logical X operator X_L. (b), (c) Move the Z-defect by six plaquettes along a U-shape path, following the procedure outlined in Figure 8.30, resulting in an intermediate logical X operator X_L'. (d) and (e) Move the Z-defect to its left by two plaquettes, returning it to its original position. The intermediate logical X operator X_L'' is an X-string operator with a loop. (f) Annihilate the loop in X_L'' by the action of four X-stabilizers $X_{p_1}^f$ through $X_{p_4}^f$. The final Z-defect has the logical X operator $X_L^{(3)}$. The superscript f means final. The bottom row shows the number of surface-code cycles needed during the moves.

which may differ from the original logical operators by sign factors. These factors should have been tracked throughout the entire procedure of moving the defect.

It is worth noting that the direction of the move is not a concern in the procedure described above. Additionally, moving an X-defect is completely analogous to moving a Z-defect, so we will not repeat the discussion here.

With these concepts and tools, we are now prepared to study braiding two defects and understanding how this braiding realizes the CNOT gate.

8.3.5 Logical CNOT

We warm ourselves up by braiding a Z-defect with itself and study the consequence on the logical X operator X_L of the Z-defect qubit, to echo our focus in the Heisenberg picture. We consider the concrete example in Figure 8.31 and enumerate the necessary steps as follows.

1. The original Z-defect is shown as an unfilled square in Figure 8.31(a), which is linked to the forgotten Z-defect (filled square) by its logical X operator X_L.
2. The entire move cannot be completed in one go but must be broken down into two paths. The first is the U-shaped path in Figure 8.31(b), along which the Z-defect is moved by six plaquettes. The second is the horizontal path consisting of two plaquettes in Figure 8.31(d). This division of the loop path is not unique but is a convenient choice. The loop path must be divided because otherwise, the plaquette at the center of the loop would be completely isolated from the rest of the surface code structure when all the Z-stabilizers along the loop are stopped simultaneously, effectively dividing the original lattice of data qubits into two uncorrelated lattices, thereby disrupting the surface code structure.

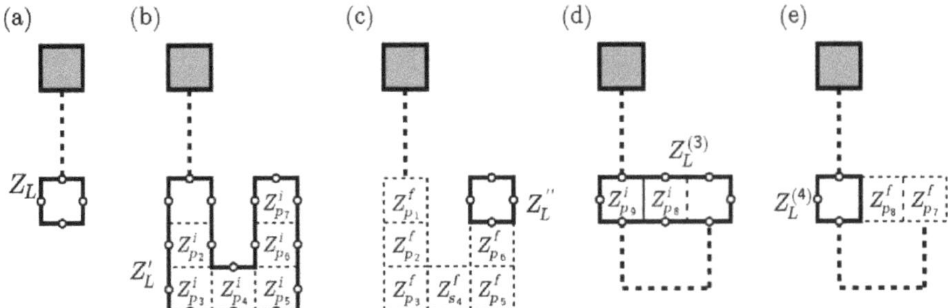

Figure 8.32 Braid a Z-defect (unfilled square) with itself by moving it along a closed loop path. (a) The original Z-defect and the logical Z operator Z_L. (b), (c) Move the Z-defect by six plaquettes along a U-shape path, following the procedure outlined in Figure 8.30, resulting in an intermediate logical Z operator Z_L''. (d), (e) Move the Z-defect to its left by two plaquettes, returning it to its original position. The final Z-defect has the logical Z operator $Z_L^{(4)}$. The superscript f means final.

3. Completing the U-shaped path requires $d + 1$ surface-code cycles to stabilize the surface code, as before. After completing the horizontal path, the Z-defect returns to its initial position, as seen in Figure 8.31(e); however, the logical X operator of the Z-defect qubit becomes X_L'', which contains a loop. This loop can be eliminated by the action of the four X-stabilizers $X_{P_1}^f$ through $X_{P_4}^f$ enclosed within the loop. (Recall how we eliminated a loop in the case of the $\mathbb{Z}_2$ toric code model?) After contracting the loop, another d surface-code cycles are executed to stabilize the Z-defect qubit, resulting in Figure 8.31(f).
4. The final logical X operator becomes $X_L^{(3)}$, which may differ from the original X_L by the phase factor

$$\left\langle \prod_{j=1}^{4} X_{P_j}^f \right\rangle = \pm 1, \tag{8.75}$$

which is the outcome of measuring all the X-stabilizers enclosed by the loop in Figure 8.31(e).

Similarly, we can assess how self-braiding a Z-defect affects its logical Z operator. This process can be examined using the same loop path as in Figure 8.31, but this time focusing on the logical Z operator. The procedure, depicted in Figure 8.32, mirrors that in Figure 8.31. In this case, however, we do not encounter a loop as seen in Figure 8.31(e) that needs to be contracted before the final stabilizing process. The final logical Z operator $Z_L^{(4)}$ may also differ from the original Z_L by a sign:

$$\left\langle \prod_{j=7}^{8} Z_{P_j}^f \right\rangle \left\langle \prod_{j=8}^{9} Z_{P_j}^i \right\rangle \left\langle \prod_{j=1}^{6} Z_{P_j}^f \right\rangle \left\langle \prod_{j=2}^{7} Z_{P_j}^i \right\rangle = \pm 1, \tag{8.76}$$

where on the LHS, the expectation values are respectively from right to left of the Z-stabilizers measured in Figure 8.32(b) through (e).

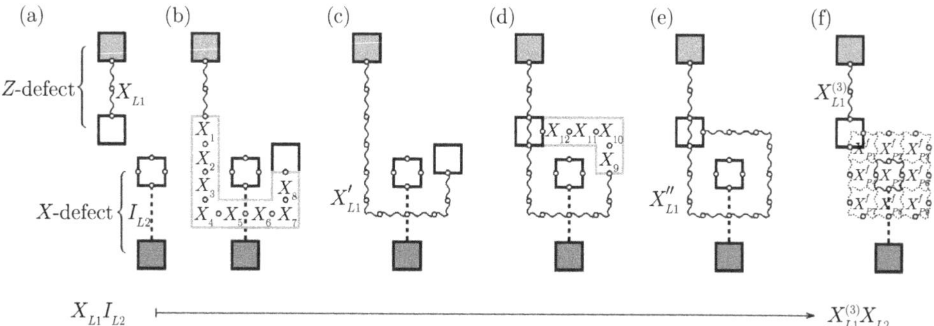

Figure 8.33 Braiding a Z-defect around an X-defect. (a) Original configuration: a Z-defect (unfilled square) with its logical X operator X_{L_1} acted on and an X-defect (unfilled square) with the trivial logical operator $\mathbb{1}_{L_2}$. (b)–(e) Braiding the Z-defect around the X-defect along a closed loop. (f) Shrinking the loop in (e) by the action of the eight X-stabilizers $X_{p_1}, X_{p_2}, X_{p_3}, X_{p_4}, X_{p_6}, X_{p_7}, X_{p_8}$, and X_{p_9}. Filled squares depict the forgotten defects. The long arrow at bottom indicates that the braiding turns the Pauli operator $X_{L_1} \otimes \mathbb{1}_{L_2}$ into $X_{L_1}^{(3)} \otimes X_{L_2}$.

Realizing CNOT

Having warmed up, we now proceed to braid a Z-defect around another Z-defect, aiming to compare the outcome of this braiding with the effects of a CNOT gate in the Heisenberg picture. It is essential to verify if the braiding process corresponds to the transformations outlined in Equation (8.72).

To facilitate this comparison, we will generate a pair of Z-defects and a pair of X-defects. In each pair, one defect will be active, while the other will be disregarded. We will then braid the active Z-defect around the active X-defect, examining how this braiding affects the logical Pauli operators. We will explore this in four distinct scenarios, each corresponding to a row in Equation (8.72). Nonetheless, having already discussed the procedure for moving a defect, our focus will primarily be on the specific details involved in maneuvering one defect around another.

- Effect on $X \otimes \mathbb{1}$:
  - The effect is displayed step by step in Figure 8.33. In Figure 8.33(a), The (functioning) forgotten defects are drawn as (unfilled) filled squares. The functioning Z-defect qubit is acted on by its logical X operator X_{L_1}, whereas the functioning Z-defect qubit is not acted on by the trivial operator $\mathbb{1}_{L_2}$. That is, the original Pauli operator on the two defect qubits is

$$X_{L_1} \otimes \mathbb{1}_{L_2}. \tag{8.77}$$

    We shall move the functioning Z-defect around the X-defect along a closed loop path indicated in the figure and see how this braiding may alter the Pauli operator above.
  - As in the case shown in Figure 8.31, the braiding here has to be done in two steps: first completing the path of moving in Figure 8.33(b), resulting in the configuration in (c), with the logical X operator extended to X'_{L_1}, then completing the path in (d) and arriving at (e), where the logical X operator becomes

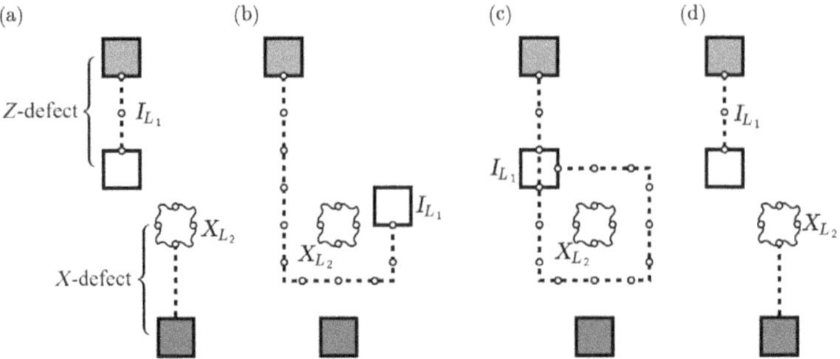

Figure 8.34 Braiding a Z-defect around an X-defect. (a) Original configuration: a Z-defect (unfilled square) with no nontrivial logical operators acted on and an X-defect acted on by its logical X operator X_{L_2}. (b)–(d) Moving the Z-defect around the X-defect along a closed path. No actual loop to be contracted in (c). Filled squares are the forgotten defects.

X''_{L_1}, which contains a closed loop. As opposed to the case in Figure 8.31(e), where the loop in the logical X could be contracted, the loop in X''_{L_1} here cannot be because it encloses the functioning X-defect, a hole on the lattice. Nevertheless, since $X^2 = \mathbb{1}$, exerting the eight X-stabilizers,

$$X_{p_1}, X_{p_2}, X_{p_3}, X_{p_4}, X_{p_6}, X_{p_7}, X_{p_8}, X_{p_9}, \tag{8.78}$$

in between the loop and the X-defect can effectively shrink the loop to the operator X_{L_2} consisting of four Pauli X matrices acting on the four data qubits on the boundary of the X-defect,[18] as in Figure 8.33(f). This operator X_{L_2} is a logical X operator of the X-defect qubit. The logical operator X''_{L_1} is simultaneously shortened to be $X_{L_1}^{(3)}$, the final logical X operator of the returned Z-defect qubit. Note that $X_{L_1}^{(3)}$ may differ from the original X_{L_1} by a sign, which should be tracked down by the software.

– Hence, the effect of the braiding in Figure 8.33 is, up to a sign factor:

$$X_{L_1} \otimes \mathbb{1}_{L_2} \xrightarrow{\text{braiding a } Z\text{-defect around an } X\text{-defect}} X_{L_1} \otimes X_{L_2}, \tag{8.79}$$

in agreement with the first row of Equation (8.72).

• Effect on $\mathbb{1} \otimes X$:

In this case, as seen in Figure 8.34, the effect of braiding a Z-defect around an X-defect is simpler because the functioning Z-defect is not attached to an X-string but a trivial one – the trivial operator $\mathbb{1}_{L_1}$. Consequently, after moving the Z-defect around the X-defect along a closed loop path, there is no actual loop to contract. Thus, the original Pauli operator $\mathbb{1}_{L_1} \otimes X_{L_2}$ remains invariant under the braiding. That is,

$$\mathbb{1}_{L_1} \otimes X_{L_2} \xrightarrow{\text{braiding a } Z\text{-defect around an } X\text{-defect}} \mathbb{1}_{L_1} \otimes X_{L_2}, \tag{8.80}$$

consistent with the second row of Equation (8.72).

---

[18] In topology, this shrinking is possible because the loop is homotopic to the boundary of the defect.

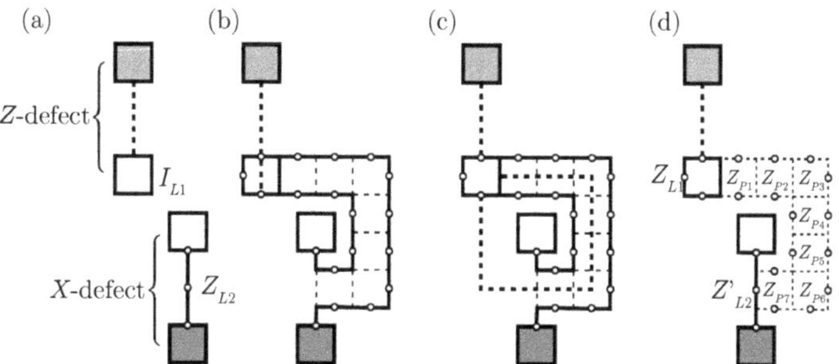

Figure 8.35 Braiding a Z-defect around an X-defect. (a) Original configuration: a Z-defect (unfilled square) with no nontrivial logical operators acted on and an X-defect acted on by its logical Z operator Z_{L_2}. (b)–(d) Moving the Z-defect around the X-defect along a closed path. The moving Z-defect pushes the logical operator Z_{L_2} from a straight string operator in A to a curvy string operator in (c), which is equivalent to the loop operator Z_{L_1} together with the straight string operator Z'_{L_2} in (d). Filled squares are the forgotten defects.

- Effect on $Z \otimes \mathbb{1}$:

This case is similar to the first case in that there is a nontrivial string to take care of in the end of the braiding. Figure 8.35 shows the braiding process. The original Pauli operator on the two defect qubits is

$$\mathbb{1}_{L_1} \otimes Z_{L_2}. \tag{8.81}$$

The braiding is performed as usual but a special care is needed: When moving the Z-defect across the logical Z operator Z_{L_2} of the X-defect qubit, the Z-string that defines Z_{L_2} is deformed due to the stopped Z-stabilizers along the way, as in Figure 8.35(b) to (c). The seven stopped Z-stabilizers Z_{p_1} through Z_{p_7} must be restored, resulting in (d), where the Z-string is broken into a closed Z-loop, namely the logical Z operator on the returned Z-defect, and again a Z-string Z'_{L_2} attached to the X-defect. This Z'_{L_2} may differ from the original logical Z operator Z_{L_2} by a sign due to the restored Z-stabilizers, which should be tracked by the software. As a result,

$$\mathbb{1}_{L_2} \otimes Z_{L_1} \xrightarrow{\text{braiding a } Z\text{-defect around an } X\text{-defect}} Z_{L_1} \otimes Z_{L_2}, \tag{8.82}$$

mirroring the third row of Equation (8.72).

- Effect on $\mathbb{1} \otimes Z$:

> **Exercise 8.12** *This effect is even simpler to check. Try it as an exercise to show that*
> $$Z_{L_1} \otimes \mathbb{1}_{L_2} \xrightarrow{\text{braiding a } Z\text{-defect around an } X\text{-defect}} Z_{L_1} \otimes \mathbb{1}_{L_2}. \tag{8.83}$$

Therefore, in view of Equation (8.72), we have shown that

$$\text{braiding a } Z\text{-defect around an } X\text{-defect} = \text{CNOT on the two defect qubits.} \tag{8.84}$$

Astute readers may ponder the question: Do we have to braid a Z-defect with an X-defect to realize the CNOT gate on defect qubits? No, we don't. In fact, we can realize the CNOT gate on two defect qubits of the same kind. Nevertheless, to achieve for example the CNOT gate on two Z-defect qubits, we would need an ancilla X-defect and an ancilla Z-defect and resort to the braiding between a Z-defect and an X-defect. Exercises 8.12 and 8.13, you are invited to explore such cases on your own.

Exercise 8.13 *This exercise aims to construct the CNOT gate on two Z-defect qubits. To this end, you need to create three pairs of Z-defects and use one in each pair as a functioning defect while forgetting the other one; you also need to create a pair of X-defects and use only one in the pair.*

1. *Convince yourself in the Heisenberg picture that the quantum circuit in Figure 8.36 taking the defect qubits as input indeed behaves as the CNOT gate on two Z-defect qubits:*

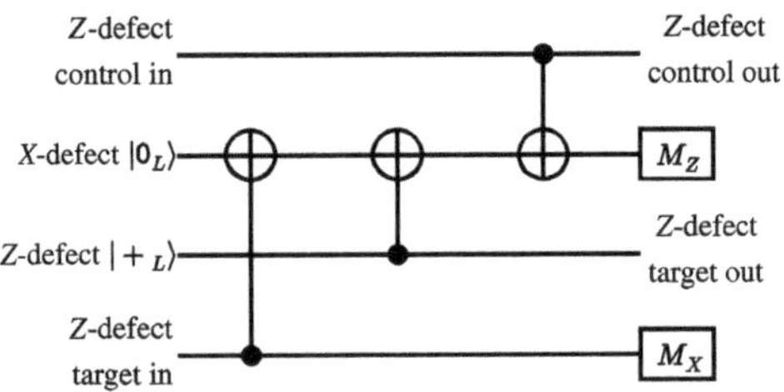

Figure 8.36 Quantum circuit realizing the CNOT gate on two Z-defect qubits.

Here, M_Z and M_X are the measurement outcomes of the corresponding defect qubits, which are logical qubits. Explain why we need them and what information we can extract from them.

2. *Verify that the circuit above is realized by the braiding shown in Figure 8.37 among the four defects:*

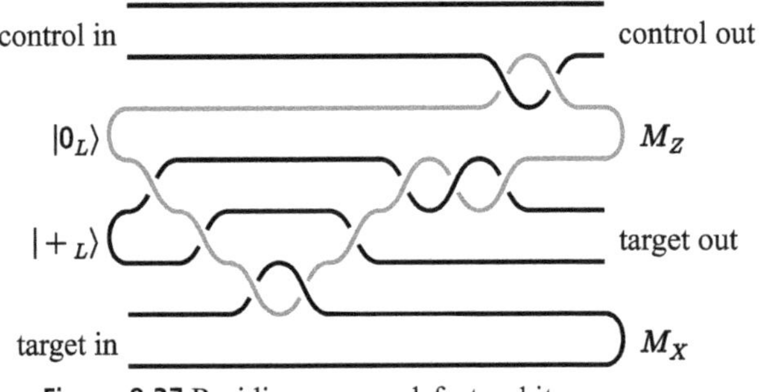

Figure 8.37 Braiding among defect qubits.

Recall that each defect qubit is devised by a pair of defects. So, here, each defect qubit is represented by two lines, one for each defect in the pair. You should keep track of how the initial Pauli operators on the control and target defect qubits evolve through the braiding. The braiding above can also be drawn as shown in Figure 8.38.

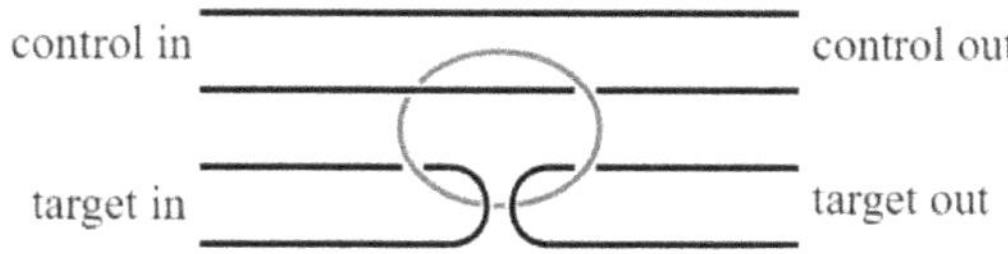

Figure 8.38 Redrawn braiding for defect qubits.

Exercise 8.14 *Following a logic similar to that in the previous exercise, construct the quantum circuit that realizes the CNOT gate on two X-defect qubits. Then, draw the corresponding braiding diagram of the circuit.*

Single-Qubit Logical Gates

The single-qubit logical gates on the defect qubits are less topologically intuitive and involve lots of quantum gate operations on the data qubits on the surface code lattice. We shall not dwell on them here, being constrained by the length of this book. The reader is encouraged to read the references cited at the beginning of this section.

8.4 Overview of Error Correction in Surface Code Quantum Computation

Surface code quantum computation is a fault-tolerant way of realizing the quantum circuit model. Along the way of any computational process on the surface code lattice, errors of the data qubits have been diagnosed and corrected. We outline here how error correction is conducted in surface code quantum computation but leave the detail to the literature.

1. **Syndrome extraction process:**
   - Measurement qubits are entangled with data qubits to measure the Z- and X-stabilizers using controlled-NOT (CNOT) and Hadamard gates.
   - For Z-stabilizer measurements, measurement qubits are initialized in $|0\rangle$, entangled, and measured. The results indicate parity errors (odd number of qubits in $|1\rangle$ state).
   - For X-stabilizer measurements, measurement qubits start in $|+\rangle$ state, are entangled with adjacent data qubits, and measured in the X basis.

2. **Collecting error syndromes:**
   - Syndrome bits from measurement qubits are recorded after each round, forming a pattern indicating potential errors.
   - Multiple rounds of measurements provide a temporal syndrome sequence, essential for identifying error patterns, especially small chains of errors.
3. **Error correction and decoding:**
   - The decoding algorithm, such as minimum weight perfect matching, analyzes syndrome patterns to infer likely error locations and types.
   - The decoder traces error paths across the lattice and determines corrective quantum operations (usually X or Z gates).
   - These operations are applied to data qubits to reverse the effects of errors, restoring the quantum information.

Through these steps, the surface code systematically detects and corrects errors across the quantum computing lattice, maintaining the integrity of quantum information despite the occurrence of errors. This process underpins the robustness of quantum computation using surface codes, enabling computations to proceed over many cycles with high fidelity.

8.5   Remarks on Surface Code and Outlook

The surface code quantum computation promises significant advancements in achieving low logical error rates. A noteworthy point is that a logical qubit consisting of approximately 10^3 physical qubits can achieve an error rate of 10^{-17}, provided the physical qubit operations maintain an error rate of 10^{-4}. This estimation assumes the code distance being 3. Recent calculations [57], however, suggest a more optimized scenario: Approximately 64 physical qubits may suffice to construct a single logical qubit with an error rate below 10^{-17}, assuming a physical error rate of 10^{-4} and an optimal code distance of 8. This refinement in the estimation indicates that with enhancements in the code distance d, the number of required physical qubits can be substantially reduced. Nonetheless, increasing d introduces greater complexity in controlling physical qubits.

Substantial progress in quantum error correction using surface codes has been observed in various aspects of quantum computers. These advancements encompass increased qubit relaxation and dephasing lifetimes, minimized cross-talk between qubits, optimized control fidelity, and sophisticated readout and reset operations. These developments are crucial for maintaining low error rates, which are essential for effective QEC in large-scale qubit systems. In 2023, a pivotal achievement was the demonstration of a logical qubit prototype, a fundamental element of an error-corrected quantum computer [254]. This breakthrough involved expanding the system from 17 to 49 physical qubits in a surface code, significantly reducing the logical qubit's error rate. Experimental results indicated that the logical error rate of a distance-5 code surpassed that of a distance-3 code, affirming the effectiveness of quantum error correction.

A more astonishing area of progress in surface code is being achieved in neutral atom arrays [255]. A programmable quantum processor with up to 280 physical qubits, based on encoded logical qubits has showcased high two-qubit gate fidelities, arbitrary connectivity, fully programmable single-qubit rotations, and mid-circuit readout. Researchers have advanced two-qubit logic gates by scaling surface code distance from 5 to 7, thereby enhancing fault-tolerance. Additionally, the realization [255] of three-dimensional codes [256, 257] marked a significant leap in error detection algorithmic performance, surpassing physical qubit fidelities in cross-entropy benchmarking and quantum simulations of fast scrambling. It is still uncertain how these results will influence the future development of QEC, but the advancements are undoubtedly exciting to witness.

In the realm of QEC, color code stands out as a notable contender alongside surface codes. Color codes are structured on 2D and 3D lattices, with qubits positioned at the vertices. Error correction in these codes is facilitated through syndrome measurements from the lattice faces and volumes, potentially offering a more efficient error correction mechanism, particularly for specific quantum gate operations. In 2023, two significant advancements in 2D color codes emerged. The first study [258] focused on enhancing the practicality of fault-tolerant quantum computing with color codes. It introduced innovative approaches like Pauli-based computation (PBC), transforming gate-based quantum circuits into a series of Pauli operator measurements on initial states. This technique enables parallelism in measurement layers, effectively reducing the overall measurement duration. Additionally, the study dwelt on magic state distillation, tackling the constraints of the Eastin–Knill theorem [259], which states that no QEC code can have both a universal and a transversal logical gate set. A groundbreaking development was the integration of decoding graphs with error-rate-dependent weights, improving the thresholds of triangular color codes under standard circuit-level noise models and potentially extending their applicability to other types of color codes. The second study [260] proposed a novel approach for decoding quantum color codes using the MaxSAT formulation, analogous to solving a variation of the LightsOut puzzle on the color code lattice. This strategy aids in identifying minimum-weight solutions, crucial for efficient decoding in quantum color codes. Furthermore, researchers successfully operated a 40-qubit color code processor, achieving high-fidelity state preparation and feedforward entanglement teleportation [255].

The recent advancements in QEC herald the beginning of a new era in practical QEC, marking a substantial shift from existing paradigms in quantum computing. These developments are driven by enhancements in the fabrication processes of quantum computers, reduction of environmental noise, optimization of quantum processor circuit designs, and the advancement of dynamical decoupling protocols. A pivotal goal in this journey is to attain logical error rates of 1 in 10^6 or lower per QEC cycle, a crucial benchmark for the effective use of quantum computers in impactful applications. This progress points towards a bright future in quantum computing, where the amalgamation of sophisticated QEC techniques is expected to yield more robust and scalable quantum systems, pushing the frontiers of computational capabilities and applications.

Conclusion

In this chapter, we explored the foundational principles of topological orders and the surface code, two concepts central to the development of fault-tolerant quantum computation. We began by discussing topological phases of matter, which introduced a way to characterize states of matter not by symmetry but through their topological properties. This perspective, which includes robust ground-state degeneracies and anyonic excitations, provides a framework for understanding phases such as the quantum Hall states.

The $\mathbb{Z}_2$ toric code model serves as a concrete example of a topological phase in a two-dimensional qubit lattice, illustrating key features such as ground-state degeneracy, stabilizer operations, and the role of noncontractible loops in defining logical qubits. This model is not only theoretically significant but also practically relevant for designing surface codes, which leverage topological protection to provide a resilient approach to quantum error correction. The toric code's use of anyons and their braiding statistics illustrated a path toward topological quantum computation, a paradigm distinct from the quantum circuit model.

The next chapter will build on these concepts by exploring topological quantum computation, examining how the properties of non-Abelian anyons enable computational operations through braiding and fusion. This transition marks an exciting step from the protection offered by topological orders to the computational potential of topological quantum systems.

9   Topological Quantum Computation

Our journey through the $\mathbb{Z}_2$ toric code model and surface code quantum computation in the previous chapter, particularly our examination of braiding the $\mathbb{Z}_2$ toric code anyons and the surface code defect, has paved us a way into the domain of topological quantum computation (TQC) [18, 241]. TQC stands distinct from the conventional quantum circuit model: While the latter operates with physical qubits, TQC leverages the Hilbert space of multiple non-Abelian anyons for encoding logical qubits. The quantum gates in TQC are realized through the process of braiding these non-Abelian anyons. Despite these differences, TQC also parallels the quantum circuit model, serving as an approximation where exact quantum gates from the circuit model are approached through anyon braiding with near-perfect precision.

This chapter explores the intricate world of non-Abelian anyons, with a focus on their fusion and braiding properties. An elegant proof will be presented for a theorem asserting that the number of anyon species in a topological phase is equal to the ground-state degeneracy (GSD) of the phase on the torus. Our discussion then shifts to the Fibonacci anyon [261], a quintessential example of a non-Abelian anyon that facilitates TQC. The Fibonacci anyon is noteworthy for its simplicity and power in TQC. We will expound on the defining characteristics of Fibonacci anyons, the Hilbert space constituted by three such anyons, and how their braiding can implement all single-qubit gates on a logical qubit encoded within this space.

As a historical note, although anyons can be realized as elementary excitations in topologically ordered matter phases, the concept of anyons was proposed independently of and much earlier than the concept of topological order. The possibility of particles with fractional statistics in two dimensions was first suggested by Leinaas and Myrheim (1977) [262], who showed that two-dimensional systems can admit more general particle statistics. Wilczek later coined the term "anyons" in the early 1980s [227, 263, 264, 265, 266], demonstrating how braiding in two dimensions leads to these exotic quasiparticles. Yong-Shi Wu [226, 267, 268, 269] and others [270, 271, 272, 273] subsequently expanded the theoretical framework, clarifying how anyons arise as topological excitations in low-dimensional quantum systems and further establishing their fundamental properties. In particular, Wu introduced a systematic way to treat the braid group for more than two anyons, going beyond the pairwise exchange picture and emphasizing the role of multi-anyon configuration spaces. His formulations underscored how braiding phases derive from the topological properties of worldlines in 2 + 1 dimensions, thereby solidifying the concept of anyons as genuinely new types of particles governed by nontrivial statistics.

9.1 Fusion and Braiding of Anyons

Previous chapters have shed light on the fusion of anyons in contexts like $v = 1/m$ FQH states and the $\mathbb{Z}_2$ toric code phase, which host merely Abelian anyons with a trivial Hilbert space. We also touched upon surface code defects, whose braiding properties bear resemblance to non-Abelian anyons. This section aims to provide a comprehensive understanding of the fusion and braiding of anyons, with a particular emphasis on non-Abelian types. While the terms "fusion category" and "braided fusion category" might seem daunting, they will be introduced and discussed in an approachable and practical manner, making these complex concepts accessible. This inviting introduction to categorical concepts will unfortunately lose a certain rigor that can be dispensed with for now but as necessary in more advanced studies beyond the scope of this book.

9.1.1 Fusion Rules

We have previously encountered examples of fusion rules in our study, such as $e \otimes m = \epsilon$ within the $\mathbb{Z}_2$ toric code phase. It's important to note that in fusion rules, we only consider anyon types, and a fusion process always involves two anyon types, which may or may not be distinct (e.g., $e \otimes e = 1$). In any given topological phase, there is a finite set of anyon types, which we can denote with a finite set of labels $\{a, b, c, \cdots\}$.

One of the remarkable aspects of studying anyon fusion and braiding is that reference to a specific topological phase is not necessary. The consistency conditions of fusion rules alone can determine whether a given set of labels represents a valid set of anyon types. A set of labels that satisfies these conditions forms a mathematical structure known as a fusion category. Each fusion category corresponding to anyon types is associated with a topological phase that contains exactly these anyon types. Conversely, the anyon types within a topological phase constitute a fusion category. This deep mathematical structure and its implications will be explored and unfolded in the forthcoming sections.[1]

Finiteness of Fusion Rules

For any set of anyon types

$$A = \{a, b, c, \cdots\}, \tag{9.1}$$

---

[1] More precisely speaking, the correspondence is between a topological phase and a **modular tensor category**. For simplicity and the level of this introductory chapter, we shall not bother with such mathematical depth. Having understood this chapter, interested readers are encouraged to dive deeper into the area.

which must include precisely one trivial anyon type, their fusion rules form a finite set. A typical fusion rule is expressed as

$$a \otimes b = \bigoplus_c N_{ab}^c c, \tag{9.2}$$

where $N_{ab}^c \in \mathbb{N}$, termed **fusion coefficients**, quantify the outcomes of the fusion process. In the context of anyons, fusion is inherently commutative,[2] implying $a \otimes b = b \otimes a$. Often, for a given anyon type a, the relevant fusion coefficients N_{ab}^c, considering all b and c in A, can be organized into a matrix

$$(N_a)_{\ c}^b, \quad b, c \in A, \tag{9.3}$$

known as the **fusion matrix** of a.

As mentioned earlier, the fusion rule in Equation (9.2) can be likened to spin coupling, for instance

$$\frac{1}{2} \otimes \frac{1}{2} = 0 \oplus 1, \tag{9.4}$$

where the spin $\frac{1}{2}$, spin 0, and spin 1 correspond to the lowest three irreducible representations of $SU(2)$.[3] It's important to note that $SU(2)$ has an infinite number of irreducible representations labeled by half-integers, so they do not constitute a legitimate set of anyon types. Each spin state possesses an intrinsic Hilbert space, which is nontrivial, except for spin 0. Thus, coupling two spins typically yields multiple spin states, ensuring that the dimensions on both sides of a spin coupling equation are equal. This analogy leads us to categorize anyons into two distinct classes: Abelian and non-Abelian, as defined here.

Definition 9.1. *Given a set of anyon types $A = \{a, b, c, \cdots\}$, an anyon type $a \in A$ is deemed **non-Abelian** if it meets at least one of the following two criteria for any $b \in A$:*

- *$N_{ab}^c > 1$ for some $c \in A$.*
- *There exist at least two distinct types $c, d \in A$ such that $N_{ab}^c, N_{ab}^d > 0$.*

*In contrast, a is **Abelian** if neither condition is met. It's important to note that a, b, c, and d in such fusion rules are not necessarily different. For example, in the fusion rule (8.2) of the Fibonacci anyon, $a = b = c = \tau$.*

*Consequently, a topological phase is labeled **Abelian** if it solely comprises Abelian anyons; otherwise, it is classified as **non-Abelian**.*

Exercise 9.1 *Confirm that the $\mathbb{Z}_2$ toric code anyons satisfy the criteria of being Abelian according to Definition 9.1.*

---

[2] In a general fusion category, fusion may not be commutative.

[3] Fusion coefficients are analogous to the **Kronecker coefficients** in coupling the irreducible representations of a group, also known as multiplicities. In the case of $SU(2)$, Kronecker coefficients for coupling any two irreducible representations are either 0 or 1. For larger groups like $SU(3)$, these coefficients can be greater than 1.

The definition above abstracts away the details of the internal[4] spaces of anyons. From our analysis in Example 8.1, we know that Abelian anyons must possess trivial internal spaces due to the finite nature of the set of anyon types and the uniqueness of their fusion products. Conversely, non-Abelian anyons can have nontrivial internal spaces. Nevertheless, these spaces are markedly different from a spin's internal space, which is integer-dimensional. In the spin coupling equation (9.4), we can represent the basis states of the two spin $\frac{1}{2}$ entities as superpositions of the basis states of spin 0 and spin 1, and vice versa. Nonetheless, probing the internal space of a non-Abelian anyon is generally challenging,[5] leading to its characterization as a nontrivial yet somewhat structureless space. In a fusion rule, the dimension of this internal space is of particular interest and warrants careful consideration.

Quantum Dimension: Definition

The notion of quantum dimension is pivotal in understanding the internal structure of anyons. This dimension encapsulates the size of the internal space of an anyon, but how do we define such a dimension for spaces that are elusive to direct examination? The answer lies in considering the asymptotic behavior in the thermodynamic limit:

Definition 9.2. *The **quantum dimension** d_a of an anyon type a is defined as the asymptotic dimension of the internal space of a, when the number of type a anyons on a sphere tends towards infinity. Specifically, if there are n anyons of type a, the total Hilbert space V_n formed by their fusion is a tensor product of the individual internal spaces of these n anyons. Hence, the quantum dimension is given by*

$$d_a := \lim_{n \to \infty} (\dim V_n)^{\frac{1}{n}}. \tag{9.5}$$

For a set of anyon types A, the quantity

$$D = \sqrt{\sum_{a \in A} d_a^2} \tag{9.6}$$

*is termed the **total quantum dimension** of A and is a crucial factor in the definition of topological entanglement entropy, as seen in Equation (8.49).*

The specific choice of the sphere topology in this definition will become apparent in subsequent discussions. Before probing deeper into the nature of the space V_n, it is instructive to note that the fusion rules of anyon types facilitate the straightforward computation of their respective quantum dimensions.

---

4   Referring to this space as "internal" is somewhat imprecise. As will be discussed, this space encapsulates the nonlocal nature of non-Abelian anyons and incorporates contributions from all non-Abelian anyons of the same type in the thermodynamic limit. For simplicity, we will continue to use the term "internal space."

5   In certain lattice models of non-Abelian topological phases, such as the string-net model [233, 274, 275, 276], this becomes feasible [277] by embedding an anyon's internal space in the Hilbert space of the model.

The conservation of dimensions in fusion processes, as indicated by Equation (9.2), leads to the relation

$$d_a d_b = \sum_c N_{ab}^c d_c. \tag{9.7}$$

Example 9.1

The fusion rules in Equation (8.57) for the $\mathbb{Z}_2$ toric code anyons, combined with Equation (9.7), yield

$$d_1 = d_e = d_m = d_\epsilon = 1, \tag{9.8}$$

corroborating the earlier notion that Abelian anyons have a one-dimensional internal Hilbert space. Accordingly, the total quantum dimension is $D = 2$.

Example 9.2

Consider the set of anyon types $\{1, \sigma, \psi\}$, governed by the fusion rules

$$\sigma \otimes \sigma = 1 \oplus \psi, \quad \sigma \otimes \psi = \sigma, \quad \psi \otimes \psi = 1, \tag{9.9}$$

with 1 denoting the trivial anyon type. The last fusion rule suggests that $d_\psi = 1$, as per Equation (9.7). Then, the first rule implies

$$d_\sigma^2 = d_1 + d_\psi = 2 \implies d_\sigma = \sqrt{2}, \tag{9.10}$$

an intriguingly irrational value. This characteristic underscores the non-Abelian nature of the Ising anyon type σ, with a total quantum dimension of $D = 2$.

Exercise 9.2 *Formulate the fusion coefficients for the $\mathbb{Z}_2$ toric code anyons as fusion matrices. Apply the same process to the anyon types discussed in Example 9.2.*

In Example 9.2, the concept of an irrational dimension emerges, a notion foreign to the traditional context of quantum mechanics. This dimension is not representative of a local Hilbert space but rather reflects the nonlocal essence of non-Abelian anyons. To elucidate this further, let us study the fusion space in greater detail.

9.1.2 Fusion Space and a Theorem

In this subsection, we aim to gain a comprehensive understanding of the quantum dimension, approached from both physical and mathematical perspectives, by looking deeply into the concept of fusion space. In the course of this investigation, we will prove a significant theorem mentioned earlier. This theorem establishes a profound equality between the number of distinct anyon types present in a topological phase and the GSD of that phase on the torus.

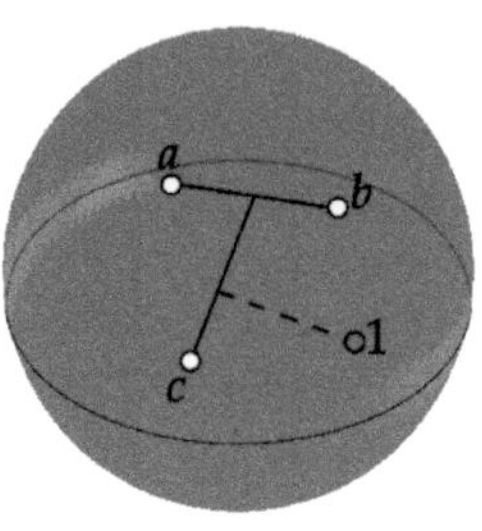

Figure 9.1 Three-punctured sphere illustrating the fusion space $V^c_{ab} = V^1_{abc}$. Each puncture, depicted as a white hole, represents an anyon. The solid lines show the fusion process of anyons a and b into c. The dashed line symbolizes a virtual fusion of a, b, and c into the trivial anyon 1, represented by a circle at the end of the line.

Fusion Space as Punctured Sphere

In discussing the fusion space of anyons, it is often useful to consider the sphere topology. This choice simplifies our analysis since every topological phase on the sphere has a unique ground state, eliminating the need to specify a particular basis ground state above which anyons are excited. This rationale underlies the mention of "sphere" in Definition 9.2 and naturally leads to the representation of anyon fusion spaces by punctured spheres.

Remember that in the $\mathbb{Z}_2$ toric code model, anyons are excited at locations where local vertex or plaquette constraints are violated. Thus, an anyon can be viewed as a singularity in the ground-state topology, analogous to a hole or puncture. When we adopt the sphere topology, an anyon is thus represented as a puncture on the sphere, making the sphere effectively a **punctured sphere**.[6]

Consider now the fusion of two anyons, a and b, which may not be of the same type. While they can fuse to produce several anyons of different types, we can concentrate on a specific subspace V^c_{ab}, corresponding to a particular fusion product c. Following the general fusion rule in Equation (9.2), the complete fusion space V_{ab} of a and b is the direct sum of the individual fusion spaces V^c_{ab} for each distinct anyon type c. Formally, this is expressed as

$$V_{ab} = \bigoplus_{c \in a \otimes b} V^c_{ab}, \tag{9.11}$$

where $c \in a \otimes b$ signifies the anyon type c appearing on the right-hand side of Equation (9.2). The space V^c_{ab} is graphically depicted as a **3-punctured sphere**, as in Figure 9.1. In this representation, a puncture where an anyon of type a is located is referred to as an a-**puncture**.

On this 3-punctured sphere, we identify exactly three nontrivial anyons, denoted as types a, b, and c. This configuration suggests that the fusion space V^c_{ab} is analogous to the space V^1_{abc}, which describes the fusion of a, b, and c into the trivial anyon 1, representing the ground state. This fusion process can be conceptualized as initially merging a and b into an intermediate anyon type, which then fuses with c to become the trivial anyon "1." Considering that the sphere only contains a, b, and c, the intermediate anyon is logically identified as c. Hence, the equivalence of $V^c_{ab} = V^1_{abc}$ is depicted by a dashed line in the figure, linking the trivial anyon "1" to the fusion path terminating at c.

---

[6] Punctured spheres are a standard tool in topological field theory.

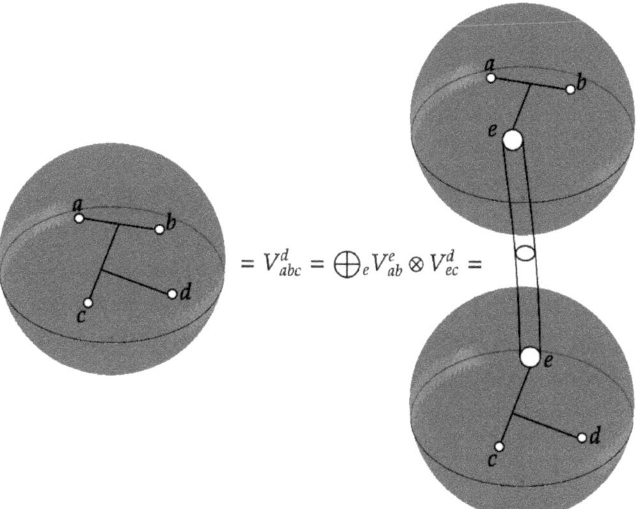

Figure 9.2 Fusion space V^d_{abc} represented as a 4-punctured sphere, corresponding to $\bigoplus_e V^e_{ab} \otimes V^d_{ec}$, visualized as the connected sum of two 3-punctured spheres.

In the context of topological phases, an anyon is often referred to as a **topological charge**, a term that generalizes the concept of charge, like e in the $\mathbb{Z}_2$ toric code phase. In this framework, a flux is also considered a topological charge. The equivalence $V^c_{ab} = V^1_{abc}$ implies that the total topological charge on a 3-punctured sphere is trivial. Similarly, two anyons, not necessarily of different types, can be the only anyons on a sphere if and only if their fusion results in the trivial anyon 1. If this is not the case, there must be at least a third anyon on the sphere. But what about having exactly one anyon on the sphere?

Suppose there is precisely one anyon, say, a, on the sphere. Since the trivial anyon 1 is always present, it must fuse with a. Nonetheless, since $1 \otimes a = a$, there must be another anyon of type a on the sphere to satisfy the fusion rule. Therefore, a 1-punctured sphere is not possible, indicating that the total topological charge of a topological phase on the sphere must be trivial!

> **Exercise 9.3** *The total topological charges on other topologies do not have to be trivial. Demonstrate that exactly one anyon can exist on the torus. Consider a specific set of anyons, such as $\{1, \sigma, \psi\}$ in Example 9.2, and think about noncontractible loops.*

Now, let's consider the dimension of V^c_{ab}. As aforementioned, a non-Abelian anyon possesses a nontrivial, yet structureless, internal space. Thus, the dimension of V^c_{ab} is determined solely by the number of ways a and b can fuse into c. Formally,

$$\dim V^c_{ab} := N^c_{ab}. \tag{9.12}$$

For spaces involving more than two anyons fusing into one type, like V^d_{abc}, we can analyze the fusion in stages. For instance, let a and b fuse into an intermediate anyon type e, which then fuses with c to form d. While a, b, c, and d are specified, e can vary according to the fusion rules of a and b. Therefore, by Equation (9.11), we have

$$V^d_{abc} = \bigoplus_{e \in a \otimes b} V^e_{ab} \otimes V^d_{ec}. \tag{9.13}$$

The fusion space V^d_{abc} can be visualized as the 4-punctured sphere shown on the left in Figure 9.2. The decomposition of V^d_{abc} in Equation (9.13) is represented by joining the two 3-punctured spheres, each symbolizing the fusion spaces V^e_{ab} and V^d_{ec}. Imagine connecting the two e-punctures with a tube and then shortening it until it disappears, resulting in the 4-punctured sphere depicted.

Exercise 9.4 *In topology, joining two punctured spheres along the boundaries of their respective punctures is known as a connected sum. To perform a connected sum of two topological spaces, one removes a disk from each space and then glues them along the boundaries of the resultant holes. Determine the connected sum of two tori. What about the connected sum of a sphere and a torus?*

GSD and Anyon Types

We are now prepared to establish a fundamental theorem relating the ground-state degeneracy (GSD) of a topological phase on the torus to the number of anyon types present in the phase.

Theorem 9.1. *The GSD of a topological phase on the torus is equal to the number of anyon types existing in the phase.*

Proof Consider Figure 9.3 illustrating our approach.

The proof employs punctured spheres. A 2-punctured sphere, with each puncture hosting an anyon a, can be seen as representing the fusion spaces V^a_{1a} and V^a_{a1}. By gluing two identical 2-punctured spheres as shown in Figure 9.3, we form a composite space:

$$\bigoplus_a V^a_{1a} \otimes V^a_{a1}. \tag{9.14}$$

On the other hand, this gluing process transforms the original spheres into a torus. Notably, on each sphere, the two anyons a are connected by a string, indicating their fusion into the trivial anyon 1. Thus, the gluing produces a noncontractible loop on the torus, each corresponding to an anyon a. This loop serves as a basis ground state in the anyon basis, as discussed in Exercise 8.7. Since the sum is taken over all possible a (direct sum), the torus with the noncontractible loop represents the ground-state Hilbert space $\mathcal{H}^0_{T^2}$ of the topological phase on the torus. Consequently,

$$\bigoplus_a V^a_{1a} \otimes V^a_{a1} \cong \mathcal{H}^0_{T^2}. \tag{9.15}$$

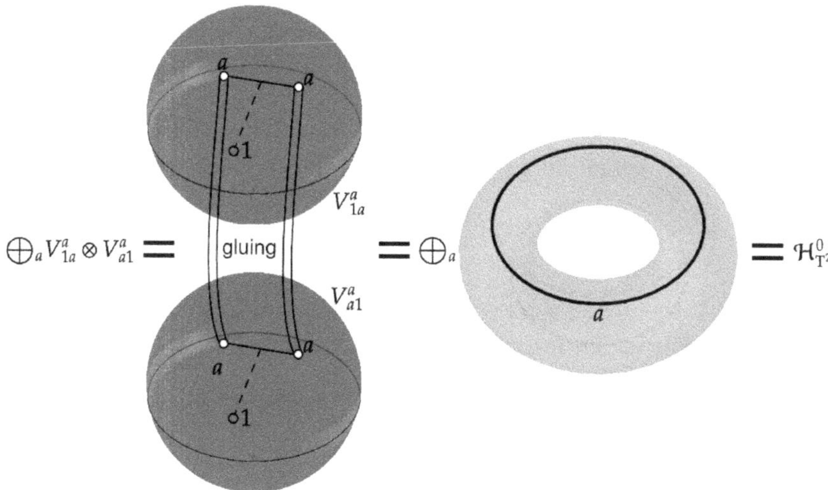

Figure 9.3 Two 2-punctured spheres, each with an anyon a, are glued together, resulting in a torus with a noncontractible loop corresponding to anyon a.

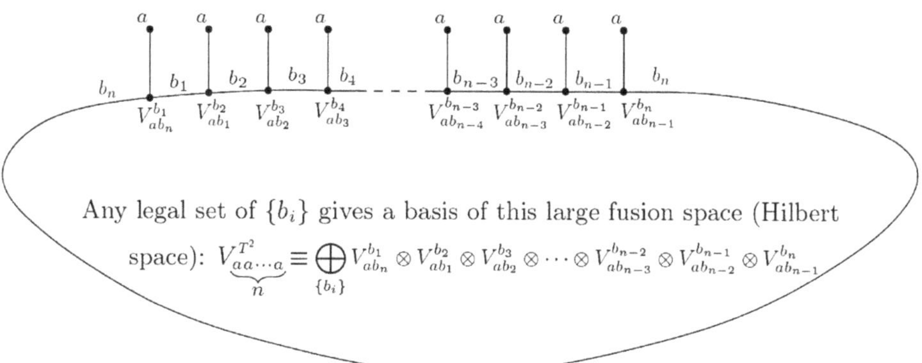

Figure 9.4 Fusion space of n anyons a on the torus, visualized as an n-punctured torus. A noncontractible loop consists of intermediate anyons b_1 through b_n collectively denoted by $\{b_i\}$.

Therefore, the dimension of the ground-state Hilbert space is

$$\dim \mathcal{H}_{T^2}^0 = \dim \bigoplus_a V_{1a}^a \otimes V_{a1}^a = \sum_a N_{1a}^a N_{a1}^a = \sum_a 1 = \# \text{ anyon types}, \qquad (9.16)$$

where $N_{1a}^a = N_{a1}^a$ by definition, and $\#$ denotes "the number of." $\qquad\square$

Quantum Dimension: Physical Meaning

The discussion of fusion spaces and punctured spheres offers insightful perspectives into the physical meaning of quantum dimension. To illustrate this more concretely, instead of an n-punctured sphere, let's consider an n-punctured torus where each puncture hosts an anyon a, as depicted in Figure 9.4.

In this configuration, a large noncontractible loop wraps around a hole of the torus. The n-punctured torus represents the space $V^{T^2}_{aa\cdots a}$, where the fusion product of n anyons a is unspecified, labeled by b_n. This label represents any outcome allowed by the fusion rules. Since only two anyons can fuse at a time, the fusion space is conceptualized as the direct sum of numerous tensor product spaces:

$$V^{T^2}_{aa\cdots a} = \bigoplus_{\{b_i\}} V^{b_1}_{ab_n} \otimes V^{b_2}_{ab_1} \otimes V^{b_3}_{ab_2} \otimes \cdots \otimes V^{b_{n-2}}_{ab_{n-3}} \otimes V^{b_{n-1}}_{ab_{n-2}} \otimes V^{b_n}_{ab_{n-1}}, \tag{9.17}$$

with b_1 through b_n representing intermediate, unspecified anyons, summed over the set $\{b_i\}$. The dimension of this space is calculated as follows:

$$\begin{aligned}
\dim V^{T^2}_{aa\cdots a} &= \sum_{\{b_i\}} \dim V^{b_1}_{ab_n} \times \dim V^{b_2}_{ab_1} \times \dim V^{b_3}_{ab_2} \times \cdots \times \dim V^{b_{n-2}}_{ab_{n-3}} \\
&\quad \times \dim V^{b_{n-1}}_{ab_{n-2}} \times \dim V^{b_n}_{ab_{n-1}} \\
&= \sum_{\{b_i\}} N^{b_1}_{ab_n} \times N^{b_2}_{ab_1} \times N^{b_3}_{ab_2} \times \cdots \times N^{b_{n-2}}_{ab_{n-3}} \times N^{b_{n-1}}_{ab_{n-2}} \times N^{b_n}_{ab_{n-1}}.
\end{aligned} \tag{9.18}$$

In this formulation, the dimensions of spaces in a direct sum are added, while those in a tensor product are multiplied. This gives us a sum of products of fusion coefficients over the intermediate anyons $\{b_i\}$. Although initially complex, this can be simplified by reinterpreting the fusion coefficients as matrix elements of fusion matrices:

$$\dim V^{T^2}_{aa\cdots a} = \sum_{\{b_i\}} (N_a)^{b_1}_{b_n} (N_a)^{b_2}_{b_1} (N_a)^{b_3}_{b_2} \cdots (N_a)^{b_{n-2}}_{b_{n-3}} (N_a)^{b_{n-1}}_{b_{n-2}} (N_a)^{b_n}_{b_{n-1}} = \mathrm{Tr}(N_a)^n, \tag{9.19}$$

where the result is the trace of the n-th power of the fusion matrix N_a. In the thermodynamical limit ($n \to \infty$), the trace is dominated by the largest eigenvalue of N_a, denoted as d_a. Hence, in the limit as $n \to \infty$,

$$\dim V^{T^2}_{aa\cdots a} \to d^n_a. \tag{9.20}$$

The LHS is an integer, being the sum of products of integer fusion coefficients. Consequently, the RHS matches this integer, although d_a itself may not be an integer. It is precisely the quantum dimension d_a, as defined in (9.5).

Exercise 9.5 *Using the results from Exercise 9.2, demonstrate that $d_\sigma = \sqrt{2}$, as the largest eigenvalue of the fusion matrix N_σ.*

Now you can see that although the total Hilbert space is integer-dimensional, it cannot be divided as n local Hilbert spaces respectively attributable to the n anyons a, as a local Hilbert space must be integer-dimensional. Therefore, the quantum dimension d_a indeed reflects the nonlocal nature of the non-Abelian anyon a; it is the asymptotic degeneracy per anyon a in the thermodynamical limit.

Exercise 9.6 *One can also identify the largest eigenvalue of the fusion matrix of a non-Abelian anyon a with its quantum dimension d_a by examining the fusion space of n anyons a on the sphere. Let's consider the set of anyon types $\{1, \tau\}$, with the fusion rules*

$$1 \otimes \tau = \tau, \quad \tau \otimes \tau = 1 \oplus \tau. \tag{9.21}$$

First, write down the fusion matrix N_τ of τ. Compute the largest eigenvalue d_τ of N_τ. Then study the fusion space $V^1_{\tau\tau\cdots\tau}$ represented by the n-punctured sphere in Figure 9.5 and show that in the limit $n \to \infty$,

$$\lim_{n\to\infty} \dim V^1_{\tau\tau\cdots\tau} = d_\tau^n. \tag{9.22}$$

Note that unlike the case of the torus, $\dim V_{\tau\tau\cdots\tau}$ is not a trace!

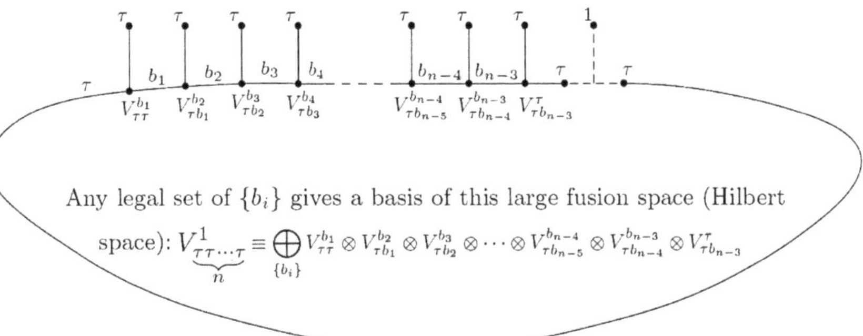

Figure 9.5 Fusion space of n anyons a on the sphere, represented as an n-punctured sphere. Intermediate anyons b_1 through b_{n-3} are collectively denoted by $\{b_i\}$.

Exercise 9.7 *As an analogy, consider the coupling of spin halves. For example,*

$$\frac{1}{2} \otimes \frac{1}{2} = 0 \oplus 1. \tag{9.23}$$

*The total tensor product space is four-dimensional. The one-dimensional subspace $V^0_{\frac{1}{2}\frac{1}{2}}$ is termed an **intertwiner space**. What's the dimension of the intertwiner space $V^0_{\frac{1}{2}\frac{1}{2}\frac{1}{2}}$? What about $V^0_{\frac{1}{2}\frac{1}{2}\frac{1}{2}\frac{1}{2}}$? Can you generalize your result to the intertwiner space of n spin halves?*

9.1.3 Fusion Category

Discerning readers might have wondered whether fusion is associative, an aspect we have implicitly assumed so far. Fusion, indeed, is associative. Given a set A of anyon types, for any $a, b, c \in A$, the associativity can be expressed as

$$(a \otimes b) \otimes c \cong a \otimes (b \otimes c), \tag{9.24}$$

where the symbol $\cong$ denotes an isomorphism, rather than an equality $=$. This distinction is crucial because both sides of the equation are linear spaces. Two isomorphic linear spaces allow for linear transformations that map a basis of one space to a basis of the other.

To define these basis transformations, we introduce a graphical method called the **fusion-tree basis**. This is a visual and intuitive means to represent the bases of fusion spaces, elucidating the underlying associative property of fusion in a more accessible manner. The fusion-tree basis is detailed as follows.

By the generic fusion rule (9.2), the basis states of the fusion space V_{ab} can be depicted as

$$\tag{9.25}$$

where c is an anyon type resulting from the fusion of a and b, denoted as $c \in a \otimes b$. We then denote the space $(a \otimes b) \otimes c$ as $V_{(ab)c}$, where the parentheses indicate the order of fusion. Similarly, $V_{a(bc)}$ represents the space for $a \otimes (b \otimes c)$. Since $V_{(ab)c}$ is a direct sum,

$$V_{(ab)c} = \bigoplus_d V^d_{(ab)c}, \tag{9.26}$$

we can consider each subspace $V^d_{(ab)c}$ individually. Within a $V^d_{(ab)c}$, the fusion-tree basis states are illustrated as

$$\tag{9.27}$$

where m is an intermediate anyon type in the fusion of a and b, indicated as $m \in a \otimes b$. Correspondingly,

$$\tag{9.28}$$

where $n \in b \otimes c$, represents the fusion-tree basis states of $V^d_{a(bc)}$. The isomorphism (9.24) implies that there is a linear transformation between the bases (9.27) and (9.28):

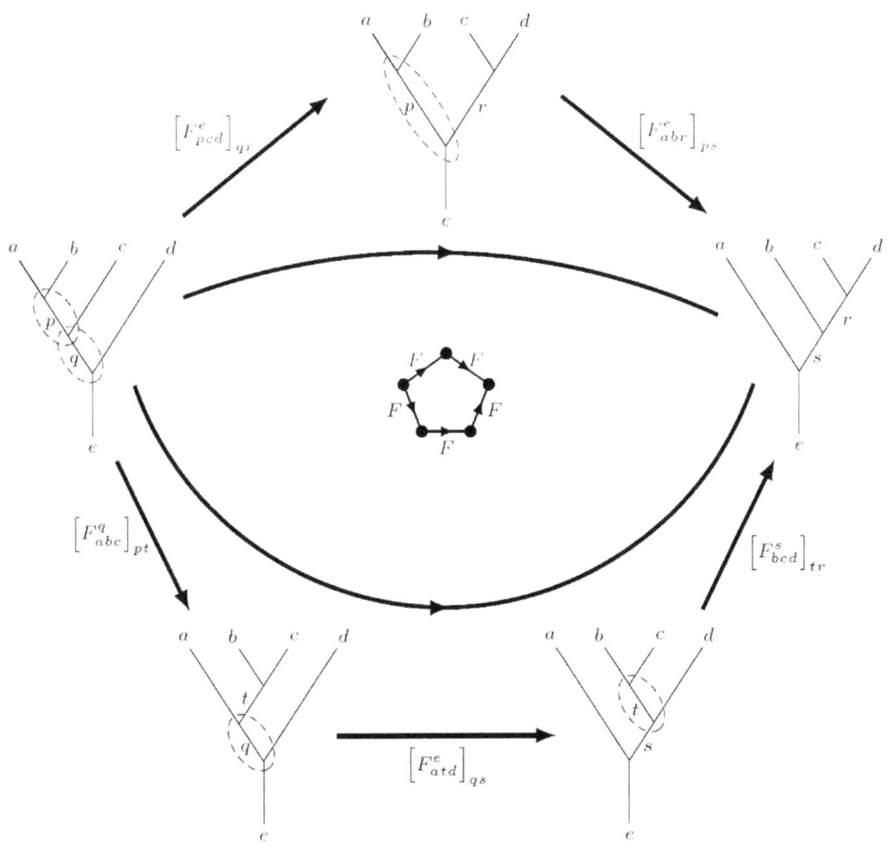

Figure 9.6 Graphical representation of the pentagon identity.

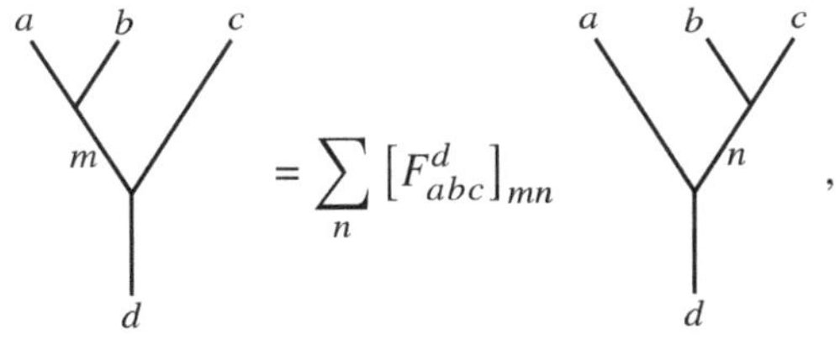

(9.29)

for any given $m \in a \otimes b$. The coefficients $[F^d_{abc}]_{mn}$ in this transformation are known as **F-symbols** and can be thought of as the elements of the matrix $[F^d_{abc}]$. This matrix should be unitary, as required by the isomorphism.

You see, in defining the basis transformation (9.29), the actual values of the F-symbols are initially unknown; they are placeholders for the coefficients that will be determined based on a given set A of anyon types and their fusion rules. Even so, it's important to note that such a basis transformation may not always be possible for arbitrarily chosen A and fusion rules. This limitation arises because the associative property (9.24) imposes stringent constraints on the permissible fusion rules for A.

To elucidate these constraints, consider the fusion of four anyons of types a, b, c, and d. Associativity dictates that

$$((a \otimes b) \otimes c) \otimes d \cong a \otimes (b \otimes (c \otimes d)). \quad (9.30)$$

Specifically, the fusion space of a, b, c, and d is a direct sum of spaces, each with a distinct fusion product. Consequently, the isomorphism (9.24) implies an isomorphism

$$V^e_{((ab)c)d} \cong V^e_{a(b(cd))},\tag{9.31}$$

where the subscripts indicate the order of fusion and e is any allowable fusion product of a, b, c, and d. Therefore, a fusion-tree basis state in $V^e_{((ab)c)d}$ can be transformed into a basis state in $V^e_{a(b(cd))}$, as depicted in Figure 9.6. Intriguingly, there are two different routes for transforming a basis state from $V^e_{a(b(cd))}$ to $V^e_{((ab)c)d}$, as illustrated in the figure. These two transformation routes should yield the same result, given their common start and end points. This leads to a crucial equality, known as the **pentagon identity**:

$$\left[F^e_{pcd}\right]_{qr}\left[F^e_{abr}\right]_{ps} = \sum_t \left[F^q_{abc}\right]_{pt}\left[F^e_{atd}\right]_{qs}\left[F^s_{bcd}\right]_{tr}.\tag{9.32}$$

The pentagon identity is named for the pentagon shape formed by the two routes of basis transformations, as clearly seen in the figure. This identity represents the sought-after constraints imposed by the associativity (9.24).

> **Exercise 9.8** *Observe that the direct comparison of the two transformation routes in Figure 9.6 involves more summations than the single summation in Equation (9.32). By eliminating the identical summations on both sides of the equality, we arrive at the pentagon identity (9.32). Confirm that this process of removing summations indeed results in (9.32).*

It is crucial to recognize that the pentagon identity is not a singular equation but a collection of nonlinear equations. Each equation corresponds to a different choice of intermediate anyon types. Therefore, when given a set A of anyon types and their corresponding fusion rules, one can evaluate whether they satisfy the pentagon identity. If the identity can be resolved with nonvanishing F-symbols, then A, along with its fusion rules, constitutes a legitimate **fusion category**.[7] If they fail to fulfill the pentagon identity, they do not form a fusion category, indicating that the fusion rules are ill-defined. It's important to note that for a given set A and associated fusion rules, there may be multiple sets of F-symbols that satisfy the pentagon identity. A set of fusion rules is considered consistent if there is at least one solution to the pentagon identity – a set of nonzero F-symbols. In summary:

$$\left.\begin{array}{l} \text{A set of anyon types} \\ + \text{ a set of consistent fusion rules} \\ + \text{ a solution to the pentagon identity} \end{array}\right\} = \quad \text{a fusion category.}\tag{9.33}$$

---

[7] In mathematical terms, the anyon types in A are referred to as the **representative simple objects** of the fusion category. In addition, the rigorous definition of fusion category involves a few extra mathematical conditions, which are not important for our purposes here.

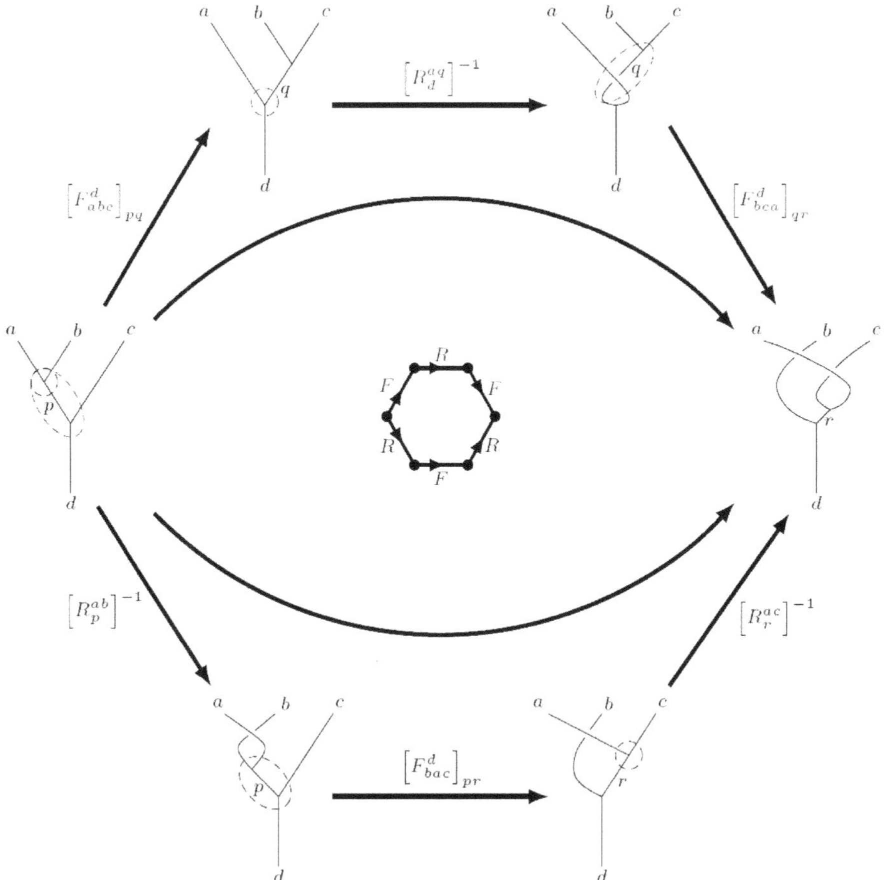

Figure 9.7 Graphical representation of the hexagon identity.

hexagon identity thus serves as a compatibility condition between the R-symbols and F-symbols. In other words, if the hexagon identity can be satisfied with a given set of F-symbols from the fusion category, then the braiding structure is compatible with it. When this compatibility is achieved, the fusion category combined with the compatible braiding structure forms a **braided fusion category**:

$$\left.\begin{array}{l} \text{A fusion category} \\ + \text{ a compatible braiding structure} \end{array}\right\} = \text{ a braided fusion category.} \quad (9.41)$$

> **Exercise 9.10** *Confirm that the diagram in Figure 9.7 indeed results in the hexagon identity* (9.40).

In Equation (9.38), which defines the R-symbol, if we depict the worldline of anyon a passing beneath that of anyon b, we conventionally have

 $= R_c^{ab*}$ 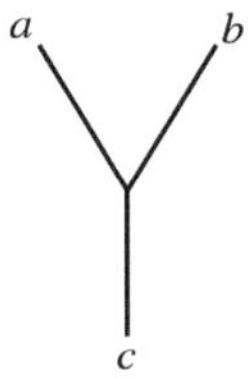 $= R_c^{ab^{-1}}$ 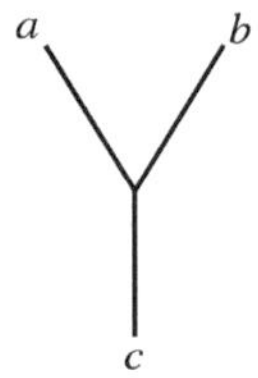 . (9.42)

This introductory exploration of braided fusion categories sets the stage for further discourse on topological quantum computation in the subsequent sections.

> **Exercise 9.11** *For each fusion category you found in Exercise 9.9, check if it can be promoted to be a braided fusion category by solving the hexagon identity* (9.40).

9.2 Fibonacci Anyons and Universal TQC

Universal topological quantum computation can be achieved using various types of non-Abelian anyons. Among these, Fibonacci anyons stand out as the simplest viable candidates for this purpose.

9.2.1 Fibonacci Anyons

As explored in Exercise 9.9, the fusion category of Fibonacci anyons, henceforth denoted by τ, consists of the anyon types $\{1, \tau\}$. The fusion rules for these anyons are given by

$$\tau \otimes \tau = 1 \oplus \tau, \quad 1 \otimes \tau = \tau, \tag{9.43}$$

where 1 represents the trivial anyon. These fusion rules confirm the non-Abelian nature of the Fibonacci anyon.

According to fusion rules (9.43), the Hilbert space associated with n Fibonacci anyons (τ) on the torus grows as follows:

$$\begin{aligned} \tau \qquad & \dim \mathcal{H}_\tau = 1 \\ \tau \otimes \tau = 1 \oplus \tau \qquad & \dim \mathcal{H}_{2\tau} = 3 \\ \tau \otimes \tau \otimes \tau = 1 \oplus \tau \oplus \tau \qquad & \dim \mathcal{H}_{3\tau} = 4 \\ \tau \otimes \tau \otimes \tau \otimes \tau = 1 \oplus 1 \oplus \tau \oplus \tau \oplus \tau \qquad & \dim \mathcal{H}_{4\tau} = 7 \\ \vdots \qquad & \\ & \dim \mathcal{H}_{n\tau} = L(n), \end{aligned} \tag{9.44}$$

where $L(n)$ is the n-th **Lucas number**. It is known that as $n \to \infty$, $L(n) \to (\sqrt{5} + 1/2)^n$.

$$L(n) \to \left(\frac{\sqrt{5}+1}{2} \right)^n . \tag{9.45}$$

The irrational number on the RHS is the quantum dimension of the Fibonacci anyon. If you have completed Exercise 9.6, you should have found that

$$d_\tau = \frac{\sqrt{5}+1}{2},\tag{9.46}$$

which is the largest eigenvalue of the fusion matrix N_τ.

Since the total topological charge on the sphere must be trivial (1), as indicated in Exercise 9.6, the space of n Fibonacci anyons on the sphere, denoted as $V_{n\tau}^1$, is merely a subspace of $V_{n\tau}$, the total space for n τ's on the torus. Therefore, for any finite n, $\dim V_{n\tau}^1 < \dim V_{n\tau}$; however, in the limit as $n \to \infty$, both the total dimension on the sphere and that on the torus are dominated by the quantum dimension d_τ.

> **Exercise 9.12** *Enhance your understanding of the space of multiple non-Abelian anyons by considering a sphere on which the total topological charge of multiple Fibonacci anyons can be τ. This scenario is equivalent to the plane, where a Fibonacci anyon is placed at infinity. Calculate the dimensions for $n = 1, 2, 3, 4,$ and 5 Fibonacci anyons (excluding the one at infinity). Can you derive a closed-form formula for the dimension as a function of n, the number of Fibonacci anyons on the plane? This result relates to the concept of the **generalized Pauli exclusion principle** [272, 269, 278, 279].*

9.2.2 The Fibonacci Category

The Fibonacci category is a braided fusion category whose set of simple objects is $\{1, \tau\}$. Fusion rules (9.43) are associative and thus should comply with the pentagon identity (9.32), whose solution is a set of F-symbols.

F-Symbols of the Fibonacci Anyon

Since we will use the fusion-basis states of the Fibonacci anyon often in what follows, let's denote them in a more convenient way: For two Fibonacci anyons,

$$\longmapsto |(\tau\ \tau)_1\rangle,\tag{9.47}$$

$$\longmapsto |(\tau\ \tau)_\tau\rangle,\tag{9.48}$$

where the subscripts indicate the fusion product of the two Fibonacci anyons. For three Fibonacci anyons, likewise,

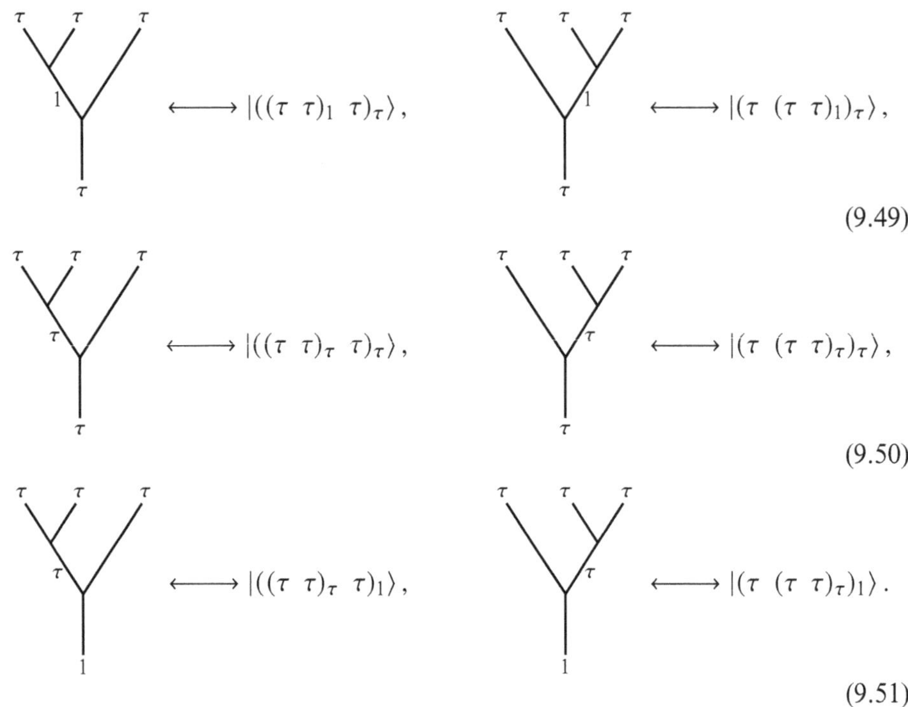

$$|((\tau\ \tau)_1\ \tau)_\tau\rangle, \qquad\qquad |(\tau\ (\tau\ \tau)_1)_\tau\rangle, \tag{9.49}$$

$$|((\tau\ \tau)_\tau\ \tau)_\tau\rangle, \qquad\qquad |(\tau\ (\tau\ \tau)_\tau)_\tau\rangle, \tag{9.50}$$

$$|((\tau\ \tau)_\tau\ \tau)_1\rangle, \qquad\qquad |(\tau\ (\tau\ \tau)_\tau)_1\rangle. \tag{9.51}$$

The transformation between the two sets of fusion-tree basis states above define the F-symbols:

$$|(\tau\ (\tau\ \tau)_i)_k\rangle = \sum_j \left[F^k_{\tau\tau\tau}\right]_{ij} |((\tau\ \tau)_j\ \tau)_k\rangle, \tag{9.52}$$

where $i, j, k \in \{1, \tau\}$. In Exercise 9.9, you should have obtained the nonzero F-symbols

$$F^1_{\tau\tau\tau} = 1,$$

$$\left[F^\tau_{\tau\tau\tau}\right] = \begin{pmatrix} F_{11} & F_{1\tau} \\ F_{\tau 1} & F_{\tau\tau} \end{pmatrix} = \begin{pmatrix} d_\tau^{-1} & \sqrt{d_\tau^{-1}} \\ \sqrt{d_\tau^{-1}} & -d_\tau^{-1} \end{pmatrix}. \tag{9.53}$$

Here, the first row is merely a number rather than a matrix because when $k = 1$ in Equation (9.52), only $i = j = \tau$ is possible.

R-Symbols of the Fibonacci Anyon

The braiding structure of the Fibonacci category is defined by

$$= R \,|(\tau\ \tau)_1\rangle = R_1^{\tau\tau}\,|(\tau\ \tau)_1\rangle,$$

$$\tag{9.54}$$

$$= R \,|(\tau\ \tau)_\tau\rangle = R_1^{\tau\tau}\,|(\tau\ \tau)_\tau\rangle.$$

The hexagon identity (9.40) in the case of the Fibonacci anyon reduces to

$$R_p[F]_{pr}R_r = [F]_{p1}[F]_{1r} + [F]_{p\tau}R_\tau[F]_{\tau r},\tag{9.55}$$

where some fixed indices are omitted to reduce clutter. Solving this hexagon identity results in

$$R = \begin{pmatrix} \mathrm{e}^{-\mathrm{i}\frac{4\pi}{5}} & 0 \\ 0 & -\mathrm{e}^{-\mathrm{i}\frac{2\pi}{5}} \end{pmatrix},\tag{9.56}$$

which acts on the basis states

$$\begin{pmatrix} |(\tau\ \tau)_1\rangle \\ |(\tau\ \tau)_\tau\rangle \end{pmatrix}.\tag{9.57}$$

> **Exercise 9.13** *Verify the reduced hexagon identity (9.55) and the solution (9.56).*

For your information, the topological spin of the Fibonacci anyon can be obtained from the R-symbols as

$$h_\tau = \frac{2}{5},\tag{9.58}$$

indicating that τ is neither a fermion nor a boson.

9.2.3 Braid Groups

As we prepare to exploit the braiding of Fibonacci anyons for quantum computation, it is pertinent to introduce the underlying mathematical principles of braids, especially for those who may be unfamiliar with this topic. Braids, commonly known through their practical application in hairstyles (e.g., in Figure 9.8), possess a rich mathematical structure known as the braid group. Although braids have been a part of human culture for millennia, their formal mathematical study was initiated by Emil Artin in 1925, when he published a seminal paper on the subject [280].

Figure 9.8 Example of a hair braid.

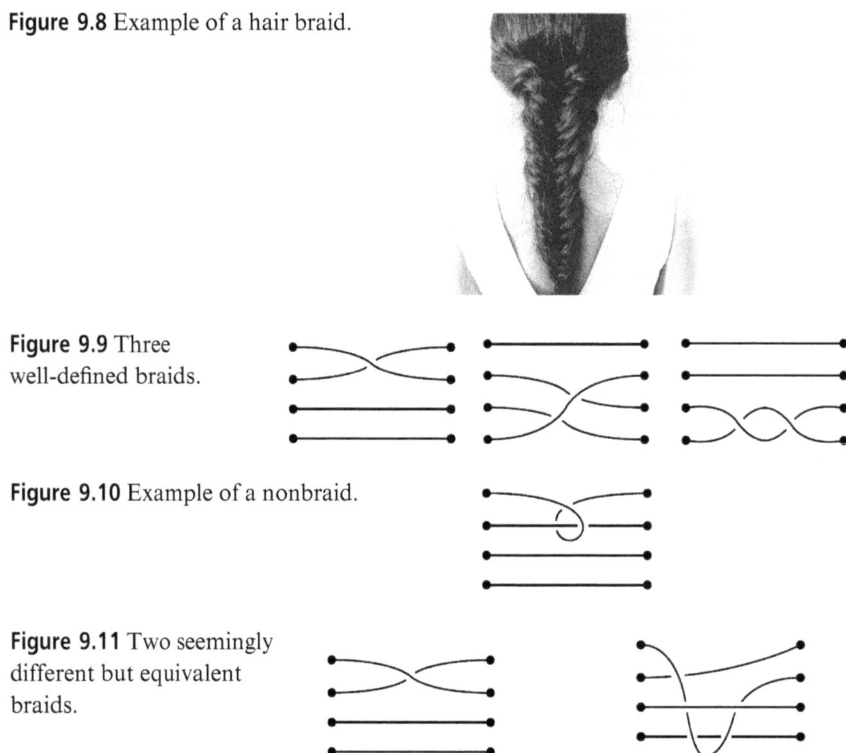

Figure 9.9 Three well-defined braids.

Figure 9.10 Example of a nonbraid.

Figure 9.11 Two seemingly different but equivalent braids.

The set of all possible braids with a fixed number of strands constitutes a group in mathematical terms. To understand this braid group, let's first define what constitutes a braid. We will approach this in a practical, rather than abstract, manner. An n-strand braid is composed of n strands that start parallel to each other and extend in one direction (either horizontally or vertically). These strands are allowed to cross over each other but must avoid self-crossing. For example, the three braids in Figure 9.9 are well-defined, whereas the braid in Figure 9.10 is not a valid braid due to the self-crossing of the top strand.

It's important to note that some braids may look different but are, in fact, equivalent. Consider the two braids in Figure 9.11. In the left braid, if you pull the strand connecting the top left point to the second-to-top right point, making it pass between the third and fourth strands, you end up with the braid on the right. This pulling action, which doesn't introduce any new actual crossings, results in a braid that looks different but is essentially the same. Such manipulations are referred to as braid **isotopies**, an equivalence relation. Therefore, n-strand braids are defined up to braid isotopy, allowing for a certain degree of flexibility in their presentation.

It becomes clear upon examination that all n-strand braids, considered up to isotopy, constitute a group known as the **braid group** B_n. This group is characterized by the following three properties:

- **Group multiplication (concatenation)**: Concatenating any two n-strand braids results in another n-strand braid. Consider, for example, the concatenation of two four-strand braids in Figure 9.12. This operation, as the group multiplica-

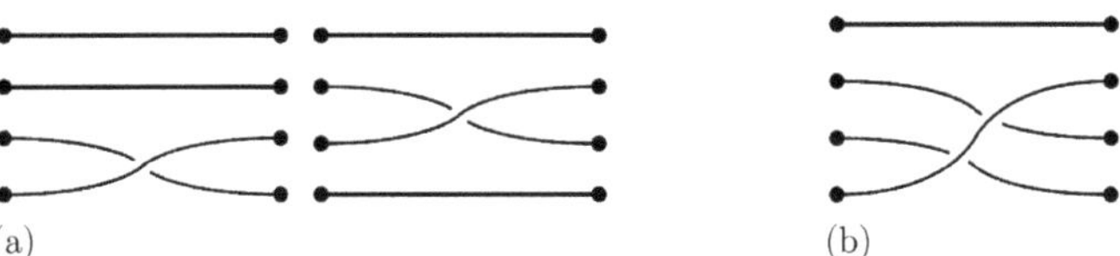

Figure 9.12 (a) Concatenating two braids results in (b) another braid.

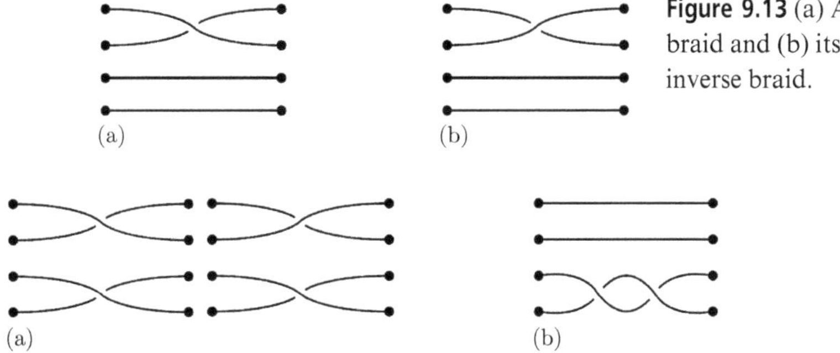

Figure 9.13 (a) A braid and (b) its inverse braid.

Figure 9.14 (a) Concatenating the two braidings and (b) the resulting braid. Two crossings inverse to each other are cancelled via concatenation.

tion, is evidently noncommutative. Hence, B_n is a non-Abelian group.

- **Identity element**: The identity element in B_n is represented by the **trivial braid**, consisting of n parallel strands without any crossings.
- **Inverse elements**: The inverse of a crossing is realized by swapping the two crossed strands. For example, The two braids in Figure 9.13 are inverses of each other: This is because their concatenation results in the trivial four-strand braid. Figure 9.14 is an illustration: The top two crossings in part (a) are inverses of each other, while the lower two crossings are identical.

The structure and properties of the braid group B_n are crucial for understanding the dynamics of braiding, especially in the realm of quantum computation. This understanding is vital when manipulating anyons and exploring their topological properties.

It should be apparent that B_n is a discrete yet infinite group for any $n > 2$. This leads to an interesting consideration:

Exercise 9.14 *What about the two-strand braid group B_2?*

To rigorously define the n-strand braid group, one must consider braid isotopy. Since each crossing involves only two neighboring strands, any n-strand braid can be constructed as a concatenation of several basic n-strand braids, each containing exactly one crossing. These basic elements are known as **braid generators**.

For n strands, let's arrange the strands vertically and number them from left to right as 1 through n, representing the identity or trivial braid, depicted in Figure 9.15.

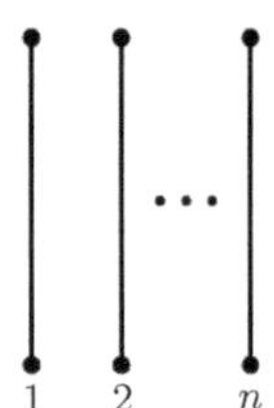

Figure 9.15 Trivial/identity braid.

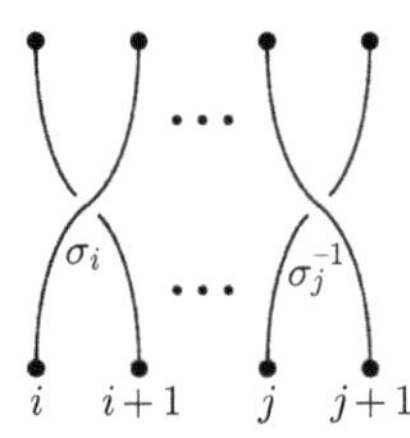

Figure 9.16 A braid group generator and its inverse.

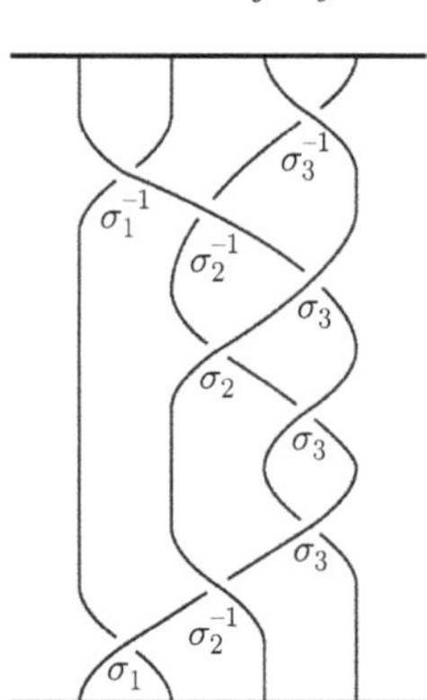

Figure 9.17 Example of a four-strand braid with nine crossings.

There are $n - 1$ braid generators, labeled as σ_1 through σ_{n-1}. The generator σ_i signifies the crossing of the i-th strand over the $(i + 1)$-th strand, with its inverse involving the $(i + 1)$-th strand crossing over the i-th strand, as seen in Figure 9.16.

By vertically concatenating these generators and their inverses, we can construct any n-strand braid. In such a braid, it is important to note that no two crossings occur at the same height. That is, at the height of any given crossing, there are no other crossings horizontally. For instance, consider the four-strand braid in Figure 9.17.

This braid can be interpreted as a sequence of generators and their inverses:

$$\sigma_1\sigma_2^{-1}\sigma_3\sigma_3\sigma_2\sigma_3\sigma_2\sigma_2^{-1}\sigma_1^{-1}\sigma_3^{-1}. \tag{9.59}$$

The exploration of the braid group B_n and its generators lays a critical foundation for understanding braiding processes in quantum computation, particularly in the manipulation and interaction of anyons. The concept of braid isotopy can be elucidated through two key relations.

First, if two crossings σ_i and σ_j involve distinct strands (i.e., $|i - j| > 1$), their relative vertical order does not affect the braid, implying they commute, as shown in Figure 9.18.

Second, one may freely pull a strand if this action does not create any nontrivial crossings. Essentially, pulling the strand back merely reverses the crossings created.

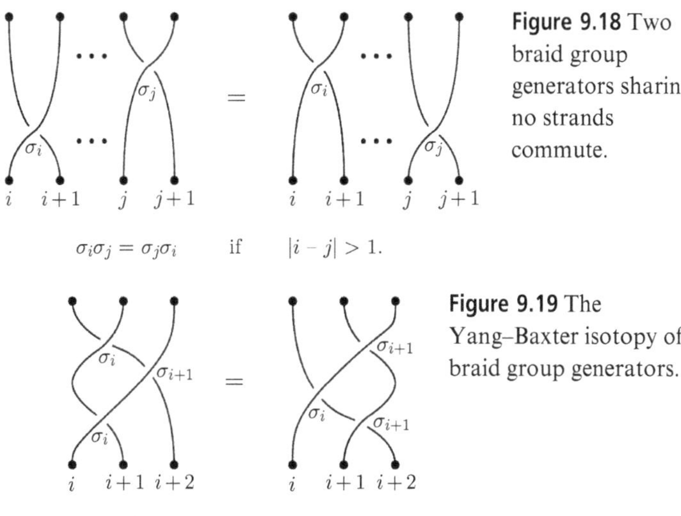

Figure 9.18 Two braid group generators sharing no strands commute.

$$\sigma_i \sigma_j = \sigma_j \sigma_i \quad \text{if} \quad |i - j| > 1.$$

Figure 9.19 The Yang–Baxter isotopy of braid group generators.

$$\sigma_i \sigma_{i+1} \sigma_i = \sigma_{i+1} \sigma_i \sigma_{i+1}.$$

This principle is best illustrated with a three-strand scenario in Figure 9.19. The pivotal relation derived from this isotopy, as shown in the figure, is

$$\sigma_j \sigma_{i+1} \sigma_i = \sigma_{i+1} \sigma_i \sigma_{i+1}, \tag{9.60}$$

which is a specific case of the renowned **Yang–Baxter equation** [281, 282].

Exercise 9.15 *Utilize braid isotopy to simplify the four-strand braid in Figure 9.17, effectively reducing the number of crossings.*

With these principles, we can formally define the braid group B_n using its presentation:

$$B_n := \left(\sigma_1, \sigma_2, \cdots, \sigma_{n-1} \mid \sigma_i \sigma_j = \sigma_j \sigma_i, \forall |i - j| > 1; \ \sigma_i \sigma_{i+1} \sigma_i = \sigma_{i+1} \sigma_i \sigma_{i+1}, \forall i \right). \tag{9.61}$$

In essence, B_n is generated by arbitrary products of the $n-1$ braid generators and/or their inverses, subject to two fundamental relations: commutativity and the Yang–Baxter equation.

Physicists often focus not just on a group itself, but on its representations, particularly within linear (Hilbert) spaces. A question then arises: On which Hilbert spaces should we represent braid groups? In the context of braiding Fibonacci anyons, which form a fusion space, the action of braiding induces linear transformations on this space. These transformations, as suggested by the hexagon identity (9.40), encode the representation of the braiding. Thus, braids and braid groups act on Fibonacci anyon spaces via their representations.

In subsequent discussions, we will focus on braiding three Fibonacci anyons. As three Fibonacci anyons create a three-dimensional Hilbert space, we aim to derive the three-dimensional representation of the braid group B_3 in this space. It will be shown that quantum computation emerges from the braiding within this three-dimensional setting.

9.2.4 Braiding Three Fibonacci Anyons

To proceed, we will utilize the 3τ basis states as defined in the left part of Equation (9.52), but with further renaming for clarity:

$$\begin{aligned}
|0\rangle &:= |((\tau\ \tau)_1\ \tau)_\tau\rangle, \\
|1\rangle &:= |((\tau\ \tau)_\tau\ \tau)_\tau\rangle, \\
|N\rangle &:= |((\tau\ \tau)_\tau\ \tau)_1\rangle.
\end{aligned} \tag{9.62}$$

The third basis state is denoted as $|N\rangle$ for a specific reason. Can you guess why?

The first two basis states, $|0\rangle$ and $|1\rangle$, span a two-dimensional subspace $V_{3\tau}^\tau$ of the total three-dimensional space $V_{3\tau}$, whereas $|N\rangle$ spans the one-dimensional subspace $V_{3\tau}^1$. Astute readers may recognize that the two subspaces $V_{3\tau}^\tau$ and $V_{3\tau}^1$ are orthogonal due to their correspondence to different fusion products, τ and 1, of the three Fibonacci anyons τ (recall the use of $\oplus$ in the fusion rules). This orthogonality inspires us to consider $V_{3\tau}^\tau$ as the Hilbert space of a logical qubit, with $|0\rangle$ and $|1\rangle$ as the computational basis states. Therefore, $|N\rangle$ represents a noncomputational state. Did your guess align with this explanation?

Yet, its mere orthogonality to $V_{3\tau}^1$ does not fully justify $V_{3\tau}^\tau$ as a well-defined logical space. We must also ensure that this orthogonality is preserved under single-qubit quantum gate operations. As braiding is our primary means of manipulating Fibonacci anyons, we seek to realize single-qubit gates on the logical qubit $V_{3\tau}^\tau$ by braiding the three Fibonacci anyons. This is feasible only if the subspace $V_{3\tau}^\tau$ remains invariant under the braiding of the three Fibonacci anyons; in other words, the braiding should not cause any mixing of the subspaces $V_{3\tau}^\tau$ and $V_{3\tau}^1$. To verify whether this expectation holds true, we need to represent the braid group B_3 on the three-dimensional space $V_{3\tau}$ and determine if this three-dimensional representation can be decomposed into a two-dimensional representation on $V_{3\tau}^\tau$ and a one-dimensional representation on $V_{3\tau}^1$.

To this end, we need to derive the representation matrices for the two generators of B_3. As we arrange the three Fibonacci anyons horizontally on a plane from left to right, the two generators σ_1 and σ_2 can be depicted as in Figure 9.20.

For the case of three Fibonacci anyons, the braid group B_3 is presented by

$$B_3 := \langle \sigma_1, \sigma_2 \mid \sigma_1\sigma_2\sigma_1 = \sigma_2\sigma_1\sigma_2 \rangle, \tag{9.63}$$

where the commutativity relation is not applicable, given that there are only two generators.

It's important to remember that in our chosen basis for $V_{3\tau}$, the leftmost two Fibonacci anyons fuse first, as indicated in Figure 9.20. Also, recall that the R-matrix for the Fibonacci category acts on the fusion of two Fibonacci anyons. Consequently, in this chosen basis, the action of braid generator σ_1 on $V_{3\tau}$ aligns with that of the R-matrix. This insight, combined with Equation (9.56), yields the

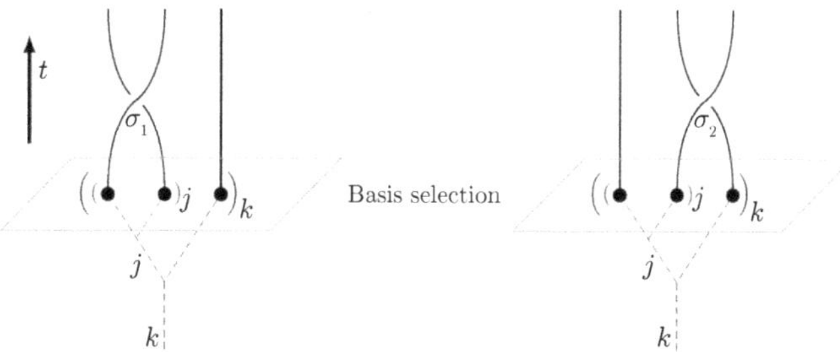

Figure 9.20 Two generators σ_1 and σ_2 of the braid group involving the worldlines of three Fibonacci anyons, along with the chosen basis. The three bullets denote the three Fibonacci anyons.

representation matrix $\rho(\sigma_1)$ for generator σ_1, along with its action on the three basis states:

$$\rho(\sigma_1)\begin{pmatrix}|0\rangle := & |((\tau\ \tau)_1\ \tau)_\tau\rangle \\ |1\rangle := & |((\tau\ \tau)_\tau\ \tau)_\tau\rangle \\ |N\rangle := & |((\tau\ \tau)_\tau\ \tau)_1\rangle\end{pmatrix} = \begin{pmatrix} e^{-\hat{\imath}\frac{4\pi}{5}} & 0 & 0 \\ 0 & -e^{-\hat{\imath}\frac{2\pi}{5}} & 0 \\ 0 & 0 & -e^{-\hat{\imath}\frac{2\pi}{5}}\end{pmatrix}\begin{pmatrix}|0\rangle \\ |1\rangle \\ |N\rangle\end{pmatrix}. \tag{9.64}$$

To derive the representation matrix $\rho(\sigma_2)$ of the second generator, σ_2, we face a more complex task. This is because σ_2 braids the worldlines of the right two Fibonacci anyons in Figure 9.20, which do not directly fuse in the chosen basis, as illustrated in Figure 9.20. To resolve this, a dedicated reader might deduce the following approach: Transform the basis using F-symbols to a basis where the right two Fibonacci anyons fuse first, perform the braiding, and then transform back to the original basis. Formally, this is represented as

$$\rho(\sigma_2) = F^{-1}RF. \tag{9.65}$$

To illustrate this process, let's consider the basis state $|0\rangle$. Utilizing Equations (9.52) and (9.53), we find:

$$|0\rangle = \sum_{j=1}^{\tau} \left[F^\tau_{\tau\tau\tau}\right]_{1j} |(\tau\ (\tau\ \tau)_j)_\tau\rangle$$
$$= F_{11}|(\tau\ (\tau\ \tau)_1)_\tau\rangle + F_{1\tau}|(\tau\ (\tau\ \tau)_\tau)_\tau\rangle. \tag{9.66}$$

After expressing $|0\rangle$ as a superposition of states with the right two Fibonacci anyons fused first, we can apply the braiding:

$$R|0\rangle = F_{11}R_1^{\tau\tau}|(\tau\ (\tau\ \tau)_1)_\tau\rangle + F_{1\tau}R_\tau^{\tau\tau}|(\tau\ (\tau\ \tau)_\tau)_\tau\rangle$$
$$= e^{-\hat{\imath}\frac{4\pi}{5}}F_{11}|(\tau\ (\tau\ \tau)_1)_\tau\rangle - e^{-\hat{\imath}\frac{2\pi}{5}}F_{1\tau}|(\tau\ (\tau\ \tau)_\tau)_\tau\rangle. \tag{9.67}$$

Inserting the R-symbols from Equation (9.56) and then applying the inverse basis transformation F^{-1}, we obtain

$$F^{-1}R|0\rangle = \left([F^{-1}]_{11}e^{-\hat{\imath}\frac{4\pi}{5}}F_{11} - [F^{-1}]_{1\tau}e^{-\hat{\imath}\frac{2\pi}{5}}F_{1\tau}\right)|0\rangle$$

$$+ \left([F^{-1}]_{\tau 1} e^{-i\frac{4\pi}{5}} F_{11} - [F^{-1}]_{\tau\tau} e^{-i\frac{2\pi}{5}} F_{1\tau} \right) |1\rangle$$

$$= -e^{i\frac{\pi}{5}} d_\tau^{-1} |0\rangle - i e^{-i\frac{\pi}{10}} \sqrt{d_\tau^{-1}} |1\rangle . \tag{9.68}$$

Thus, we have

$$\rho(\sigma_2) |0\rangle = -e^{i\frac{\pi}{5}} d_\tau^{-1} |0\rangle - i e^{-i\frac{\pi}{10}} \sqrt{d_\tau^{-1}} |1\rangle . \tag{9.69}$$

Exercise 9.16 *Determine $\rho(\sigma_2) |1\rangle$ and $\rho(\sigma_2) |N\rangle$.*

Combining Equation (9.69) with the results from Exercise 9.16, we arrive at

$$\rho(\sigma_2) \begin{pmatrix} |0\rangle := |((\tau\ \tau)_1\ \tau)_\tau\rangle \\ |1\rangle := |((\tau\ \tau)_\tau\ \tau)_\tau\rangle \\ |N\rangle := |((\tau\ \tau)_\tau\ \tau)_1\rangle \end{pmatrix} = \begin{pmatrix} -e^{i\frac{\pi}{5}} d_\tau^{-1} & -i e^{-i\frac{\pi}{10}} \sqrt{d_\tau^{-1}} & 0 \\ -i e^{-i\frac{\pi}{10}} \sqrt{d_\tau^{-1}} & -d_\tau^{-1} & 0 \\ 0 & 0 & -e^{-i\frac{2\pi}{5}} \end{pmatrix} \begin{pmatrix} |0\rangle \\ |1\rangle \\ |N\rangle \end{pmatrix} . \tag{9.70}$$

Remarkably, both $\rho(\sigma_1)$ in Equation (9.64) and $\rho(\sigma_2)$ above are block-diagonal with a 2×2 block and a 1×1 block. The 2×2 blocks act solely on the subspace $V_{3\tau}^\tau$. Since σ_1 and σ_2 generate the entire three-strand braid group B_3 (9.63), the representation of B_3 on $V_{3\tau}$ is effectively reduced to a two-dimensional representation on $V_{3\tau}^\tau$ and a 1-dimensional one on $V_{3\tau}^1$.

Therefore, $V_{3\tau}^\tau$ is indeed an invariant two-dimensional subspace of $V_{3\tau}$ under braiding the three Fibonacci anyons and qualifies as a logical qubit! Focusing on $V_{3\tau}^\tau$, we can take the reduced two-dimensional representation matrices:

$$\tilde{\rho}(\sigma_1) := \begin{pmatrix} e^{-i\frac{4\pi}{5}} & 0 \\ 0 & -e^{-i\frac{2\pi}{5}} \end{pmatrix} , \qquad \tilde{\rho}(\sigma_2) := \begin{pmatrix} e^{-i\frac{2\pi}{5}} d_\tau^{-1} & -i e^{-i\frac{\pi}{10}} \sqrt{d_\tau^{-1}} \\ -i e^{-i\frac{\pi}{10}} \sqrt{d_\tau^{-1}} & -d_\tau^{-1} \end{pmatrix} . \tag{9.71}$$

One can see that these two matrices cannot be diagonalized simultaneously, implying that B_3 act on $V_{3\tau}^\tau$ nontrivially. But how nontrivial is it?

In fact, as can be proven [283], $\tilde{\rho}(\sigma_1)$ and $\tilde{\rho}(\sigma_2)$ generate a dense subset of $SU(2)$. To physicists, one can say that $\tilde{\rho}(B_3)$ is essentially the entire group $SU(2)$. In our familiar words, braiding the three Fibonacci anyons realizes all possible single qubit quantum gates on the logical qubit $V_{3\tau}^\tau$ spanned by the computational basis states $|0\rangle$ and $|1\rangle$, as illustrated in Figure 9.21.

Exercise 9.17 *Try to find a braid of three Fibonacci anyons that realizes the Hadamard gate with a gate fidelity of 0.99. For a unitary single-qubit gate U_{braid} that approximates a certain unitary single-qubit gate U, its **gate fidelity** is defined by*

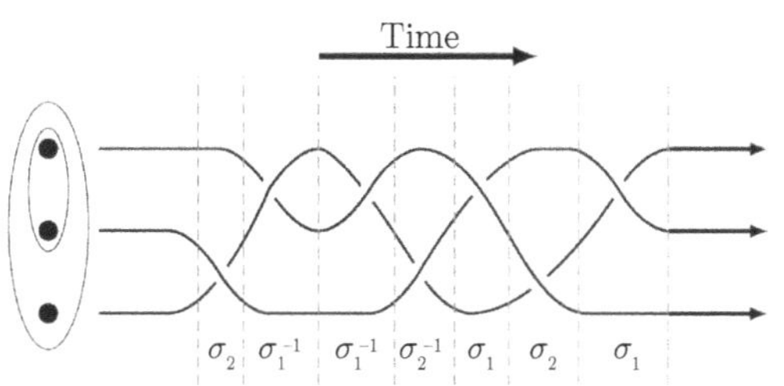

Figure 9.21
Braiding three Fibonacci anyons to realize single qubit logical gates.

$$F = \frac{1}{2}\left|\mathrm{Tr}(U^{\dagger}U_{braid})\right|,$$

where $F = 1$ is the highest fidelity. Note that such a braid may not be unique, and you should try to find a braid as short as possible, i.e., with fewest crossings. You may use softwares such as Mathematica.

9.2.5 Robust Universal Quantum Computation

We have seen how braiding three Fibonacci anyons enables the realization of arbitrary single-qubit logical gates on a logical qubit encoded in their three-dimensional Hilbert space. To achieve universal quantum computation through braiding Fibonacci anyons, one critical component remains: demonstrating that such braiding can realize a two-qubit entangling gate, such as any controlled gate. Indeed, this is feasible. A straightforward approach involves creating six Fibonacci anyons, divided into two groups of three. This configuration allows for two logical qubits, each encoded in the three-dimensional Hilbert spaces of the respective sets of three Fibonacci anyons. By intricately braiding the anyons within one set with those in the other, it's possible to implement a specific two-qubit entangling gate on these logical qubits.

The detailed methodology for such braiding, due to its complexity, is beyond the scope of this book. Nevertheless, readers interested in exploring these processes can find comprehensive discussions in the literature [284, 285, 286, 287]. For an illustrative example of how such an entangling gate might be realized by braiding six Fibonacci anyons [288], refer to Figure 9.22.

This method of realizing universal quantum computation is known as topological quantum computation. It derives its name from the fact that the quantum gates are determined solely by the topology of the braiding process. To conclude this chapter, four crucial remarks are necessary:

Relation to the Quantum Circuit Model

The exploration of single-qubit gates in the context of topological quantum computation (TQC), particularly through braiding three Fibonacci anyons, reveals intriguing contrasts with the quantum circuit model. Notably, the representation matrices $\tilde{\rho}(\sigma_1)$ and $\tilde{\rho}(\sigma_2)$ for the generators of B_3, as specified in (9.71), consist of irrational number elements. This characteristic introduces an inherent

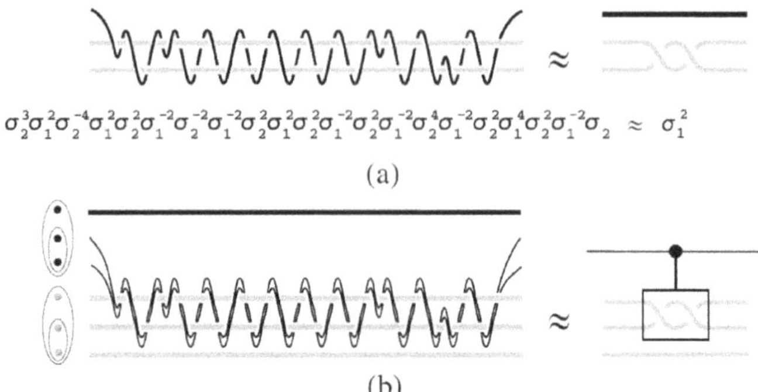

$$\sigma_2^3\sigma_1^2\sigma_2^{-4}\sigma_1^2\sigma_2^2\sigma_1^{-2}\sigma_2^{-2}\sigma_1^{-2}\sigma_2^2\sigma_1^2\sigma_2^2\sigma_1^{-2}\sigma_2^2\sigma_1^{-2}\sigma_2^4\sigma_1^{-2}\sigma_2^2\sigma_1^4\sigma_2^2\sigma_1^{-2}\sigma_2 \approx \sigma_1^2$$

(a)

(b)

Figure 9.22 An example of braiding two sets of three Fibonacci anyons to realize a two-qubit controlled gate (a reproduction of Figure 3 in [288]).

approximation when these matrices, or their inverses, are used to synthesize specific quantum gates, such as the Hadamard gate (4.62). The implication is profound: TQC, by its nature, approximates the quantum circuit model rather than replicating it exactly.

This approximation stems from the topological nature of TQC, where quantum gates are realized through the braiding of anyons, and the gate operations depend on the topological properties of these braids rather than on precise numerical transformations. Consequently, achieving perfect precision in gate implementation is inherently challenging in TQC. A quantum gate in TQC is typically approximated by a series of braids, each offering varying degrees of precision. Yet, identifying the optimal braid sequence to approximate a specific quantum gate is not straightforward. In contrast to the quantum circuit model, where gates are represented by exact unitary matrices, TQC requires a more heuristic approach to discover efficient braiding patterns that closely approximate the desired gates.

The challenge is further compounded by the lack of a general algorithm known to efficiently search for such optimal braids. Each gate approximation is a unique problem, requiring case-by-case analysis and often extensive computational resources. This aspect of TQC highlights a key difference from traditional quantum computing models: While TQC promises inherent fault tolerance and robustness against certain types of errors, it also demands innovative solutions to address the approximation problem in gate synthesis.

Hence, while TQC presents a novel and potentially powerful framework for quantum computation, it also poses unique challenges in gate implementation. The field continues to evolve, with ongoing research aimed at developing more efficient methods for approximating quantum gates within the TQC paradigm and understanding the trade-offs between fault tolerance and operational feasibility in this emerging model of quantum computation.

Robustness of TQC

The fault tolerance of TQC is a widely recognized feature, attributed to its robustness against errors caused by local perturbations. This robustness can be understood from two perspectives:

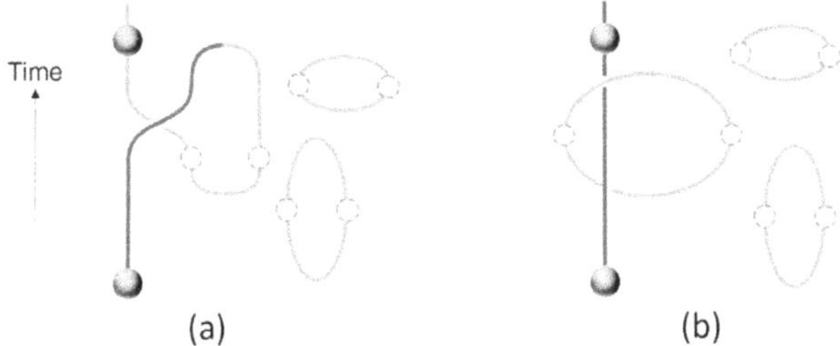

Figure 9.23 (a) A pair of Fibonacci anyons created from the vacuum; one in the pair may annihilate with a tracked Fibonacci anyon, while the other replaces the tracked anyon. (b) A pair of Fibonacci anyons created from the vacuum; one in the pair may braid with a tracked Fibonacci anyon and then annihilate with the other in the pair.

First, TQC typically operates within a two-dimensional material that is in a topological phase, such as the Fibonacci phase. Topological phases are inherently robust against local perturbations. The influence of such perturbations is exponentially suppressed by the size of the system. In practical terms, any macroscopic material's size is essentially infinite compared to the scale of microscopic local perturbations, leading to a high degree of inherent stability.

Second, for effective TQC operation, conditions near absolute zero temperature ($T = 0$) are essential. Under such conditions, only a fixed number of Fibonacci anyons are deliberately excited and maintained at sufficient separation from each other. These anyons, which are actively involved in the computational process, are referred to as *tracked anyons*. At these low temperatures, the dominant source of thermal fluctuation in the system is the creation of ephemeral Fibonacci anyon pair states, representing the most significant potential error source in a realistic topological quantum computer.

Despite their short lifetimes, these thermally created anyon pairs can potentially interfere with the computation by interacting with the tracked anyons. That said, given the considerable separation between the tracked anyons, there are only two feasible ways for such interference to occur, as depicted in Figure 9.23: (a) A pair of Fibonacci anyons emerges from the vacuum; one of them may annihilate with a tracked Fibonacci anyon, with the other taking its place; (b) a pair of Fibonacci anyons emerges from the vacuum; one of them may braid with a tracked Fibonacci anyon before annihilating with its pair. Although these scenarios involve interference with the braiding processes of the tracked anyons, it can be demonstrated that the overall computation remains unaffected. This robustness against interference has been corroborated through experimental quantum simulations [289], further emphasizing the resilience of TQC in practical applications.

Leakage Error

As discussed in Section 9.2.4, three Fibonacci anyons can encode a logical qubit within their three-dimensional Hilbert space. Braiding these anyons preserves the integrity of this logical qubit, ensuring that it does not mix with the noncomputational state $|N\rangle$. Nonetheless, when extending this principle to devise a two-qubit gate, the situation becomes more complex. Typically, six Fibonacci anyons are needed, which together occupy a 13-dimensional Hilbert space. Of this space, only a four-dimensional subspace is utilized for the two logical qubits. Consequently, in such a scenario, braiding six Fibonacci anyons is unlikely to preserve the two-qubit logical space. This leads to the possibility of logical states mixing with the remaining nine-dimensional noncomputational subspace, resulting in what are known as **leakage errors**.

A notable aspect of TQC is that multiple braids can implement the same quantum gate, but it raises an intriguing question: For any given two-qubit gate, do there exist six-strand braids that can perfectly preserve the logical space, thereby avoiding leakage? Currently, the answer to this question remains elusive. This uncertainty also applies to configurations with more logical qubits and correspondingly more Fibonacci anyons.

Despite these challenges, it has been proven that the leakage error associated with any logical gate can be made arbitrarily small by using sufficiently long braids to implement the gate [286]. While this finding offers some promise for the feasibility of TQC, it does not completely resolve the problem. The algorithmic difficulty in identifying such long braids for a specific logical gate remains a significant hurdle. In recent developments, machine learning techniques have been employed to assist in the search for braids as single-qubit gates [287].

How Fast Can It Be?

Increasing computational speed has been a constant objective in technological advancement. While it might be premature to focus on the speed of quantum gates in the current stage of TQC development, it is still valuable to contemplate the potential speed of TQC operations, at least from a qualitative standpoint. Anticipating and preparing for future challenges is always a prudent approach.

Hypothetically, in a functional topological quantum computer, the operational environment is maintained at a temperature extremely close to absolute zero ($T = 0$). This is essential to suppress thermal fluctuations that could interfere with computation. Within this setup, tracked Fibonacci anyons are generated at specific locations, initiated by finely tuned probe heads positioned very close to the desired points. The process of braiding these anyons involves moving them around each other, a task facilitated by the movement of the probe heads.

This movement must be carefully controlled in terms of speed, however, as rapid movements can lead to excessive thermal exchange between the probe heads and the material, potentially generating unwanted anyons. These extraneous anyons could inadvertently interact with the tracked anyons, thereby disrupting the computational process. Furthermore, excessively fast movements risk destabilizing the very topological phase that underpins the material, thus compromising the entire computation.

The challenge of maintaining precision in the movement of anyons becomes even more pronounced when dealing with longer braids. Finding ways to increase the speed of braiding, while still preserving the integrity of the computation and the material's topological phase, presents an exciting yet formidable challenge. As TQC continues to evolve, addressing this challenge will be crucial for enhancing the efficiency and practicality of quantum computations in this paradigm.

Exercise 9.18 *Consider a scenario where topological quantum computation is realized using Fibonacci anyons on the boundary of a disk. In this exercise, you will analyze a single-logical-qubit toy model. For this model, a logical qubit requires at least three boundary Fibonacci anyons. The basis states of the minimal Hilbert space of these anyons can be visualized using the triangle in Figure 9.24.*

*In Figure 9.24, each open (boundary) edge hosts a Fibonacci anyon and is assigned the value τ. The lower internal edge is fixed with the value τ, while the other two internal edges j_1 and j_2 can take values in $\{1, \tau\}$, ensuring that the **fusion rules are satisfied at each vertex**. The basis vectors of the Hilbert space can be denoted by $|j_1, j_2\rangle$. Braiding of the Fibonacci anyons corresponds to braiding the open edges, which is generated by the operations σ_{12} and σ_{23} as depicted in Figure 9.24.*

1. Determine the dimension of the Hilbert space.
2. The logical qubit is a two-dimensional subspace of this Hilbert space, invariant under the operations σ_{12} and σ_{23}. Utilize the F-matrix to identify the logical basis.
3. Derive the representation matrices of the braiding generators σ_{12} and σ_{23} over the logical space. Hint: These matrices should be two-dimensional.
4. Calculate the representation matrices of σ_{12} and σ_{23} in the basis $|j_1, j_2\rangle$. Hint: The dimension of these matrices is the answer to Question 1.

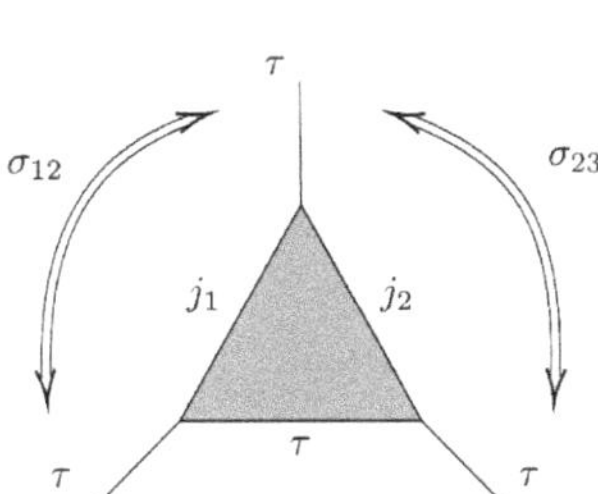

Figure 9.24
Representation of three boundary Fibonacci anyons in a minimal Hilbert space configuration.

Conclusion

This chapter guided us through the framework of topological quantum computation (TQC), examining how it leverages the unique properties of non-Abelian anyons for quantum processing. Starting with a foundation in the fusion and braiding of anyons, we saw how specific non-Abelian anyons, such as Fibonacci anyons, provide a basis for encoding logical qubits. Through braiding operations, these anyons can implement unitary transformations that approximate quantum gates, offering a robust approach to quantum computation with inherent fault tolerance due to topological protection.

We further examined the Fibonacci anyons in detail, illustrating how their fusion rules and braiding properties facilitate universal single-qubit and entangling two-qubit gates, key elements in any computational model. The braid group representation within the space of these anyons provided a powerful tool for realizing the gate operations needed for quantum algorithms, demonstrating TQC's potential to approximate the quantum circuit model, albeit with challenges in achieving exact gate precision.

As we close this chapter and in fact the entire book, we recognize both the promise and the obstacles within TQC: its robustness against local perturbations makes it a compelling model for fault-tolerant quantum computing, but the challenges in implementing precise gate approximations and managing leakage errors highlight areas for ongoing research and development. TQC continues to be an exciting frontier, with significant implications for scalable, error-resistant quantum computation.

References

[1] Ince, D. *The computer: A very short introduction*. No. 292 in Very Short Introductions (Oxford University Press, Oxford, 2011). OCLC: ocn761374747.

[2] Stokes, J. *Inside the machine: An illustrated introduction to microprocessors and computer architecture* (No Starch Press, 2015). OCLC: 913792823.

[3] Ifrah, G. *The universal history of computing: From the abacus to the quantum computer* (Wiley, New York, 2001).

[4] Kurzweil, R. *How to create a mind: The secret of human thought revealed* (Penguin Books, New York, 2013).

[5] Hyman, A. *Charles Babbage: Pioneer of the computer* (Oxford University Press, Oxford, 1982).

[6] Menabrea, L. F. Sketch of the Analytical Engine invented by Charles Babbage, Esq. In *Ada's legacy: Cultures of computing from the Victorian to the digital age* (Association for Computing Machinery and Morgan & Claypool, 2015). https://doi.org/10.1145/2809523.2809528.

[7] Liu, C. *The three-body problem*. Book I in Three-Body trilogy (Tor Books, New York, 2014).

[8] Patterson, D. A., Hennessy, J. L. & Alexander, P. *Computer organization and design: The hardware/software interface*, 5th edn. (Elsevier Morgan Kaufmann, Amsterdam, Heidelberg, 2014).

[9] Silberschatz, A., Galvin, P. B. & Gagne, G. *Operating system concepts*, 9th edn. (Wiley, Hoboken, NJ, 2013).

[10] MacCartney, S. *ENIAC: The triumphs and tragedies of the world's first computer* (Walker, New York, 1999).

[11] Moore, G. E. Cramming more components onto integrated circuits. Reprinted from Electronics, **38**(8), April 19, 1965, pp.114 ff. *IEEE Solid-State Circuits Society Newsletter* **11**(3), 33–35 (2006). http://ieeexplore.ieee.org/document/4785860/.

[12] Brynjolfsson, E. & McAfee, A. *The second machine age: Work, progress, and prosperity in a time of brilliant technologies* (W. W. Norton & Company, New York, London, 2016).

[13] Nielsen, M. A. & Chuang, I. L. *Quantum computation and quantum information*, 10th Anniversary edn. (Cambridge University Press, Cambridge, 2010).

[14] Georgescu, I., Ashhab, S. & Nori, F. Quantum simulation. *Reviews of Modern Physics* **86**, 153–185 (2014). https://link.aps.org/doi/10.1103/RevModPhys.86.153.

[15] Farhi, E., Goldstone, J., Gutmann, S. & Sipser, M. Quantum computation by adiabatic evolution (2000). http://arxiv.org/abs/quant-ph/0001106. ArXiv:quant-ph/0001106.

[16] Zanardi, P. & Rasetti, M. Holonomic quantum computation. *Physics Letters A* **264**, 94–99 (1999). https://linkinghub.elsevier.com/retrieve/pii/S0375960199008038.

[17] Kitaev, A. Fault-tolerant quantum computation by anyons. *Annals of Physics* **303**, 2–30 (2003). https://linkinghub.elsevier.com/retrieve/pii/S0003491602000180.

[18] Nayak, C., Simon, S. H., Stern, A., Freedman, M. & Das Sarma, S. Non-Abelian anyons and topological quantum computation. *Reviews of Modern Physics* **80**, 1083–1159 (2008). https://link.aps.org/doi/10.1103/RevModPhys.80.1083.

[19] Devoret, M. H. & Schoelkopf, R. J. Superconducting circuits for quantum information: An outlook. *Science* **339**, 1169–1174 (2013). https://www.science.org/doi/10.1126/science.1231930.

[20] Monroe, C. & Kim, J. Scaling the ion trap quantum processor. *Science* **339**, 1164–1169 (2013). https://www.science.org/doi/10.1126/science.1231298.

[21] Nakahara, M. & Ohmi, T. *Quantum computing: From linear algebra to physical realizations* (CRC Press, Boca Raton, FL, 2008). OCLC: ocn177825572.

[22] Bloch, I., Dalibard, J. & Nascimbene, S. Quantum simulations with ultracold quantum gases. *Nature Physics* **8**, 267–276 (2012). https://www.nature.com/articles/nphys2259.

[23] Vandersypen, L. M. K. & Chuang, I. L. NMR techniques for quantum control and computation. *Reviews of Modern Physics* **76**, 1037–1069 (2005). https://link.aps.org/doi/10.1103/RevModPhys.76.1037.

[24] Waldrop, M. M. The chips are down for Moore's law. *Nature* **530**, 144–147 (2016). https://www.nature.com/articles/530144a.

[25] Rivest, R. L., Shamir, A. & Adleman, L. A method for obtaining digital signatures and public-key cryptosystems. *Communications of the ACM* **21**, 120–126 (1978). https://dl.acm.org/doi/10.1145/359340.359342.

[26] Wardlaw, W. P. The RSA public key cryptosystem. In Joyner, D. (ed.), *Coding theory and cryptography*, 101–123 (Springer, Berlin, Heidelberg, 2000).

[27] Zhou, X. & Tang, X. Research and implementation of RSA algorithm for encryption and decryption. In *Proceedings of 2011 6th International Forum on Strategic Technology*, vol. 2, 1118–1121 (2011).

[28] NaQi *et al.* Analysis and research of the RSA algorithm. *Information Technology Journal* **12**, 1818–1824 (2013). https://scialert.net/fulltext/?doi=itj.2013.1818.1824.

[29] Al-Kaabi, S. S. & Belhaouari, S. B. Methods toward enhancing RSA algorithm: A survey (2019). https://papers.ssrn.com/abstract=3412776.

[30] Manin, U. I. & Dyson, F. J. *Mathematics as metaphor* (American Mathematical Society, Providence, RI, 2007).

[31] Feynman, R. P. Simulating physics with computers. *International Journal of Theoretical Physics* **21**, 467–488 (1982). https://doi.org/10.1007/BF02650179.

[32] Landauer, R. Irreversibility and heat generation in the computing process. *IBM Journal of Research and Development* **5**, 183–191 (1961). http://ieeexplore.ieee.org/document/5392446/.

[33] Ikonen, J., Salmilehto, J. & Mottonen, M. Energy-efficient quantum computing. *npj Quantum Information* **3**, 17 (2017). https://www.nature.com/articles/s41534-017-0015-5.

[34] Deutsch, D. Quantum theory, the Church–Turing principle and the universal quantum computer. *Proceedings of the Royal Society of London A* **400**, 97–117 (1985). https://royalsocietypublishing.org/doi/10.1098/rspa.1985.0070.

[35] Shor, P. W. Polynomial-time algorithms for prime factorization and discrete logarithms on a quantum computer. *SIAM Journal on Computing* **26**, 1484–1509 (1997). http://epubs.siam.org/doi/10.1137/S0097539795293172.

[36] Grover, L. K. A fast quantum mechanical algorithm for database search. In *Proceedings of the 28th Annual ACM Symposium on Theory of Computing: STOC '96*, 212–219 (ACM Press, Philadelphia, PA, 1996). http://portal.acm.org/citation.cfm?doid=237814.237866.

[37] Zhao, Y. *et al.* Realization of an error-correcting surface code with superconducting qubits. *Physical Review Letters* **129**, 030501 (2022). https://link.aps.org/doi/10.1103/PhysRevLett.129.030501.

[38] Ni, Z. *et al.* Beating the break-even point with a discrete-variable-encoded logical qubit. *Nature* **616**, 56–60 (2023). https://www.nature.com/articles/s41586-023-05784-4.

[39] Arute, F. *et al.* Quantum supremacy using a programmable superconducting processor. *Nature* **574**, 505–510 (2019). https://www.nature.com/articles/s41586-019-1666-5.

[40] Kelly, J. A preview of Bristlecone, Google's new quantum processor (2018). https://blog.research.google/2018/03/a-preview-of-bristlecone-googles-new.html.

[41] Gambetta, J. The hardware and software for the era of quantum utility is here (2023). https://www.ibm.com/quantum/blog/quantum-roadmap-2033.

[42] Neven, H. Meet Willow, our state-of-the-art quantum chip (2024). https://blog.google/technology/research/google-willow-quantum-chip/.

[43] Lin, Rui *et al.* AI-Enabled Parallel Assembly of Thousands of Defect-Free Neutral Atom Arrays (2025). Phys. Rev. Lett. **135**, 06060. https://doi.org/10.1103/2ym8-vs82.

[44] Chiu, NC *et al.* Continuous operation of a coherent 3,000-qubit system (2025). *Nature*, (2025). https://doi.org/10.1038/s41586-025-09596-6.

[45] Quantinuum. Quantinuum Launches Industry-First, Trapped-Ion 56-Qubit Quantum Computer, Breaking Key Benchmark Record. https://www.quantinuum.com/press-releases/quantinuum-launches-industry-first-trapped-ion-56-qubit-quantum-computer-that-challenges-the-worlds-best-supercomputers.

[46] Intel. Introducing Tunnel Falls. https://www.intel.com/content/www/us/en/research/quantum-computing.html.

[47] Petit, L. *et al.* Universal quantum logic in hot silicon qubits. *Nature* **580**, 355–359 (2020). https://www.nature.com/articles/s41586-020-2170-7.

[48] Bolgar, C. Microsoft's Majorana 1 chip carves new path for quantum computing. https://news.microsoft.com/source/features/innovation/microsofts-majorana-1-chip-carves-new-path-for-quantum-computing.

[49] Zhu, Q. *et al.* Quantum computational advantage via 60-qubit 24-cycle random circuit sampling. *Science Bulletin* **67**, 240–245 (2022). https://linkinghub.elsevier.com/retrieve/pii/S2095927321006733.

[50] Xinhua. China's 176-qubit quantum computing platform goes online(2023). https://english.news.cn/20230531/0946675301284c1786b4ee27251c89a3/c.html.

[51] BAQIS Quafu Group. Quafu-RL: The Cloud quantum computers based quantum reinforcement Learning (sic; 2023). https://arxiv.org/abs/2305.17966. Version Number: 2.

[52] BAQIS Quafu Group. Quafu-Qcover: Explore combinatorial optimization problems on Cloud-based quantum computers (2023). https://arxiv.org/abs/2305.17979.

[53] BAQIS. BAQIS Unveiled "Quafu" quantum Cloud platform at 2023 ZGC Forum (2023). http://en.baqis.ac.cn/news/detail/?cid=1699.

[54] Pikulin, D. I. *et al.* Protocol to identify a topological superconducting phase in a three-terminal device (2021). https://arxiv.org/abs/2103.12217.

[55] Aaronson, S. Is quantum mechanics an island in theoryspace? (2004). http://arxiv.org/abs/quant-ph/0401062. ArXiv: quant-ph/0401062.

[56] Watrous, J. *The theory of quantum information* (Cambridge University Press, Cambridge, 2018). https://www.cambridge.org/core/books/theory-of-quantum-information/AE4AA5638F808D2CFEB070C55431D897.

[57] Fowler, A. G., Mariantoni, M., Martinis, J. M. & Cleland, A. N. Surface codes: Towards practical large-scale quantum computation. *Physical Review* **86**, 032324 (2012). https://link.aps.org/doi/10.1103/PhysRevA.86.032324.

[58] Preskill, J. Quantum computing in the NISQ era and beyond. *Quantum* **2**, 79

(2018). https://quantum-journal.org/papers/q-2018-08-06-79/.

[59] Gottesman, D. An introduction to quantum error correction and fault-tolerant quantum computation (2009). http://arxiv.org/abs/0904.2557. ArXiv:0904.2557 [quant-ph].

[60] Bacon, D. M. *Decoherence, control, and symmetry in quantum computers.* Doctoral thesis, University of California, Berkeley (2001).

[61] Bohm, G. Quantum computers (2018). https://group.mercedes-benz.com/company/magazine/technology-innovation/quantum-computers-future-daimler-google-ibm-technology.html.

[62] ExxonMobil. ExxonMobil and IBM to advance energy sector application of quantum computing (2019). https://corporate.exxonmobil.com/news/news-releases/2019/0108exxonmobil-and-ibm-to-advance-energy-sector-application-of-quantum-computing.

[63] CERN. CERN Quantum Technology Initiative unveils strategic roadmap shaping CERN's role in next quantum revolution (2021). https://home.cern/news/press-release/knowledge-sharing/cern-quantum-technology-initiative-unveils-strategic-roadmap.

[64] Mitsubishi. Establishment of quantum strategic industry alliance for revolution (2021). https://www.m-chemical.co.jp/en/news/2021/_icsFiles/afieldfile/2021/09/29/210901qstar.pdf.

[65] Yin, H.-L. *et al.* Measurement-device-independent quantum key distribution over a 404 km optical fiber. *Physical Review Letters* **117**, 190501 (2016). https://link.aps.org/doi/10.1103/PhysRevLett.117.190501.

[66] Gibney, E. Mini-satellite paves the way for quantum messaging anywhere on Earth. https://doi.org/10.1038/d41586-025-00581-7.

[67] Lu, C.-Y., Cao, Y., Peng, C.-Z. & Pan, J.-W. Micius quantum experiments in space. *Reviews of Modern Physics* **94**, 035001 (2022). https://link.aps.org/doi/10.1103/RevModPhys.94.035001.

[68] Ursin, R. *et al.* Entanglement-based quantum communication over 144 km. *Nature Physics* **3**, 481–486 (2007). https://www.nature.com/articles/nphys629.

[69] Hu, X.-M., Guo, Y., Liu, B.-H., Li, C.-F. & Guo, G.-C. Progress in quantum teleportation. *Nature Reviews Physics* **5**, 339–353 (2023). https://www.nature.com/articles/s42254-023-00588-x.

[70] Heshami, K. *et al.* Quantum memories: Emerging applications and recent advances. *Journal of Modern Optics* **63**, 2005–2028 (2016). https://www.tandfonline.com/doi/full/10.1080/09500340.2016.1148212.

[71] Ma, L. *et al.* High-performance cavity-enhanced quantum memory with warm atomic cell. *Nature Communications* **13**, 2368 (2022). https://www.nature.com/articles/s41467-022-30077-1.

[72] Wang, Y., Craddock, A. N., Sekelsky, R., Flament, M. & Namazi, M. Field-deployable quantum memory for quantum networking. *Physical Review Applied* **18**, 044058 (2022). https://link.aps.org/doi/10.1103/PhysRevApplied.18.044058.

[73] Hammerer, K., Sørensen, A. S. & Polzik, E. S. Quantum interface between light and atomic ensembles. *Reviews of Modern Physics* **82**, 1041–1093 (2010). https://link.aps.org/doi/10.1103/RevModPhys.82.1041.

[74] Zhong, T., Kindem, J. M., Miyazono, E. & Faraon, A. Nanophotonic coherent light–matter interfaces based on rare-earth-doped crystals. *Nature Communications* **6**, 8206 (2015). https://www.nature.com/articles/ncomms9206.

[75] Manninen, J., Blick, R. H. & Massel, F. Hybrid optomechanical superconducting qubit system. *Physical Review Research* **6**, 023029 (2024). https://link.aps.org/doi/10.1103/PhysRevResearch.6.023029.

[76] Azuma, K. *et al.* Quantum repeaters: From quantum networks to the quantum internet (2022). https://arxiv.org/abs/2212.10820. arXiv Version Number: 2.

[77] QuantumXC. Quantum Xchange launches "Phio" the first commercial QKD network for quantum communications in the U.S. https://quantumxc.com/blog/quantum-launches-phio/.

[78] Hawking, S. W. Breakdown of predictability in gravitational collapse. *Physical Review D* **14**, 2460–2473 (1976). https://link.aps.org/doi/10.1103/PhysRevD.14.2460.

[79] Harlow, D. Jerusalem lectures on black holes and quantum information. *Reviews of Modern Physics* **88**, 015002 (2016). https://link.aps.org/doi/10.1103/RevModPhys.88.015002.

[80] Maldacena, J. The large-N limit of superconformal field theories and supergravity. *International Journal of Theoretical Physics* **38**, 1113–1133 (1999). http://link.springer.com/10.1023/A:1026654312961.

[81] Ryu, S. & Takayanagi, T. Holographic derivation of entanglement entropy from the anti-de Sitter space/conformal field theory correspondence. *Physical Review Letters* **96**, 181602 (2006). https://link.aps.org/doi/10.1103/PhysRevLett.96.181602.

[82] Swingle, B. Entanglement renormalization and holography. *Physical Review D* **86**, 065007 (2012). https://link.aps.org/doi/10.1103/PhysRevD.86.065007.

[83] Verlinde, E. & Verlinde, H. Black hole entanglement and quantum error correction. *Journal of High Energy Physics* **2013**, 107 (2013). http://link.springer.com/10.1007/JHEP10(2013)107.

[84] Baker, O. K. Quantum information science in high energy physics. In *Topics on Quantum Information Science* (IntechOpen, 2021). https://www.intechopen.com/chapters/77398.

[85] Almheiri, A., Dong, X. & Harlow, D. Bulk locality and quantum error correction in AdS/CFT. *Journal of High Energy Physics* **2015**, 163 (2015). http://link.springer.com/10.1007/JHEP04(2015)163.

[86] Harlow, D. The Ryu–Takayanagi formula from quantum error correction. *Communications in Mathematical Physics* **354**, 865–912 (2017). http://link.springer.com/10.1007/s00220-017-2904-z.

[87] Akers, C. & Rath, P. Holographic Rényi entropy from quantum error correction. *Journal of High Energy Physics* **2019**, 52 (2019). https://link.springer.com/10.1007/JHEP05(2019)052.

[88] Kharzeev, D. E. & Levin, E. Deep inelastic scattering as a probe of entanglement: Confronting experimental data. *Physical Review D* **104**, L031503 (2021). https://link.aps.org/doi/10.1103/PhysRevD.104.L031503.

[89] Susskind, L. Computational complexity and black hole horizons (2014). https://arxiv.org/abs/1402.5674.

[90] Stanford, D. & Susskind, L. Complexity and shock wave geometries. *Physical Review D* **90**, 126007 (2014). https://link.aps.org/doi/10.1103/PhysRevD.90.126007.

[91] Pastawski, F., Yoshida, B., Harlow, D. & Preskill, J. Holographic quantum error-correcting codes: Toy models for the bulk/boundary correspondence. *Journal of High Energy Physics* **2015**, 149 (2015). http://link.springer.com/10.1007/JHEP06(2015)149.

[92] Bennett, C. H. & Landauer, R. The fundamental physical limits of computation. *Scientific American* **253**, 48–56 (1985). https://www.scientificamerican.com/article/the-fundamental-physical-limits-of.

[93] Lloyd, S. Ultimate physical limits to computation. *Nature* **406**, 1047–1054 (2000). https://www.nature.com/articles/35023282.

[94] Thomas, H. Physics of computation: From classical to quantum. In Skjeltorp, A. T. & Vicsek, T. (eds.), *Complexity from microscopic to macroscopic scales: Coherence and large deviations*, NATO Science Series, 1–20 (Springer

Netherlands, Dordrecht, 2002). https://doi.org/10.1007/978-94-010-0419-0_1.

[95] Sipser, M. *Introduction to the theory of computation* (PWS Pub. Co., Boston, 1997).

[96] Petzold, C. *The annotated Turing: A guided tour through Alan Turing's historic paper on computability and the Turing machine* (Wiley, Hoboken, NJ, 2008).

[97] Hilbert, D. Mathematical problems. *Bulletin of the American Mathematical Society* **8**, 437–479 (1902). https://www.ams.org/bull/1902-08-10/S0002-9904-1902-00923-3/.

[98] Godel, K. Uber formal unentscheidbare Sätze der Principia mathematica und verwandter Systeme I. *Monatshefte für Mathematik und Physik* **37**, 173–198 (1931). http://link.springer.com/10.1007/BF01700692.

[99] Church, A. An unsolvable problem of elementary number theory. *American Journal of Mathematics* **58**, 345 (1936). https://www.jstor.org/stable/2371045?origin=crossref.

[100] Turing, A. M. On computable numbers, with an application to the Entscheidungsproblem. *Proceedings of the London Mathematical Society* **s2-42**, 230–265 (1937). http://doi.wiley.com/10.1112/plms/s2-42.1.230.

[101] Motwani, R. & Raghavan, P. *Randomized algorithms* (Cambridge University Press, Cambridge, 1995).

[102] Kirkpatrick, S., Gelatt, C. D. & Vecchi, M. P. Optimization by simulated annealing. *Science* **220**, 671–680 (1983). https://www.science.org/doi/10.1126/science.220.4598.671.

[103] Arora, S. & Barak, B. *Computational complexity: A modern approach* (Cambridge University Press, Cambridge, 2009). OCLC: ocn286431654.

[104] Goldreich, O. *Computational complexity: A conceptual perspective* (Cambridge University Press, Cambridge, 2008). OCLC: ocn192050142.

[105] Papadimitriou, C. H. *Computational complexity* (Addison-Wesley, Reading, MA, 1994).

[106] Jukna, S. *Boolean function complexity: Advances and frontiers*, vol. 27 of *Algorithms and combinatorics* (Springer, Berlin, Heidelberg, 2012). https://link.springer.com/10.1007/978-3-642-24508-4.

[107] Wegener, I. *Complexity theory: Exploring the limits of efficient algorithms* (Springer, Berlin, Heidelberg, 2005). http://link.springer.com/10.1007/3-540-27477-4.

[108] Lenstra, A. K., Lenstra, H. W., Manasse, M. S. & Pollard, J. M. The number field sieve. In Lenstra, A. K. & Lenstra, H. W. (eds.), *The development of the number field sieve*, Lecture Notes in Mathematics, 11–42 (Springer, Berlin, Heidelberg, 1993).

[109] Boudot, F. *et al.* Comparing the difficulty of factorization and discrete logarithm: A 240-digit experiment. In Micciancio, D. & Ristenpart, T. (eds.), *Advances in Cryptology – CRYPTO 2020*, Lecture Notes in Computer Science, 62–91 (Springer International Publishing, Cham, Switzerland, 2020). https://link.springer.com/10.1007/978-3-030-56880-1_3.

[110] West, D. B. *Introduction to graph theory*, 2nd edn. (Pearson, Upper Saddle River, NJ, 2000).

[111] Kruskal, J. B. On the shortest spanning subtree of a graph and the traveling salesman problem. *Proceedings of the American Mathematical Society* **7**, 48–50 (1956). https://www.ams.org/proc/1956-007-01/S0002-9939-1956-0078686-%7/.

[112] Prim, R. C. Shortest connection networks and some generalizations. *Bell System Technical Journal* **36**, 1389–1401 (1957). https://ieeexplore.ieee.org/document/6773228.

[113] Held, M. & Karp, R. M. A dynamic programming approach to sequencing problems. *Journal of the Society for Industrial and Applied Mathematics* **10**, 196–210 (1962). https://epubs.siam.org/doi/10.1137/0110015.

[114] Euler, L. Solutio problematis ad geometriam situs pertinentis. *Commentarii academiae scientiarum Petropolitanae* **8**, 128–140 (1741). https://scholarlycommons.pacific.edu/euler-works/53.

[115] Hierholzer, C. & Wiener, C. Ueber die Moglichkeit, einen Linienzug ohne Wiederholung und ohne Unterbrechung zu umfahren. *Mathematischen Annalen* **6**, 30–32 (1873). https://doi.org/10.1007/BF01442866.

[116] Kwan, M.-K. Graphic programming using odd or even points. *Chinese Mathematics 1* **1962**, 237–277 (1962).

[117] Edmonds, J. & Johnson, E. L. Matching, Euler tours and the Chinese postman. *Mathematical Programming* **5**, 88–124 (1973).

[118] Pevzner, P. A., Tang, H. & Waterman, M. S. An Eulerian path approach to DNA fragment assembly. *Proceedings of the National Academy of Sciences of the United States of America* **98**, 9748–9753 (2001). https://pnas.org/doi/full/10.1073/pnas.171285098.

[119] Tarjan, R. E. & Vishkin, U. An efficient parallel biconnectivity algorithm. *SIAM Journal on Computing* **14**, 862–874 (1985). http://epubs.siam.org/doi/10.1137/0214061.

[120] Berkman, O. & Vishkin, U. Finding level-ancestors in trees. *Journal of Computer and System Sciences* **48**, 214–230 (1994). https://linkinghub.elsevier.com/retrieve/pii/S002200000580002%9.

[121] Savage, C. A survey of combinatorial gray codes. *SIAM Review* **39**, 605–629 (1997). http://epubs.siam.org/doi/10.1137/S0036144595295272.

[122] Roy, K. Optimum gate ordering of CMOS logic gates using Euler path approach: Some insights and explanations. *Journal of Computing and Information Technology* **15**, 85 (2007). http://cit.srce.unizg.hr/index.php/CIT/article/view/1629.

[123] Achlioptas, D. *et al.* Random constraint satisfaction: A more accurate picture. *Constraints* **6**, 329–344 (2001). https://doi.org/10.1023/A:1011402324562.

[124] Mezard, M., Parisi, G. & Zecchina, R. Analytic and algorithmic solution of random satisfiability problems. *Science* **297**, 812–815 (2002). https://www.science.org/doi/10.1126/science.1073287.

[125] Lame, G. Note sur la limite du nombre des divisions dans la recherche du plus grand commun diviseur entre deux nombres entiers. *Comptes rendus des seances de l'Academie des Sciences* 867–870 (1844).

[126] Dijkstra, E. W. A note on two problems in connexion with graphs. *Numerische Mathematik* **1**, 269–271 (1959). http://link.springer.com/10.1007/BF01386390.

[127] Floyd, R. W. Algorithm 97: Shortest path. *Communications of the ACM* **5**, 345 (1962). https://dl.acm.org/doi/10.1145/367766.368168.

[128] Moore, E. The shortest path through a maze. *In Proceedings of the International Symposium on the Theory of Switching* 285–292 (Harvard University Press, Cambridge, MA, 1959).

[129] Schoning, U. & Torán, J. *The satisfiability problem: Algorithms and analyses.* (corrected reprint edn.), vol. 3 of Mathematik für Anwendungen (Lehmanns Media, Berlin, 2020).

[130] Mathews, G. B. On the partition of numbers. *Proceedings of the London Mathematical Society* **s1-28**, 486–490 (1896). http://doi.wiley.com/10.1112/plms/s1-28.1.486.

[131] Chernoff, H. A measure of asymptotic efficiency for tests of a hypothesis based on the sum of observations. *Annals of Mathematical Statistics* **23**, 493–507 (1952). http://projecteuclid.org/euclid.aoms/1177729330.

[132] Hoare, C. A. R. Algorithm 64: Quicksort. *Communications of the ACM* **4**, 321 (1961). https://dl.acm.org/doi/10.1145/366622.366644.

[133] Calude, C. *Theories of computational complexity*. Vol. 35 in Annals of Discrete Mathematics (North-Holland/Elsevier, Amsterdam, New York, 1988).

[134] Leeuwen, J. v. (ed.) *Handbook of theoretical computer science* (Elsevier/MIT Press, Amsterdam, New York, Cambridge, MA, 1990).

[135] Gershenfeld, N. A. *The physics of information technology*. Cambridge Series on Information and the Natural Sciences (Cambridge University Press, Cambridge, 2011).

[136] Leff, H. S. & Rex, A. F. (eds.) *Maxwell's demon: Entropy, information, computing* (Adam Hilger, Bristol, 1990).

[137] Leff, H. S. & Rex, A. F. (eds.) *Maxwell's demon 2: Entropy, classical and quantum information, computing* (Institute of Physics, Bristol; Philadelphia, 2003). OCLC: ocm51569169.

[138] Berut, A. *et al.* Experimental verification of Landauer's principle linking information and thermodynamics. *Nature* **483**, 187–189 (2012). https:// www.nature.com/articles/nature10872.

[139] Bennett, C. H. Logical reversibility of computation. *IBM Journal of Research and Development* **17**, 525–532 (1973). http://ieeexplore.ieee.org/ document/5391327/.

[140] Fredkin, E., Landauer, R. & Toffoli, T. Physics of computation. *International Journal of Theoretical Physics* **21**, 903–903 (1982). http://link.springer.com/ 10.1007/BF02084157.

[141] Fredkin, E. & Toffoli, T. Conservative logic. *International Journal of Theoretical Physics* **21**, 219–253 (1982). http://link .springer.com/10.1007/BF01857727.

[142] Toffoli, T. & Margolus, N. *Cellular automata machines: A new environment for modeling* (MIT Press, Cambridge, MA, 1987). https://direct.mit.edu/books/ book/4258/Cellular-Automata-Mach %inesA-New-Environment-for.

[143] Toffoli, T. & Margolus, N. H. Invertible cellular automata: A review. *Physica D: Nonlinear Phenomena* **45**, 229–253 (1990).

https://linkinghub.elsevier.com/retrieve/ pii/016727899090185R%.

[144] Axelsen, H. B., Glück, R. & Yokoyama, T. Reversible machine code and its abstract processor architecture. In Diekert, V., Volkov, M. V. & Voronkov, A. (eds.), *Computer Science – Theory and Applications*, vol. 4649, 56–69 (Springer, Berlin, Heidelberg, 2007). http://link .springer.com/10.1007/978-3-540-74510 -5_9.

[145] Gerlach, W. & Stern, O. Der experimentelle Nachweis der Richtungsquantelung im Magnetfeld. *Zeitschrift für Physik* **9**, 349–352 (1922). http://link.springer.com/ 10.1007/BF01326983.

[146] Talagrand, M. *What is a quantum field theory?* (Cambridge University Press, Cambridge, 2022).

[147] Bennett, C. H. Quantum cryptography using any two nonorthogonal states. *Physical Review Letters* **68**, 3121–3124 (1992). https://link.aps.org/doi/10.1103/ PhysRevLett.68.3121.

[148] Hou, J.-M. & Chen, W. Hidden antiunitary symmetry behind "accidental" degeneracy and its protection of degeneracy. *Frontiers in Physics* **13**, 130301 (2018). http://link.s pringer.com/10.1007/s11467-017-0712-8.

[149] Hung, L.-Y. & Wan, Y. Symmetry-enriched phases obtained via pseudo anyon condensation. *International Journal of Modern Physics* **28**, 1450172 (2014). https://www .worldscientific.com/doi/abs/10.1142/ S021797921450%1720.

[150] Gaiotto, D., Kapustin, A., Seiberg, N. & Willett, B. Generalized global symmetries. *Journal of High Energy Physics* **2015**, 172 (2015). http://link.springer.com/ 10.1007/JHEP02(2015)172.

[151] Kong, L., Lan, T., Wen, X.-G., Zhang, Z.-H. & Zheng, H. Algebraic higher symmetry and categorical symmetry: A holographic and entanglement view of symmetry. *Physical Review Research* **2**,

043086 (2020). https://link.aps.org/doi/10.1103/PhysRevResearch.2.043086.

[152] Ji, W. & Wen, X.-G. Categorical symmetry and noninvertible anomaly in symmetry-breaking and topological phase transitions. *Physical Review Research* **2**, 033417 (2020). https://link.aps.org/doi/10.1103/PhysRevResearch.2.033417.

[153] Zhu, G., Sikander, S., Portnoy, E., Cross, A. W. & Brown, B. J. Non-Clifford and parallelizable fault-tolerant logical gates on constant and almost-constant rate homological quantum LDPC codes via higher symmetries (2023). https://arxiv.org/abs/2310.16982.

[154] Jozsa, R. Fidelity for mixed quantum states. *Journal of Modern Optics* **41**, 2315–2323 (1994). http://www.tandfonline.com/doi/abs/10.1080/09500349414552171.

[155] Schrodinger, E. Discussion of probability relations between separated systems. *Mathematical Proceedings of the Cambridge Philosophical Society* **31**, 555–563 (1935). https://www.cambridge.org/core/product/identifier/S0305004100013554/type/journal_article.

[156] Einstein, A., Podolsky, B. & Rosen, N. Can quantum-mechanical description of Physical reality be considered complete? *Physical Review* **47**, 777–780 (1935). https://link.aps.org/doi/10.1103/PhysRev.47.777.

[157] Bohr, N. Can quantum-mechanical description of physical reality be considered Complete? *Physical Review* **48**, 696–702 (1935). https://link.aps.org/doi/10.1103/PhysRev.48.696.

[158] Bell, J. S. On the Einstein Podolsky Rosen paradox. *Physics Physique Fizika* **1**, 195–200 (1964). https://link.aps.org/doi/10.1103/PhysicsPhysiqueFizika.1.195.

[159] Clauser, J. F., Horne, M. A., Shimony, A. & Holt, R. A. Proposed experiment to test local hidden-variable theories. *Physical Review Letters* **23**, 880–884 (1969). https://link.aps.org/doi/10.1103/PhysRevLett.23.880.

[160] Page, D. N. Information in black hole radiation. *Physical Review Letters* **71**, 3743–3746 (1993). https://link.aps.org/doi/10.1103/PhysRevLett.71.3743.

[161] Shannon, C. E. A mathematical theory of communication. *Bell System Technical Journal* **27**, 379–423 (1948). https://ieeexplore.ieee.org/document/6773024.

[162] Rényi, A. On measures of entropy and information. *Proceedings of the 4th Berkeley Symposium on Mathematics, Statistics and Probability* 547–561 (1960). https://digitalassets.lib.berkeley.edu/math/ucb/text/math_s4_%v1_article-27.pdf.

[163] Dur, W., Vidal, G. & Cirac, J. I. Three qubits can be entangled in two inequivalent ways. *Physical Review A* **62**, 062314 (2000). https://link.aps.org/doi/10.1103/PhysRevA.62.062314.

[164] Verstraete, F., Dehaene, J., De Moor, B. & Verschelde, H. Four qubits can be entangled in nine different ways. *Physical Review A* **65**, 052112 (2002). https://link.aps.org/doi/10.1103/PhysRevA.65.052112.

[165] Miyake, A. Classification of multipartite entangled states by multidimensional determinants. *Physical Review A* **67**, 012108 (2003). https://link.aps.org/doi/10.1103/PhysRevA.67.012108.

[166] Hu, Y. & Wan, Y. Entanglement entropy, quantum fluctuations, and thermal entropy in topological phases. *Journal of High Energy Physics* **2019**, 110 (2019). https://link.springer.com/10.1007/JHEP05(2019)110.

[167] Bekenstein, J. D. Black holes and entropy. *Physical Review D* **7**, 2333–2346 (1973). https://link.aps.org/doi/10.1103/PhysRevD.7.2333.

[168] Kaufman, A. M. *et al.* Quantum thermalization through entanglement in an isolated many-body system. *Science* **353**, 794–800 (2016). https://www.science.org/doi/10.1126/science.aaf6725.

[169] Chen, B. & Wu, J.-q. Universal relation between thermal entropy and entanglement entropy in conformal field theories. *Physical Review D* **91**, 086012

(2015). https://link.aps.org/doi/10.1103/ PhysRevD.91.086012.

[170] Casini, H., Huerta, M. & Myers, R. C. Towards a derivation of holographic entanglement entropy. *Journal of High Energy Physics* **2011**, 36 (2011). http:// link.springer.com/10.1007/JHEP05 (2011)036.

[171] Axler, S. *Linear algebra done right.* Undergraduate Texts in Mathematics (Springer, Cham, Switzerland, 2015). https://link.springer.com/10.1007/978-3 -319-11080-6.

[172] Miatto, F. M., Di Lorenzo Pires, H., Barnett, S. M. & Van Exter, M. P. Spatial Schmidt modes generated in parametric down-conversion. *The European Physical Journal D* **66**, 263 (2012). http://link.s pringer.com/10.1140/epjd/e2012-30035-3.

[173] Ren, J., Wang, Y. & You, W.-L. Quantum phase transitions in spin-1 *XXZ* chains with rhombic single-ion anisotropy. *Physical Review A* **97**, 042318 (2018). https://link.aps.org/doi/10.1103/ PhysRevA.97.042318.

[174] Bennett, C. H. *et al.* Purification of noisy entanglement and faithful teleportation via noisy channels. *Physical Review Letters* **76**, 722–725 (1996). https:// link.aps.org/doi/10.1103/PhysRevLett .76.722.

[175] Peres, A. Neumark's theorem and quantum inseparability. *Foundations of Physics* **20**, 1441–1453 (1990). http://link.springer.com/10.1007/ BF01883517.

[176] Bloch, F. Nuclear induction. *Physical Review* **70**, 460–474 (1946). https://link.aps.org/doi/10.1103/ PhysRev.70.460.

[177] Deutsch, D., Barenco, A. & Ekert, A. Universality in quantum computation. *Proceedings of the Royal Society of London* **449**, 669–677 (1995). https:// royalsocietypublishing.org/doi/10.1098/ rspa.1995.0065%.

[178] DiVincenzo, D. P. Two-bit gates are universal for quantum computation. *Physical Review A* **51**, 1015–1022 (1995).

https://link.aps.org/doi/10.1103/ PhysRevA.51.1015.

[179] Gray, F. Pulse code communication. https://patents.google.com/patent/ US2632058A/en.

[180] Barenco, A. *et al.* Elementary gates for quantum computation. *Physical Review A* **52**, 3457–3467 (1995). https://link.aps .org/doi/10.1103/PhysRevA.52.3457.

[181] Kitaev, A. Y. Quantum computations: Algorithms and error correction. *Russian Mathematical Surveys* **52**, 1191–1249 (1997). https://iopscience.iop.org/article/ 10.1070/RM1997v052n06ABEH0 %02155.

[182] Dawson, C. M. & Nielsen, M. A. The Solovay–Kitaev algorithm. *Quantum Information & Computation* **6**, 81–95 (2006).

[183] Deutsch, D. & Jozsa, R. Rapid solution of problems by quantum computation. *Proceedings of the Royal Society of London* **439**, 553–558 (1992). https:// royalsocietypublishing.org/doi/10.1098/ rspa.1992.0167%.

[184] Shor, P. Algorithms for quantum computation: Discrete logarithms and factoring. In *Proceedings 35th Annual Symposium on Foundations of Computer Science*, 124–134 (IEEE Computer Society Press, Santa Fe, NM, 1994). http://ieeexplore.ieee.org/document/ 365700/.

[185] Simon, D. R. On the power of quantum computation. *SIAM Journal on Computing* **26**, 1474–1483 (1997). http://epubs.siam.org/doi/10.1137/ S0097539796298637.

[186] Coppersmith, D. An approximate Fourier transform useful in quantum factoring (2002). http://arxiv.org/ abs/quant-ph/0201067. ArXiv:quant-ph/0201067.

[187] Ettinger, M., Hoyer, P. & Knill, E. The quantum query complexity of the hidden subgroup problem is polynomial. *Information Processing Letters* **91**, 43–48 (2004). https://linkinghub.elsevier.com/ retrieve/pii/S002001900400084%5.

[188] Gisin, N., Ribordy, G., Tittel, W. & Zbinden, H. Quantum cryptography. *Reviews of Modern Physics* **74**, 145–195 (2002). https://link.aps.org/doi/10.1103/RevModPhys.74.145.

[189] Jozsa, R. & Linden, N. On the role of entanglement in quantum-computational speed-up. *Proceedings: Mathematical, Physical and Engineering Sciences* **459**, 2011–2032 (The Royal Society, London, 2003). https://www.jstor.org/stable/3560059.

[190] Ekert, A., Jozsa, R., Penrose, R., Ekert, A. & Jozsa, R. Quantum algorithms: Entanglement-enhanced information processing. *Philosophical Transactions of the Royal Society of London. Series A: Mathematical, Physical and Engineering Sciences* **356**, 1769–1782 (1998). https://royalsocietypublishing.org/doi/10.1098/rsta.1998.0248%.

[191] Markov, A. Extension of the limit theorems of probability theory to a sum of variables connected in a chain. In Howard, R. A., *Dynamic Probabilistic Systems, Volume I: Markov Models* (Courier Corporation, 2007/Wiley, Hoboken, NJ, 1971; reprinted in Appendix B).

[192] Kraus, K., Bohm, A., Dollard, J. D. & Wootters, W. H. (eds.) *States, effects, and operations: Fundamental notions of quantum theory*, vol. 190 of *Lecture Notes in Physics* (Springer, Berlin, Heidelberg, 1983). http://link.springer.com/10.1007/3-540-12732-1.

[193] Preskill, J. Lecture Notes for Physics 229: Quantum Information and Computation (CreateSpace Independent Publishing Platform, 2017/California Institute of Technology, 1998).

[194] Stinespring, W. F. Positive functions on C*-algebras. *Proceedings of the American Mathematical Society* **6**, 211 (1955). https://www.jstor.org/stable/2032342?origin=crossref.

[195] Choi, M.-D. Completely positive linear maps on complex matrices. *Linear Algebra and its Applications* **10**, 285–290 (1975). https://linkinghub.elsevier.com/retrieve/pii/0024379575900750%.

[196] Jamiołkowski, A. Linear transformations which preserve trace and positive semidefiniteness of operators. *Reports on Mathematical Physics* **3**, 275–278 (1972). https://linkinghub.elsevier.com/retrieve/pii/0034487772900110%.

[197] Lindblad, G. On the generators of quantum dynamical semigroups. *Communications in Mathematical Physics* **48**, 119–130 (1976). http://link.springer.com/10.1007/BF01608499.

[198] Gorini, V., Kossakowski, A. & Sudarshan, E. C. G. Completely positive dynamical semigroups of *N*-level systems. *Journal of Mathematical Physics* **17**, 821–825 (1976). https://pubs.aip.org/jmp/article/17/5/821/225427/Completely-p%ositive-dynamical-semigroups-of-N.

[199] Aharonov, Y., Albert, D. Z. & Vaidman, L. How the result of a measurement of a component of the spin of a spin-*1/2* particle can turn out to be 100. *Physical Review Letters* **60**, 1351–1354 (1988). https://link.aps.org/doi/10.1103/PhysRevLett.60.1351.

[200] Dressel, J., Malik, M., Miatto, F. M., Jordan, A. N. & Boyd, R. W. Colloquium: Understanding quantum weak values: Basics and applications. *Reviews of Modern Physics* **86**, 307–316 (2014). https://link.aps.org/doi/10.1103/RevModPhys.86.307.

[201] Peres, A. Reversible logic and quantum computers. *Physical Review A* **32**, 3266–3276 (1985). https://link.aps.org/doi/10.1103/PhysRevA.32.3266.

[202] Shor, P. W. Scheme for reducing decoherence in quantum computer memory. *Physical Review A* **52**, R2493–R2496 (1995). https://link.aps.org/doi/10.1103/PhysRevA.52.R2493.

[203] Calderbank, A. R., Rains, E. M., Shor, P. W. & Sloane, N. J. A. Quantum error correction and orthogonal geometry. *Physical Review Letters* **78**, 405–408

(1997). https://link.aps.org/doi/10.1103/PhysRevLett.78.405.

[204] Park, J. L. The concept of transition in quantum mechanics. *Foundations of Physics* **1**, 23–33 (1970). http://link.springer.com/10.1007/BF00708652.

[205] Yamaguchi, K. & Kempf, A. Encrypted qubits can be cloned (2025). http://arxiv.org/abs/2501.02757. ArXiv:2501.02757 [quant-ph].

[206] Buzek, V. & Hillery, M. Quantum copying: Beyond the no-cloning theorem. *Physical Review A* **54**, 1844–1852 (1996). https://link.aps.org/doi/10.1103/PhysRevA.54.1844.

[207] Bruss, D., Ekert, A. & Macchiavello, C. Optimal universal quantum cloning and state estimation. *Physical Review Letters* **81**, 2598–2601 (1998). https://link.aps.org/doi/10.1103/PhysRevLett.81.2598.

[208] Steane, A. Multiple-particle interference and quantum error correction. *Proceedings of the Royal Society of London* **452**, 2551–2577 (1996). https://royalsocietypublishing.org/doi/10.1098/rspa.1996.0136%.

[209] Ekert, A. & Macchiavello, C. Error correction in quantum communication (1996). https://arxiv.org/abs/quant-ph/9602022. arXiv Version Number: 1.

[210] Gottesman, D. Class of quantum error-correcting codes saturating the quantum Hamming bound. *Physical Review A* **54**, 1862–1868 (1996). https://link.aps.org/doi/10.1103/PhysRevA.54.1862.

[211] Gottesman, D. The Heisenberg representation of quantum computers (1998). https://arxiv.org/abs/quant-ph/9807006. arXiv Version Number: 1.

[212] Gallager, R. Low-density parity-check codes. *IEEE Transactions on Information Theory* **8**, 21–28 (1962). http://ieeexplore.ieee.org/document/1057683/.

[213] Gottesman, D., Kitaev, A. & Preskill, J. Encoding a qubit in an oscillator. *Physical Review A* **64**, 012310 (2001). https://link.aps.org/doi/10.1103/PhysRevA.64.012310.

[214] Zanardi, P. & Rasetti, M. Noiseless quantum codes. *Physical Review Letters* **79**, 3306–3309 (1997). https://link.aps.org/doi/10.1103/PhysRevLett.79.3306.

[215] Lidar, D. A., Chuang, I. L. & Whaley, K. B. Decoherence-free subspaces for quantum computation. *Physical Review Letters* **81**, 2594–2597 (1998). https://link.aps.org/doi/10.1103/PhysRevLett.81.2594.

[216] Duan, L.-M. & Guo, G.-C. Preserving coherence in quantum computation by pairing quantum bits. *Physical Review Letters* **79**, 1953–1956 (1997). https://link.aps.org/doi/10.1103/PhysRevLett.79.1953.

[217] Viola, L. & Lloyd, S. Dynamical suppression of decoherence in two-state quantum systems. *Physical Review A* **58**, 2733–2744 (1998). https://link.aps.org/doi/10.1103/PhysRevA.58.2733.

[218] Wen, X. G. Mean-field theory of spin-liquid states with finite energy gap and topological orders. *Physical Review B* **44**, 2664–2672 (1991). https://link.aps.org/doi/10.1103/PhysRevB.44.2664.

[219] Wen, X.-G. Topological orders and edge excitations in fractional quantum Hall states. *Advances in Physics* **44**, 405–473 (1995). http://www.tandfonline.com/doi/abs/10.1080/00018739500101566.

[220] Landau, L. On the theory of phase transitions. In *Collected Papers of L. D. Landau*, 193–216 (Elsevier, 1965). https://linkinghub.elsevier.com/retrieve/pii/B9780080105864500341.

[221] Klitzing, K. V., Dorda, G. & Pepper, M. New method for high-accuracy determination of the fine-structure constant based on quantized Hall resistance. *Physical Review Letters* **45**, 494–497 (1980). https://link.aps.org/doi/10.1103/PhysRevLett.45.494.

[222] Tsui, D. C., Stormer, H. L. & Gossard, A. C. Two-dimensional magnetotransport in the extreme quantum limit. *Physical Review Letters* **48**, 1559–1562 (1982).

https://link.aps.org/doi/10.1103/PhysRevLett.48.1559.

[223] Wen, X. G. Vacuum degeneracy of chiral spin states in compactified space. *Physical Review B* **40**, 7387–7390 (1989). https://link.aps.org/doi/10.1103/PhysRevB.40.7387.

[224] Wen, X.-G. Topological orders in rigid states. *International Journal of Modern Physics* **04**, 239–271 (1990). https://www.worldscientific.com/doi/abs/10.1142/S021797929000%0139.

[225] Laughlin, R. B. Anomalous quantum Hall effect: An incompressible quantum fluid with fractionally charged excitations. *Physical Review Letters* **50**, 1395–1398 (1983). https://link.aps.org/doi/10.1103/PhysRevLett.50.1395.

[226] Wu, Y.-S. General theory for quantum statistics in two dimensions. *Physical Review Letters* **52**, 2103–2106 (1984). https://link.aps.org/doi/10.1103/PhysRevLett.52.2103.

[227] Arovas, D., Schrieffer, J. R. & Wilczek, F. Fractional statistics and the quantum Hall effect. *Physical Review Letters* **53**, 722–723 (1984). https://link.aps.org/doi/10.1103/PhysRevLett.53.722.

[228] Arovas, D. P., Schrieffer, R., Wilczek, F. & Zee, A. Statistical mechanics of anyons. *Nuclear Physics B* **251**, 117–126 (1985). https://linkinghub.elsevier.com/retrieve/pii/0550321385902524%.

[229] Haldane, F. D. M. Geometrical interpretation of momentum and crystal momentum of classical and quantum ferromagnetic Heisenberg chains. *Physical Review Letters* **57**, 1488–1491 (1986). https://link.aps.org/doi/10.1103/PhysRevLett.57.1488.

[230] Haldane, F. D. M. & Rezayi, E. H. Spin-singlet wave function for the half-integral quantum Hall effect. *Physical Review Letters* **60**, 956–959 (1988). https://link.aps.org/doi/10.1103/PhysRevLett.60.956.

[231] Kohmoto, M. Topological invariant and the quantization of the Hall conductance. *Annals of Physics* **160**, 343–354 (1985).

https://www.sciencedirect.com/science/article/pii/00034916859%01484.

[232] Hatsugai, Y. & Kohmoto, M. Energy spectrum and the quantum Hall effect on the square lattice with next-nearest-neighbor hopping. *Physical Review B* **42**, 8282–8294 (1990). https://link.aps.org/doi/10.1103/PhysRevB.42.8282.

[233] Hung, L.-Y. & Wan, Y. String-net models with Z N fusion algebra. *Physical Review B* **86**, 235132 (2012). https://link.aps.org/doi/10.1103/PhysRevB.86.235132.

[234] Hu, Y., Huang, Z., Hung, L.-Y. & Wan, Y. Anyon condensation: Coherent states, symmetry enriched topological phases, Goldstone theorem, and dynamical rearrangement of symmetry. *Journal of High Energy Physics* **2022**, 26 (2022). https://link.springer.com/10.1007/JHEP03(2022)026.

[235] Zhao, Y. & Wan, Y. Landau-Ginzburg Paradigm of Topological Phases. arXiv:2506.05319. https://doi.org/10.48550/arXiv.2506.05319.

[236] Zhao, Y. & Wan, Y. Noninvertible gauge symmetry in (2+1)d topological orders: A string-net model realization. *Journal of High Energy Physics* **2025**, 138 (2025). https://doi.org/10.1007/JHEP11(2025)138.

[237] Kitaev, A. & Preskill, J. Topological entanglement entropy. *Physical Review Letters* **96**, 110404 (2006). https://link.aps.org/doi/10.1103/PhysRevLett.96.110404.

[238] Bravyi, S., Hastings, M. B. & Michalakis, S. Topological quantum order: Stability under local perturbations. *Journal of Mathematical Physics* **51**, 093512 (2010). https://pubs.aip.org/jmp/article/51/9/093512/896702/Topologic%al-quantum-order-Stability-under-local.

[239] Fierz, M. Uber die relativistische Theorie kraftefreier Teilchen mit beliebigem Spin. *Helvetica Physika Acta* **12**, 3–37 (1939). https://www.e-periodica.ch/digbib/view?pid=hpa-001:1939:12::6%30.

[240] Pauli, W. The connection between spin and statistics. *Physical Review* **58**,

716–722 (1940). https://link.aps .org/doi/10.1103/PhysRev.58.716.

[241] Freedman, M. H., Kitaev, A., Larsen, M. J. & Wang, Z. Topological quantum Computation (2001). https://arxiv.org/ abs/quant-ph/0101025.

[242] Kitaev, A. Anyons in an exactly solved model and beyond. *Annals of Physics* **321**, 2–111 (2006). https://linkinghub .elsevier.com/retrieve/pii/S0003 49160500238%1.

[243] Reshetikhin, N. & Turaev, V. G. Invariants of 3-manifolds via link polynomials and quantum groups. *Inventiones Mathematicae* **103**, 547–597 (1991). http://link.springer.com/10 .1007/BF01239527.

[244] Chern, S.-S. & Simons, J. Characteristic forms and geometric invariants. *The Annals of Mathematics* **99**, 48 (1974). https://www.jstor.org/stable/1971013? origin=crossref.

[245] Turaev, V. & Viro, O. State sum invariants of 3-manifolds and quantum 6j-symbols. *Topology* **31**, 865–902 (1992). https://linkinghub.elsevier.com/retrieve/ pii/004093839290015A%.

[246] Turaev, V. G. *Quantum invariants of knots and 3-manifolds* (De Gruyter, Berlin, 2016). https://www.degruyter .com/document/doi/10.1515/ 9783110435221/%html.

[247] Dijkgraaf, R. & Witten, E. Topological gauge theories and group cohomology. *Communications in Mathematical Physics* **129**, 393–429 (1990). http://link.springer .com/10.1007/BF02096988.

[248] Dennis, E., Kitaev, A., Landahl, A. & Preskill, J. Topological quantum memory. *Journal of Mathematical Physics* **43**, 4452–4505 (2002). https:// pubs.aip.org/jmp/article/43/9/4452/ 230976/Topological%-quantum-memory.

[249] Chen, X., Gu, Z.-C. & Wen, X.-G. Local unitary transformation, long-range quantum entanglement, wave function renormalization, and topological order. *Physical Review B* **82**, 155138 (2010).

https://link.aps.org/doi/10.1103/ PhysRevB.82.155138.

[250] Dehn, M. Die Gruppe der Abbildungsklassen: Das arithmetische Feld auf Flachen. *Acta Mathematica* **69**, 135–206 (1938). http://projecteuclid.org/ euclid.acta/1485888214.

[251] Mignard, M. & Schauenburg, P. Modular categories are not determined by their modular data. *Letters in Mathematical Physics* **111**, 60 (2021). https://link.springer.com/10.1007/ s11005-021-01395-0.

[252] Wen, X. & Wen, X.-G. Distinguish modular categories and 2+1D topological orders beyond modular data: Mapping class group of higher genus manifold (2019). https://arxiv.org/abs/ 1908.10381.

[253] Raussendorf, R. & Harrington, J. Fault-tolerant quantum computation with high threshold in two dimensions. *Physical Review Letters* **98**, 190504 (2007). https://link.aps.org/doi/ 10.1103/PhysRevLett.98.190504.

[254] Google Quantum AI *et al.* Suppressing quantum errors by scaling a surface code logical qubit. *Nature* **614**, 676–681 (2023). https://www.nature.com/articles/ s41586-022-05434-1.

[255] Bluvstein, D. *et al.* Logical quantum processor based on reconfigurable atom arrays. *Nature* (2023). https://www .nature.com/articles/s41586-023- 06927-3.

[256] Bombin, H. & Martin-Delgado, M. A. Topological quantum distillation. *Physical Review Letters* **97**, 180501 (2006). https://link.aps.org/doi/ 10.1103/PhysRevLett.97.180501.

[257] Bombin, H. & Martin-Delgado, M. A. Topological computation without braiding. *Physical Review Letters* **98**, 160502 (2007). https://link.aps.org/doi/ 10.1103/PhysRevLett.98.160502.

[258] Zhang, J., Wu, Y.-C. & Guo, G.-P. Facilitating practical fault-tolerant quantum computing based on color

codes (2023). https://arxiv.org/abs/
2309.05222. arXiv Version Number: 4.

[259]   Eastin, B. & Knill, E. Restrictions on
transversal encoded quantum gate sets.
Physical Review Letters **102**, 110502
(2009). https://link.aps.org/doi/10.1103/
PhysRevLett.102.110502.

[260]   Berent, L., Burgholzer, L., Derks, P.-J.
H. S., Eisert, J. & Wille, R. Decoding
quantum color codes with MaxSAT
(2023). https://arxiv.org/abs/2303.14237.
arXiv Version Number: 2.

[261]   Freedman, M. H., Larsen, M. & Wang,
Z. A modular functor which is universal
for quantum computation.
Communications in Mathematical Physics
227, 605–622 (2002). http://link.springer
.com/10.1007/s002200200645.

[262]   Leinaas, J. M. & Myrheim, J. On the
theory of identical particles. *Il Nuovo
Cimento B* **37**, 1–23 (1977). http://
link.springer.com/10.1007/BF02727953.

[263]   Wilczek, F. Magnetic flux, angular
momentum, and statistics. *Physical
Review Letters* **48**, 1144–1146 (1982).
https://link.aps.org/doi/10.1103/
PhysRevLett.48.1144.

[264]   Wilczek, F. Quantum mechanics of
fractional-spin particles. *Physical Review
Letters* **49**, 957–959 (1982). https://link
.aps.org/doi/10.1103/PhysRevLett
.49.957.

[265]   Wilczek, F. & Zee, A. Linking numbers,
spin, and statistics of solitons. *Physical
Review Letters* **51**, 2250–2252 (1983).
https://link.aps.org/doi/10.1103/PhysRev
Lett.51.2250.

[266]   Wilczek, F. *Fractional statistics and
anyon superconductivity* (World Scientific,
Singapore, Teaneck, NJ, 1990).

[267]   Wu, Y.-S. Multiparticle quantum
mechanics obeying fractional Statistics.
Physical Review Letters **53**, 111–114
(1984). https://link.aps.org/doi/
10.1103/PhysRevLett.53.111.

[268]   Wu, Y.-S. & Zee, A. Comments on the
Hopf lagrangian and fractional statistics
of solitons. *Physics Letters B* **147**,
325–329 (1984). https://www

.sciencedirect.com/science/article/pii/
03702693849%01266.

[269]   Wu, Y.-S. Statistical distribution for
generalized ideal gas of
fractional-statistics particles. *Physical
Review Letters* **73**, 922–925 (1994).
https://link.aps.org/doi/10.1103/
PhysRevLett.73.922.

[270]   Halperin, B. I. Statistics of quasiparticles
and the hierarchy of fractional quantized
Hall states. *Physical Review Letters* **52**,
1583–1586 (1984). https://link.aps
.org/doi/10.1103/PhysRevLett.52.1583.

[271]   Shizuya, K. & Tamura, H. Anyon
statistics and its variation with
wavelength in Maxwell–Chern–Simons
gauge theories. *Physics Letters B* **252**,
412–416 (1990). https://www.science
direct.com/science/article/pii/
03702693909%0561J.

[272]   Haldane, F. D. M. "Fractional statistics"
in arbitrary dimensions: A generalization
of the Pauli principle. *Physical Review
Letters* **67**, 937–940 (1991). https://link
.aps.org/doi/10.1103/PhysRevLett
.67.937.

[273]   Nayak, C. & Wilczek, F. Exclusion
statistics: Low-temperature properties,
fluctuations, duality, and applications.
Physical Review Letters **73**, 2740–2743
(1994). https://link.aps.org/doi/10.1103/
PhysRevLett.73.2740.

[274]   Levin, M. A. & Wen, X.-G. String-net
condensation: A physical mechanism for
topological phases. *Physical Review B* **71**,
045110 (2005). https://link.aps.org/doi/
10.1103/PhysRevB.71.045110.

[275]   Hu, Y., Wan, Y. & Wu, Y.-S. Boundary
Hamiltonian theory for gapped
topological orders. *Chinese Physics
Letters* **34**, 077103 (2017). https://
iopscience.iop.org/article/10.1088/
0256-307X/34/7/077%103.

[276]   Hu, Y., Luo, Z.-X., Pankovich, R., Wan,
Y. & Wu, Y.-S. Boundary Hamiltonian
theory for gapped topological phases on
an open surface. *Journal of High Energy
Physics* **2018**, 134 (2018). http://link
.springer.com/10.1007/JHEP01(2018)134.

[277] Zhao, Y., Wang, H., Hu, Y. & Wan, Y. Symmetry fractionalized (irrationalized) fusion rules and two domain-wall Verlinde formulae (2023). https://arxiv.org/abs/2304.08475. arXiv Version Number: 2.

[278] Hu, Y., Stirling, S. D. & Wu, Y.-S. Emergent exclusion statistics of quasiparticles in two-dimensional topological phases. *Physical Review B* **89**, 115133 (2014). https://link.aps.org/doi/10.1103/PhysRevB.89.115133.

[279] Li, Y., Wang, H., Hu, Y. & Wan, Y. Anyonic exclusions statistics on surfaces with gapped boundaries. *Journal of High Energy Physics* **2019**, 78 (2019). https://link.springer.com/10.1007/JHEP04(2019)078.

[280] Artin, E. Theory of braids. *The Annals of Mathematics* **48**, 101 (1947). https://www.jstor.org/stable/1969218?origin=crossref.

[281] Yang, C. N. Some exact results for the many-body problem in one dimension with repulsive delta-function interaction. *Physical Review Letters* **19**, 1312–1315 (1967). https://link.aps.org/doi/10.1103/PhysRevLett.19.1312.

[282] Baxter, R. J. Exactly solved models in statistical mechanics. In *Integrable systems in statistical mechanics*, vol. 1, 5–63 (Series on Advances in Statistical Mechanics; World Scientific, Singapore, 1985). http://www.worldscientific.com/doi/abs/10.1142/9789814415255_%0002.

[283] Kauffman, L. H. & Lomonaco, S. J. Chapter 14: Quantum computing and quantum topology. In Kauffman, L. & Lomonaco, S. J. (eds.), *Mathematics of quantum computation and quantum technology* (Chapman and Hall/CRC, 2007). https://www.taylorfrancis.com/books/9781584889007.

[284] Xu, H. & Wan, X. Constructing functional braids for low-leakage topological quantum computing. *Physical Review A* **78**, 042325 (2008). https://link.aps.org/doi/10.1103/PhysRevA.78.042325.

[285] Field, B. & Simula, T. Introduction to topological quantum computation with non-Abelian anyons. *Quantum Science and Technology* **3**, 045004 (2018). https://iopscience.iop.org/article/10.1088/2058-9565/aacad2.

[286] Cui, S. X., Tian, K. T., Vasquez, J. F., Wang, Z. & Wong, H. M. The search for leakage-free entangling Fibonacci braiding gates. *Journal of Physics A: Mathematical and Theoretical* **52**, 455301 (2019). https://iopscience.iop.org/article/10.1088/1751-8121/ab488e.

[287] Zhang, Y.-H., Zheng, P.-L., Zhang, Y. & Deng, D.-L. Topological quantum compiling with reinforcement learning. *Physical Review Letters* **125**, 170501 (2020). https://link.aps.org/doi/10.1103/PhysRevLett.125.170501.

[288] Bonesteel, N. E., Hormozi, L., Zikos, G. & Simon, S. H. Braid topologies for quantum computation. *Physical Review Letters* **95**, 140503 (2005). https://link.aps.org/doi/10.1103/PhysRevLett.95.140503.

[289] Fan, Y.-a. *et al.* Experimental quantum simulation of a topologically protected Hadamard gate via braiding Fibonacci anyons. *Innovation* **4** (2023). https://www.cell.com/the-innovation/abstract/S2666-6758(23)00%108-X.

Index

For EU product safety concerns, contact us at Calle de José Abascal, 56–1°,
28003 Madrid, Spain or eugpsr@cambridge.org.